Thermophilic Fungi
Basic Concepts and
Biotechnological Applications

Thermophilic Fungi
Basic Concepts and
Biotechnological Applications

Raj Kumar Salar

CRC Press
Taylor & Francis Group
Boca Raton London New York

CRC Press is an imprint of the
Taylor & Francis Group, an **informa** business

CRC Press
Taylor & Francis Group
6000 Broken Sound Parkway NW, Suite 300
Boca Raton, FL 33487-2742

First issued in paperback 2020

ISBN-13: 978-0-367-57189-4 (pbk)
ISBN-13: 978-0-8153-7070-3 (hbk)

Visit the Taylor & Francis Web site at
http://www.taylorandfrancis.com

and the CRC Press Web site at
http://www.crcpress.com

Dedicated to my mentor, Professor Kamal Rai Aneja.

Contents

Preface...xiii
Author..xv
Acknowledgments..xvii

SECTION I   Basic Concepts

Chapter 1   Introduction...3

    1.1   Overview of Thermophilic Fungi..3
    1.2   Defining Thermophily...10
    1.3   Historical Background...11
    1.4   Habitat Relationships..13
    1.5   Isolation and Culture of Thermophilic Fungi...........................15
        1.5.1   Isolation Techniques..16
            1.5.1.1   Dilution Plate Technique..............................16
            1.5.1.2   Warcup's Soil Plate Method........................16
            1.5.1.3   Humidified Chamber Technique...................16
            1.5.1.4   Paired Petri Plate Technique.......................17
            1.5.1.5   Waksman's Direct Inoculation Method........17
            1.5.1.6   Isolation from Air..17
        1.5.2   Culture Media for Isolation of Thermophilic Fungi....17
    1.6   Biotechnological Significance..20
    References...23

Chapter 2   Origin of Thermophily in Fungi...29

    2.1   Introduction...29
    2.2   Origin and Ecological Relationships...30
    2.3   Fungal Adaptations to Thermophily..33
    2.4   Hypothesis to Explain Thermophilism in Fungi........................34
        2.4.1   Protein Thermostability and Stabilization....................35
            2.4.1.1   Structure-Based...35
            2.4.1.2   Sequence-Based...36
        2.4.2   Heat Shock Proteins..36
        2.4.3   Proteome and Genome as Determinants
            of Thermophilic Adaptation...38
        2.4.4   Reduction in Genome Size...40
        2.4.5   Thermotolerance Genes..41
        2.4.6   Rapid Turnover of Essential Metabolites....................41
        2.4.7   Macromolecular Thermostability................................42

                    2.4.8   Ultrastructural Thermostability and Pigmentation.......43
                    2.4.9   Lipid Solubilization..44
          2.5   Acquired Thermotolerance...45
          2.6   Homeoviscous versus Homeophasic Adaptations...................46
          References..47

Chapter 3   Physiology of Thermophilic Fungi.......................................55

          3.1   Introduction...55
          3.2   Nutritional Requirements of Thermophilic Fungi.................56
          3.3   Growth and Metabolism of Thermophilic Fungi..................57
          3.4   Effects of Environmental Factors on Growth.......................59
                    3.4.1   Effect of Temperature.....................................59
                    3.4.2   Effect of pH..62
                    3.4.3   Effect of Oxygen..64
                    3.4.4   Effect of Solutes and Water Activity..................64
                    3.4.5   Hydrostatic Pressure......................................65
                    3.4.6   Effect of Light..66
                    3.4.7   Relative Humidity...66
          3.5   Complex Carbon Sources and Adaptations for Mixed
                    Substrate Utilization...67
          3.6   Nutrient Transport...68
          3.7   Protein Breakdown and Turnover......................................68
          3.8   Virulence...69
          References..70

Chapter 4   Habitat Diversity..75

          4.1   Introduction...75
          4.2   Natural Habitats..76
                    4.2.1   Soil...76
                              4.2.1.1   Desert Soils..............................80
                              4.2.1.2   Coal Mine Soils.........................81
                              4.2.1.3   Geothermal Soils........................81
                              4.2.1.4   Dead Sea Valley Soil...................82
                    4.2.2   Beach Sand...83
                    4.2.3   Nesting Material of Birds and Animals................84
                              4.2.3.1   Bird Nests and Feathers...............84
                              4.2.3.2   Alligator Nesting Material............85
                    4.2.4   Coal Spoil Tips..86
                    4.2.5   Hot Springs..86
          4.3   Man-Made Habitats..88
                    4.3.1   Hay...88
                    4.3.2   Wood Chip Piles..89
                    4.3.3   Nuclear Reactor Effluents...............................89
                    4.3.4   Manure...90

4.3.5   Stored Peat..90
4.3.6   Retting Guayule..91
4.3.7   Stored Grains..91
4.3.8   Municipal Waste...92
4.3.9   Composts..93
References...96

SECTION II   Taxonomy, Biodiversity, and Classification

Chapter 5   Bioprospecting of Thermophilic Fungi...105

5.1   Introduction...105
5.2   Biodiversity Perspective...106
5.3   Culturable Microbial Diversity..107
5.4   Bioprospecting the Uncultivable...108
5.5   Bioprospecting and Conservation of Fungal Diversity...........110
        5.5.1   Microbial Strain Data Network...................................111
        5.5.2   Classification of Microorganisms on the Basis
                  of Hazard..111
        5.5.3   International Depository Authorities.........................113
                  5.5.3.1   Responsibilities of an IDA.........................113
                  5.5.3.2   Distribution of IDAs and the Biological
                              Material Accepted......................................114
                  5.5.3.3   Guide to the Deposit of Microorganisms
                              under the Budapest Treaty.......................114
                  5.5.3.4   Code of Practice for IDAs....................114
                  5.5.3.5   Future Development of the IDA
                              Network Worldwide..................................115
        5.5.4   Culture Transportation...116
        5.5.5   The Premises before Dispatch of Cultures.................117
        5.5.6   Organizations Dealing with Microbial Cultures.........118
5.6   Future Perspectives...118
References...119

Chapter 6   Taxonomy and Molecular Phylogeny of Thermophilic Fungi........121

6.1   Introduction...121
6.2   Classification and Taxonomic Ranks.....................................122
6.3   What Is Phylogeny?..124
6.4   Phylogenetic Analysis..125
        6.4.1   Molecular Phylogeny of Thermophilic Fungi............125
        6.4.2   Constructing Phylogenetic Trees...............................129
        6.4.3   Phylogeny and Systematics.......................................134
        6.4.4   Thermophilic Fungal Genomes..................................135

      6.5   Future Prospects...139

      References..139

Chapter 7   Biodiversity and Taxonomic Descriptions....................143

      7.1   Introduction...143

      7.2   Key to the Identification of Thermophilic Fungi...................145

            7.2.1   Zygomycota..145

            7.2.2   Ascomycota..145

            7.2.3   Deuteromycetes (Anamorphic Fungi).....................147

      7.3   Taxonomic Descriptions of Thermophilic Taxa...................148

            7.3.1   Zygomycetes...148

            7.3.2   Ascomycetes..151

            7.3.3   Deuteromycetes (Anamorphic Fungi).....................166

      7.4   Nomenclatural Disagreement and Synonymies...................176

      References..180

Chapter 8   The Conflict of Name Change and Synonymies...........185

      8.1   Introduction...185

      8.2   The Conflict over Name Change...................................186

      8.3   The Conflict of One Fungus, Which Name?........................188

      8.4   Taxonomies and the Name Changes.................................190

      8.5   Classification of Uncultured Species...............................198

      8.6   Unwarranted Taxonomies...200

      References..201

SECTION III   *Biotechnological Applications*

Chapter 9   Role of Thermophilic Fungi in Composting....................209

      9.1   Introduction...209

      9.2   Bioconversion of Lignocellulosic Materials.....................211

      9.3   Physicochemical Aspects of Composts............................213

            9.3.1   Initial C:N and C:P Ratio.................................213

            9.3.2   Moisture Content..215

            9.3.3   Temperature..215

            9.3.4   Pile Size...216

            9.3.5   Initial pH Value of Compost..............................216

      9.4   Ecology of Thermophilic Fungi in Mushroom Compost.......217

      9.5   Role of Hydrolytic Enzymes of Thermophiles
            in Composting...222

      9.6   Methods of Mushroom Composting................................223

            9.6.1   Long Method of Composting..............................223

            9.6.2   Short Method of Composting..............................225

        9.6.3   Anglo-Dutch Method..226
        9.6.4   INRA Method..226
    9.7 Growth Promotion of *Agaricus bisporus*
         by Thermophilic Fungi...227
    9.8 Co-Composting..230
    9.9 Future Prospects..231
    References..232

Chapter 10 Bioremediation and Biomineralization..241

    10.1 Introduction...241
    10.2 Bioremediation...243
        10.2.1 Heavy Metals as Environmental Pollutants...............243
        10.2.2 Metals as a Precious Component of Life...................245
        10.2.3 Strategies to Control Heavy Metal Contamination....246
                10.2.3.1 Conventional Treatment Techniques.........247
                10.2.3.2 Bioaccumulation of Heavy Metals.............249
                10.2.3.3 Biosorption of Heavy Metals....................250
                10.2.3.4 Immobilized Biosorbent
                          for Bioremediation.....................................252
                10.2.3.5 Recovery of Metals and Regeneration
                          of Biomass...252
        10.2.4 Thermophilic Fungi in Bioremediation....................252
    10.3 Biomineralization...254
    References..256

Chapter 11 Biocatalysts of Thermophilic Fungi...259

    11.1 Introduction...259
    11.2 Extracellular Thermostable Enzymes Produced
          by Thermophilic Fungi...261
        11.2.1   Cellulases...262
        11.2.2   Amylases...271
        11.2.3   Glucoamylase..273
        11.2.4   Xylanases..273
        11.2.5   Lipases..276
        11.2.6   Proteases...277
        11.2.7   Pectinases...278
        11.2.8   Phytases..279
        11.2.9   Phosphatases...280
        11.2.10 Laccases..280
        11.2.11 α-D-Glucuronidase...281
        11.2.12 Cellobiose Dehydrogenase..281
        11.2.13 D-Glucosyltransferase...282
        11.2.14 DNase..282

11.3 Intracellular or Cell-Associated Thermostable Enzymes
Produced by Thermophilic Fungi..282
    11.3.1 Trehalase..283
    11.3.2 Invertase...285
    11.3.3 β-Glycosidase..286
    11.3.4 ATP Sulfurylase...287
    11.3.5 Protein Disulfide Isomerase......................................288
    11.3.6 Lipoamide Dehydrogenase..288
11.4 Bioactive Compounds from Thermophilic Fungi...................288
11.5 Single-Cell Protein Production...290
11.6 Tools for Genetic Recombination...290
11.7 Detrimental Activities...291
References..293

Chapter 12 Future Perspectives and Conclusions...311

12.1 Diversity Perspectives...312
12.2 Taxonomic Perspectives...313
12.3 Phylogenetic and Genomic Perspectives................................315
12.4 Biotechnological Perspectives...317
References..318

Index...321

Preface

It is now well recognized that microbial life can exist at elevated temperatures where most other life-forms fail to survive. Fungi growing at extraordinarily high temperatures appear to be "treasure troves" for fundamental research and biotechnological applications. The first modern comprehensive account of the biology and classification of thermophilic fungi was published by Cooney and Emerson in 1964. However, during the last five decades, a number of thermophilic fungi have been discovered and documented in the literature. The purpose of this book, *Thermophilic Fungi: Basic Concepts and Biotechnological Applications*, is to present an all-inclusive account on thermophilic fungi. It begins with the essential concepts of thermophilic fungi, covering their early history, habitat relationships, isolation and culture, and biotechnological relevance.

The book is divided into three sections aiming to comprehensively cover the basic aspects; taxonomy and classification, including molecular phylogeny; and biotechnological applications of thermophilic fungi. All the chapters include recent research and innovations carried out on a particular subtopic and provide a framework upon which students and researchers can build their knowledge of thermophilic fungi. A list of references at the end of each chapter is provided for the readers to learn more about a particular topic. Typically, these references include basic research, research papers, review articles, and articles from the popular literature.

Section I is designed to underpin the basic concepts of thermophilic fungi. Its aim is to provide a sufficient, albeit elaborate, overview of thermophilic fungi, covering their history, culture techniques, biotechnological significance, origin, physiology, and habitat diversity. Section II is devoted to the taxonomy, biodiversity, and classification of thermophilic fungi. It examines the bioprospecting of thermophilic fungi, including uncultivable genomes. Further, Chapter 7, on the taxonomy and molecular phylogeny of thermophilic fungi, provides an overview of the classification, phylogeny, systematics, and genomes of important thermophilic species. Over the last 50 years, many new thermophilic fungi have been discovered, and taxonomic descriptions of thermophilic taxa, including certain nomenclatural disagreements, are provided in Chapter 7. This section also examines the conflict over name changes and gives an overview of the implementation of the one-fungus, one-name concept by the International Code of Nomenclature for algae, fungi, and plants. Lastly, Section III covers the biotechnological applications of thermophilic fungi, such as their role in mushroom composting, bioremediation, and biomineralization. Chapter 11 investigates diverse biocatalysts, including enzymes, proteins, and antimicrobials, produced by thermophilic fungi and their various biotechnological applications. Finally, based on the study and understanding of thermophilic fungi, several future prospects for research on thermophilic fungi are proposed.

This book attempts to fill the gap in the literature about modern biotechnology, and I am sure that it will prove to be of benefit for students, scientists, and researchers working with various aspects of thermophilic fungi. Because no text is perfect, some errors may exist in this book. The readers are invited to point out mistakes and advise me whenever any weaknesses or shortcomings come to the fore. I sincerely hope to receive corrections and guidance letters from the valued readers to further improve this book.

Raj Kumar Salar
Department of Biotechnology
Chaudhary Devi Lal University, Sirsa, India

Author

Raj Kumar Salar is a professor in the Department of Biotechnology, Chaudhary Devi Lal University, Sirsa, India. Presently, he is chairperson of the Department of Biotechnology and dean, Faculty of Life Sciences, Chaudhary Devi Lal University, Sirsa, India. He earned his postdoctoral fellowship from the Ministry of Education, Slovak Republic, to pursue a postdoctorate from the Department of Biochemical Technology, Slovak University of Technology, Bratislava, Slovakia, and visited several other countries, including Japan, Norway, Austria, and Hungary, for academic pursuits. Besides thermophilic fungi, his current research interests include fermentation technology and harnessing the medicinal value of plants and microbes. He has published about 80 research papers and reviews in journals of national and international repute and edited three books published by Springer. Dr. Salar also has received several research and development projects from University Grants Commission (UGC), New Delhi, and Haryana State Council for Science and Technology (HSCST), Chandigarh, and an international travel grant from Department of Science and Technology (DST), Ministry of Science & Technology, New Delhi. He is a reviewer of several journals published by Wiley-VCH, Springer, Elsevier, and Taylor & Francis. Dr. Salar received the prestigious King Abdulaziz City for Science & Technology award for best paper published in *3 Biotech* journal. He has supervised several PhD and MPhil candidates and has more than 20 years of teaching and research experience.

Acknowledgments

I express my sincere gratitude to those who kindled, guided, or motivated my interest in the study of fungi throughout my career. I am thankful to my PhD supervisor, Professor Kamal Rai Aneja, who introduced me to this enigmatic group of thermophilic fungi and deepened my knowledge of fungi. My postdoctoral mentor, Professor Milan Certik (Slovak Technical University, Bratislava, Slovak Republic), encouraged me to write this book. I appreciate all those scientists throughout the world who have cited my work on thermophilic fungi, which helped me to seek recent literature to organize this book. Professor Jean Mouchacca of Laboratoire de Cryptogamie, Paris, deserves special mention for sending me literature on the taxonomy of thermophilic fungi, which proved to be of great help. I am extremely thankful to all the scientists and publishers who granted me permission to use their illustrations to shape this book. Among my PhD students, Dr. Suresh Kumar helped me to scan and organize some original photographs and create new figures, and Naresh Kumar helped in organizing and arranging references. Above all, I am indebted to my wife, Neelam, and son, Arnav, for coping with my long involvement in the writing of this book and helping in ways that enabled me to complete the work within the given time frame.

I am also grateful to colleagues, staff, and students in the Department of Biotechnology, Chaudhary Devi Lal University, Sirsa, India, for their courtesies, and to the university administration for allowing me sabbatical leave to complete this work.

Raj Kumar Salar

Section I

Basic Concepts

1 Introduction

1.1 OVERVIEW OF THERMOPHILIC FUNGI

Thermophilic—literally, "heat loving"—is a characteristic that appears among widely different groups of microorganisms. Organisms can be classified as thermophilic if their optimum temperature for growth is between 45°C and 80°C, as hyperthermophilic if their optimum temperature is above 80°C, and as mesophilic if their optimum temperature is below 45°C (Stetter et al., 1990; Madigan and Orient, 1999). Thermophilic organisms are found frequently in the Bacteria and Archaea domains, whereas hyperthermophiles are mainly confined to the Archaea. However, in the Eukarya domain life at high temperatures is an uncommon phenomenon. Of the estimated 600,000 fungi known (Mora et al., 2011), only a small fraction are considered thermophilic and are able to thrive at temperatures between 45 and 55°C. The first modern comprehensive account of the biology and classification of thermophilic fungi was published by Cooney and Emerson in 1964; it included the 11 species known at that time, with a few new to science. Since then, several thermophilic fungi have been discovered and documented in the literature. Despite their potential use in industrial processes, studies on thermophilic fungi have been neglected until recently. Further, their uncertain taxonomic affiliation puts them in a state of disarray, often leading to misidentification and confusion (Mouchacca, 1997).

Thermophilic fungi are a small assemblage of eukarya that have the unique ability to grow at elevated temperatures extending up to 61°C. During the last five decades, many species of thermophilic fungi sporulating at 45°C have been reported from various habitats and distributed worldwide (Table 1.1). Such fungi have been defined variously by several workers, but the most common and universally adapted definition of thermophilic fungi is that of Cooney and Emerson (1964). Therefore, in dealing with this account, we draw on Cooney and Emerson's definition of thermophilic fungi as those "that have a maximum temperature for growth at or above 50°C and a minimum temperature for growth at or above 20°C." However, thermophily in fungi is not as extreme as in bacteria or archaea, some species of which are able to grow in temperatures up to 113°C in hot springs or hydrothermal vents (Blochl et al., 1997). Moreover, it became important to understand how thermophilic organisms exist and thrive at temperatures that are lethal for most forms of life. It is speculated that thermophily in the Fungi kingdom arose as an adaptation to seasonal changes and high day temperatures, rather than as an adaptation for acquiring new thermal niches (Powell et al., 2012).

Thermophilic fungi have been playing their role in the economy of nature ever since they evolved on this earth. Their importance to man's economy has been realized from their ability to efficiently degrade organic matter, acting as biodeteriorants and natural scavengers; to produce extracellular as well as intracellular enzymes, organic acids, amino acids, antibiotics, phenolic compounds, polysaccharides, and sterols of

TABLE 1.1

Geographical Distribution of Thermophilic Fungi and Their Habitats

Sr. No.	Organisms	Habitat	Distribution	References
Zygomycetes				
1.	*Rhizomucor miehei*	Retting guayule, soil, sand, coal mines, hay, stored barley, compost	United States, India, Ghana, United Kingdom, Saudi Arabia	Cooney and Emerson, 1964
2.	*Rhizomucor nainitalensis*	Decomposed oak log	India	Joshi, 1982
3.	*Rhizomucor pusillus*	Composting and fermenting substrates like compost, municipal wastes, horse dung, composted wheat straw, guayule, hay, seeds of cacao, barley, oat, maize and wheat, groundnuts, pecans, sputum, bird nests, air, soil	United Kingdom, Chad, Czechoslovakia, South Africa, Indonesia, India, Japan, United States, Nigeria, Australia	Schipper, 1978
4.	*Rhizopus microsporus*	Soil, composting wheat straw, coal mine soils, fermenting plant materials, nesting materials, stored grains, air, cow dung	India, Sudan, Tanzania, Malaysia, Australia	–
5.	*Rhizopus rhizopodiformis*	Coal mine soils, nesting material of birds, lung of pullet, stomach of pig, bread, wooden slats, soil, seeds of *Lycopersicon esculentum*, *Cucumis melo*, breeder cow, oil palm effluents	India, United Kingdom, South Africa, China, Ghana, Hong Kong, Indonesia, Malaysia, Japan	Thakre and Johri, 1976
Ascomycetes				
6.	*Canariomyces thermophila*	Soil	Africa	Arx et al., 1988
7.	*Chaetomidium pingtungium*	Sugarcane field	Taiwan	Chen and Chen, 1996
8.	*Chaetomium britannicum*	Mushroom compost, soil	United Kingdom	Ames, 1963

(Continued)

TABLE 1.1 (CONTINUED)
Geographical Distribution of Thermophilic Fungi and Their Habitats

Sr. No.	Organisms	Habitat	Distribution	References
9.	*Chaetomium mesopotamicum*	Date palm plantation	Iraq	Abdullah and Zora, 1993
10.	*Chaetomium senegalensis*	Plant remains, seeds of *Capsicum annuum*, soil, decomposing wheat straw	Senegal, Netherlands, Kuwait, Iran, India	Ames, 1963
11.	*Chaetomium thermophile* var. *coprophile*	Decomposing wheat straw, horse dung, mushroom compost, vegetable detritus, soil	United States, United Kingdom, India, Netherlands, Ghana	Cooney and Emerson, 1964
12.	*Chaetomium thermophile* var. *dissitum*	Nesting materials, decomposing wheat straw, mushroom compost, soil	United States, United Kingdom, India, Netherlands, Ghana	Cooney and Emerson, 1964
13.	*Chaetomium virginicum*	Decomposing leaves	United States	Ames, 1963
14.	*Corynascus sepedonium*	Dung, soil, human skin, pasture soil, hay, coal spoil tips, compost, cellulose material, *Litchi sinensis* leaf, seeds of *Triticum*, *Foeniculum vulgare*, *Carpentaria acuminata*	India, Kenya, United States, Uzbekistan, United Kingdom, Ghana, Egypt, Hungary, Australia, China, Senegal	von Arx, 1975
15.	*Corynascus thermophilus*	Mushroom compost	United States	Lodha, 1978
16.	*Coonemeria aegyptiaca*	Sludge, soil	Egypt, Iraq	Ueda and Udagawa, 1983
17.	*Coonemeria crustacea*	Coal spoil tips, bagasse, soil	United States, United Kingdom, Ghana, Japan, Netherlands, Indonesia	Mouchacca, 1997
18.	*Dactylomyces thermophilus*	Wood and bark of *Pinus*, plant debris	Sweden, Norway, United Kingdom	von Arx, 1975
19.	*Melanocarpus albomyces*	Nesting material of chickens, decomposing wheat straw, soil, grass compost	United States, India, United Kingdom, Saudi Arabia	von Arx, 1975

(Continued)

TABLE 1.1 (CONTINUED)
Geographical Distribution of Thermophilic Fungi and Their Habitats

Sr. No.	Organisms	Habitat	Distribution	References
20.	*Melanocarpus thermophilus*	Forest soil	Iraq	Guarro et al., 1996
21.	*Talaromyces byssochlamydoides*	Soil	Japan, Egypt	Stolk and Samson, 1972
22.	*Talaromyces duponti*	Manure and damp hay, self-heated guayule shrub, leaf litter, soil, cigarette, *Hordeum vulgare*	United States, India, United Kingdom, France, South Africa, Netherlands, Jordan, Nigeria	Griffon and Maublanc, 1911
23.	*Talaromyces emersonii*	Compost, soil, piles of wood chips, riverbanks, grassland, municipal waste, peat, coal spoil tips, sugarcane bagasse, palm oil kernels, Blesbok dung, air, rhizosphere of *Cassia tora, Cassia occidentalis*	Italy, Netherlands, United Kingdom, United States, Sweden, Canada, Japan, India, Nigeria, South Africa, Indonesia	Stolk, 1965
24.	*Talaromyces thermophilus*	Guayule shrub, fermented straw, dung, compost, soils	United States, Netherlands, India, United Kingdom, Japan, Australia, Indonesia	Stolk, 1965
25.	*Thermoascus aurantiacus*	Heated hay, peat, cacao husks, mushroom compost, stored grains, coal mine soils, soil, air, self-heated wood chips, chaff, tobacco, sawdust	Germany, United States, India, Russia, Holland, Netherlands, South Africa, Italy, United Kingdom, Canada, Jordan, Australia, Indonesia, Egypt, Japan	Cooney and Emerson, 1964
26.	*Thielavia australiensis*	Nesting material of mallee fowl	Australia	Tansey and Jack, 1975

(Continued)

TABLE 1.1 (CONTINUED)
Geographical Distribution of Thermophilic Fungi and Their Habitats

Sr. No.	Organisms	Habitat	Distribution	References
27.	*Thielavia minor*	Coal mine soils, *Elaeis guineensis* leaf, groundnut kernels	India, Zaire, Zambia	Malloch and Cain, 1973
28.	*Thielavia terricola*	Soil, cow dung, compost, *Ficus* sp.	United States, China, India, Canada, Kenya, Australia, United Kingdom, Indonesia	Emmons, 1930
Deuteromycetes (Anamorphic Fungi)				
29.	*Acremonium alabamense*	Alluvial soil, needles of *Pinus taeda*	United States, United Kingdom	Morgan-Jones, 1974
30.	*Acremonium thermophilum*	Sugarcane bagasse	Trinidad	Gams and Lacey, 1972
31	*Arthrinium pterospermum*	Moldy hay, soil	United Kingdom, United States	von Arx, 1981
32.	*Chrysosporium tropicum*	Deteriorating woolen fabric, dung, soil, air	New Guinea, India, United States, Canada	Carmichael, 1962
33.	*Malbranchea cinnamomea*	Guayule rets, composting heaps, wheat straw compost, stacked tobacco leaves, soil, peanut kernels, coal spoil tips, feces of Cape sparrow, deer dung, cattle, henhouse litter, snuff, air silage	United States, Germany, United Kingdom, South Africa, Japan, Canada, Netherlands, Australia, Ghana, Egypt, India, Indonesia	van Oorschot and de Hoog, 1984
34.	*Myceliophthora fergusi*	Soil	India	van Oorschot, 1977
35.	*Myceliophthora hinnulea*	Cultivated soil	Japan	Awao and Udagawa, 1983

(Continued)

TABLE 1.1 (CONTINUED)
Geographical Distribution of Thermophilic Fungi and Their Habitats

Sr. No.	Organisms	Habitat	Distribution	References
36.	*Myceliophthora thermophila*	Soil	United States, Canada, India, United Kingdom, Japan, Australia	Basu, 1984
37.	*Papulospora thermophila*	Mushroom compost, soil	Switzerland, India, Japan	Fergus, 1971
38.	*Scytalidium indonesicum*	Soil, *Dipterocarp* forest soil	Indonesia, Java and Sumatra	Hedger et al., 1982
39.	*Scytalidium thermophilum*	Nesting litter of chickens, mushroom compost, soil, horse dung, wood chips	United States, Japan, Indonesia, India, United Kingdom, Netherlands	Narain et al., 1983
40.	*Remersonia thermophila*	Mushroom compost, straw bedding used for pigs, horse dung, soil	United States, United Kingdom, India	Seifert et al., 1997
41.	*Thermomyces ibadanensis*	Oil palm kernel stacks, soil	Nigeria, India	Apinis and Eggins, 1966
42.	*Thermomyces lanuginosus*	Soil, moist oats, cereal grains, coal mine soils, coal spoil tips, mushroom compost, guayule rets, hay, manure, leaf mold peat, garden compost, horse, sheep and pig dung, cow, air, various plant substances	United States, United Kingdom, Nigeria, Ghana, India, Japan, Australia, Indonesia	Mason, 1933

biotechnological importance; and to produce nutritionally enriched feeds and single cell protein (SCP); they are also suitable as agents of bioconversion, for example, their role in the preparation of mushroom compost. The fungal protein, or "mycoprotein," has attracted the attention of food and feed scientists and protein engineers. *Chaetomium cellulolyticum* and *Sprotrichum pulverulentum* are the most widely used organisms for upgrading animal feed and producing SCP from lignocellulosic wastes. Similarly, investigations into the process of composting municipal solid waste with thermophilic fungi, that is, *Chaetomium thermophile*, *Humicola lanuginosa*, *Mucor pusillus*, and *Thermoascus aurantiacus*, have revealed that the resulting compost is richer in N, P, and K. On the industrial front, the use of thermophilic strains can be an effective solution to the maintenance of optimal temperature in industrial fermentation during the entire cultivation period. It is well known that thermophilic activities of microbes are generally associated with protein and enzyme thermostability. The advantages of using thermostable proteins and enzymes for conducting biotechnological processes at high temperature include a reduced risk of contamination with mesophilic microbes, a decrease in the viscosity of the culture medium, an increase in the bioavailability and solubility of organic compounds, and an increase in the diffusion coefficient of substrates and products, resulting in a higher rate of reactions. Further, their involvement in genetic manipulations is a much more recent development. Nevertheless, because of these and many more advantages, thermophilic fungi appear to be suitable candidates in biotechnological applications.

In contrast, thermophilic fungi are also involved in the spoilage of stored agricultural produce. Ever since the dawn of agriculture, man has depended on the storage of agricultural produce for use in time of need. Fungi do not invade agricultural produce before harvest to any appreciable degree. But during storage, thermophilic fungi can cause deterioration of cereal grains, groundnuts, palm kernels, hay, wood chips, baggase, peat, and so forth. During storage, such products undergo a process of heating, which under some conditions may advance to a plateau where spontaneous combustion occurs. A number of thermophilic fungi from stored cereal grains have been reported. The commonly associated thermophilic fungi with stored grains are *Absidia* spp., *Aspergillus* spp., *Mucor pusillus*, *Thermomyces lanuginosus*, and *Thermoascus aurantiacus* (Mehrotra, 1985; Sharma, 1989). These fungi can cause a reduction in the rate of germination, discoloration, damage to the seed, and spoilage due to microbial activity. The thermophilic fungi of stored grains are receiving attention from scientists all over the world because of their toxigenic and pathogenic potential. The implication of thermophilic fungi in the spoilage of groundnuts and palm kernels is attributed to their strong lipolytic activity. A wide range of thermophilous fungi, namely, *Aspergillus fischieri*, *Aspergillus fumigatus*, *Chaetomium globosum*, *Chrysosporium thermophilum*, *Humicola lanuginosa*, *Mucor pusillus*, *Thermoascus aurantiacus*, *Penicillium dupontii*, *Paecilomyces varioti*, *Scopulariopsis fusca*, *Absidia blakesleana*, *Absidia ramosa*, and *Thermomyces ibadanensis*, are reported from moldy groundnuts and palm kernels (Kuku and Adeniji, 1976; Ogundero, 1980). All these species are able to utilize a variety of lipids as their carbon source.

Commercial wood chips stored outside in piles spontaneously generate heat, which causes the deterioration of chips and can lead to spontaneous ignition.

The thermophilic and thermotolerant fungi isolated from wood chip piles include *Chaetomium thermophile* var. *coporophile*, *Chaetomium thermophile* var. *dissitum*, *Humicola grisea* var. *thermoidea*, *Humicola lanuginosa*, *Sporotrichum thermophile*, *Thermomyces emersonii*, *Thermomyces thermophilus*, and *Thermoascus aurantiacus* (Tansey, 1971). Similarly, economic losses from the spoilage of bagasse (fibrous residue of sugarcane after the extraction of juice) and peat due to self-heating and combustion are significant. The presence of residual sugar in bagasse makes it prone to microbial spoilage. During storage, thermophilic microorganisms degrade the polysaccharides, causing spoilage of the cellulosic waste. The inhalation of spores released by the microorganisms can cause bagassosis, a disease of the respiratory tract. The chief fungal species present in stored bagasse are *Absidia corymbifera*, *Chrysosporium keratinophilum*, *Paecilomyces variotii*, and *Phialophora lignicola*, and they are thermotolerant (Sharma, 1989).

Hot springs are supposed to be a potential habitat for thermophilic microorganisms. However, there have been relatively few studies on the isolation of thermophilic fungi from hot springs. Chen et al. (2000) isolated and identified five species—*Aspergillus fumigatus*, *Thermomyces lanuginosus*, *Humicola insolens*, *Penicillium dupontii*, and *Rhizoctonia* sp.—of thermophilic and thermotolerant fungi from hot springs in northern Taiwan. Similarly, several fungi have been isolated from near-neutral and alkaline thermal springs of Tengchong Rehai National Park (Pan et al., 2010). These fungi were identified as *Rhizomucor miehei*, *Chaetomium* sp., *Talaromyces thermophilus*, *Talaromyces byssochlamydoides*, *Thermoascus aurantiacus* var. *levispora*, *Thermomyces lanuginosus*, *Scytalidium thermophilum*, *Malbranchea flava*, *Myceliophthora* sp., and *Coprinopsis* sp. using morphological analysis in conjunction with internal transcribed spacer (ITS) sequencing.

There are also numerous reports of thermophilic fungi as pathogens of humans and other warm-blooded animals (Tansey and Brock, 1978). The thermophilic *Mucor pusillus*, in particular, is a pathogen that causes a variety of mycoses (Meyer and Armstrong, 1973). Thermotolerant fungi, such as *Absidia ramosa* (Nottebrock et al., 1974) and *Aspergillus fumigatus* (Rippon, 1974), are more frequently reported as pathogens. The thermotolerant fungus *Dactylaria gallopava* has been found to be a cause of epidemics in young turkeys and chickens (Blalock et al., 1973; Waldrip et al., 1974).

1.2   DEFINING THERMOPHILY

Temperature is one of the environmental factors that plays an immensely important and often decisive role in the survival, growth, distribution, and diversity of organisms on the earth's surface because it acts directly on the structure and function of biomolecules and on the maintenance of the integrity of cellular structures. Temperature ranges for growth have often been used as a basic limiting factor to classify groups of microorganisms. The most common divisions are the *psychrophiles* (0°C–20°C), *mesophiles* (10°C–40°C), and *thermophiles* (25°C–80°C).

A thermophile is defined as an organism that grows at temperatures that are considered above those normally associated with biological systems. Defining thermophilic fungi has been a subject of confusion and arguments. According to Cooney and Emerson (1964), fungi growing at a maximum temperature of ≥50°C and a minimum temperature of ≥20°C are considered thermophilic, whereas fungi growing well below 20°C and up to about 50°C are regarded as thermotolerant. Other researchers disagree with this definition of thermophilic fungi and have proposed to classify fungi as thermophilic if their optimum temperature for growth is above 40°C (Crisan, 1964) or 45°C (Maheshwari et al., 2000). A definition for other thermophilic organisms cannot hold for thermophilic fungi, as their temperature tolerance differs. Because of this difference in temperature characteristics of different groups of organisms, thermophilic fungi must be defined on a separate basis if it is to be meaningful and useful. Some common terms used with reference to thermophily in fungi are

Thermophilic fungi: Those fungi that have an optimum temperature for growth at or above 45°C (Maheshwari et al., 2000).

Thermotolerant fungi: Those fungi that have a maximum temperature for growth extending up to 50°C but a minimum growth temperature well below 20°C (Mouchacca, 2000).

Thermophilous fungi: All fungi growing at elevated temperatures. This term includes both thermophilic and thermotolerant fungi (Apinis, 1963a).

Theromduric fungi: Those fungi whose reproductive structure can resist a temperature just approaching 80°C or above without loss of viability, but their normal growth temperature ensues at room temperature (22°C–25°C) (Apinis and Pugh, 1967).

Transitional thermophile: Group of fungi that grow below 20°C but can also withstand a temperature just approaching 40°C (Apinis and Pugh, 1967).

Stenothermal: Organisms with a narrow temperature range for survival and growth. They are generally found in habitats of relatively constant temperature (Brock and Fred, 1982).

Eurythermal: Organisms with a wider temperature range for survival and growth and usually found in environments where temperature varies considerably (Brock and Fred, 1982).

Poikilotrophic or poikilophilic: The term *poikilotrophic* or *poikilophilic* is used to describe organisms that can tolerate extreme environmental stress caused by fluctuating conditions.

Poiklothermic: Organisms that can withstand extremes of fluctuating temperatures and thermal conditions.

1.3   HISTORICAL BACKGROUND

The dating of meteorites using radioisotopes has estimated the earth's age to be 4.5 to 4.6 billion years. For the first 100 million years, conditions on earth were far too hot and harsh to sustain any life-form. The first cellular life-forms on earth appeared approximately 3.5 billion years ago. These earliest life-forms were believed to be thermophilic due to the high temperature conditions prevailing in the earth's atmosphere at that time. The first direct evidence of primitive life-forms on the earth was the discovery of a microbial fossil in 1977 from the Swartkoppie chert (granular silica), which dates back to 3.5 billion years ago. However, the first eukaryotic cells are believed to have appeared around 2.5 to 2.0 billion years ago. The multicellular eukaryotic organisms evolved about 1.5 billion years ago. Since then, microbes have evolved and diversified to occupy virtually every habitat on earth and are major contributors in every ecosystem.

The study of thermophilic fungi enjoys a century-old history now. The earliest studies on this group of fungi were descriptive and focused on recording the occurrence of thermophiles from various habitats suitable for the growth of these organisms. The first description of a known thermophilic hyphomycete, *Thermomyces lanuginosus*, incidentally encountered on a potato inoculated with garden soil, is often attributed to Tsiklinskaya (1899). Although Fresenius's (1850–1853) description of *Aspergillus fumigatus* and Lindt's (1886) description of *Rhizomucor pusillus* (as *Mucor pusillus*) were published earlier, it was the work of Tsiklinskaya that drew attention to thermophilism in fungi. Later, in 1907, Miehe focused his studies on the causes of the self-heating and spontaneous combustion of damp haystacks. He was the first person to show the relationship of the growth of thermophilic fungi to the thermogenesis of stored agricultural products, and he isolated four thermophilic fungi: *Mucor pusillus*, *Thermomyces lanuginosus*, *Thermoidium sulfureum*, and *Thermascus aurantiacus*. Further, Miehe (1930a, 1930b) also compared the heating capacities of mesophilic and thermophilic fungi to prove that the final temperature of the packed plant material depends on the maximum temperature of the growth of the kind of microorganism used. Griffon and Maublanc (1911) reported the first thermophilic *Penicillium*, *P. dupontii* (now *Talaromyces thermophilus*). Miehe's pioneering work on the thermal adaptation of microorganisms led Kurt Noack (1920) to isolate thermophilic fungi from several natural substrates. He went on to further observe that thermophilic fungi were also present in habitats where temperature for their growth occurs only occasionally, for example, soils of the temperate zone. Noack's work paved the way for investigations on the physiology of thermophilic fungi. Post–World War II, the need for finding alternative sources of rubber led to studies on the guayule shrub (*Parthenium argentatum*), a rubber-producing plant. It was vehemently shown that the extractability and physical properties of rubber improved when the shrub was chopped and further composted in a heap (Allen and Emerson, 1949). From this compost, they isolated several thermophilic fungi that were able to grow up to 60°C. They further observed that the development of thermophilic microflora in the retting guayule reduced the amount of resin in crude rubber, thereby improving its extractability and physical properties.

Between 1912 and 1950, no new thermophilic fungi were reported. Later, La Touche (1950) reported a new cellulolytic ascomycete, *Chaetomium thermophile*. Substantiated by the cellulolytic nature of the new ascomycete and owing to its industrial application, such discoveries generated much interest in this group of fungi among mycologists. Notable pioneering publications then followed on thermophilic fungi during the sixties, the golden years of thermophilic fungi research. These contributions were from soils of temperate climate (Apinis, 1963a; Eggins and Malik, 1969), tropical areas (Hedger, 1974; Gochenaur, 1975), and arid regions (Mouchacca, 1995). Other habitats rich in organic materials were also extensively investigated for the presence of thermophilic fungi by several workers, and data from relevant publications were critically reviewed by Tansey and Brock (1978). However, the monograph *Thermophilic Fungi: An Account of Their Biology, Activities, and Classification*, by Cooney and Emerson (1964), remained the sole treatise for decades that guided and motivated scientists engaged in research on thermophilic fungi and their possible industrial application. Based on their own cultures isolated from retting guayule and cultures obtained from other investigators, they provided detailed descriptions of the 13 thermophilic fungi known at that time.

1.4   HABITAT RELATIONSHIPS

Diversity is a hallmark of life on earth. Mycologists have so far identified and named about 100,000 species of fungi. Researchers identify hundreds of additional species of fungi every year. Mycologists face a major challenge in attempting to make sense of this large variety. Thermophilic fungi comprise a fraction of this innumerable mycoflora. They have been reported from a large variety of habitats, both natural and man-made (Figure 1.1).

There are several extreme environments found in nature, for example, those with high and low temperature, high salt, high alkali, high acid, drought conditions, high pressure, and high radiation. Such environments are unfavorable for the survival of most organisms; however, some microorganisms can grow and thrive in these

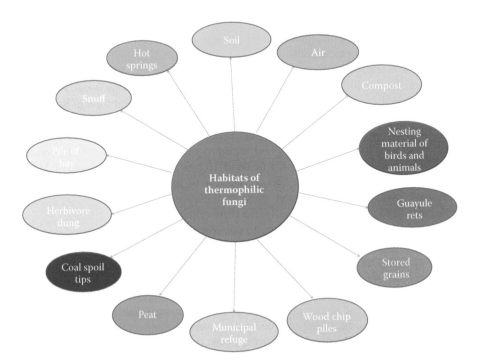

FIGURE 1.1   Some common habitats of thermophilic fungi.

conditions, mainly because of their long-term natural selection. Much is known about the occurrence of thermophilic fungi from various types of soils and in habitats where the decomposition of plant material takes place. These include composts, piles of hay, stored grains, wood chip piles, nesting materials of birds and animals, snuff, and municipal refuse, as well as other accumulations of organic matter wherein the warm, humid, and aerobic environment provides the basic physiological conditions for their growth and development. In these habitats, thermophiles may occur as either resting propagules or active mycelia, depending on the availability of nutrients and favorable environmental conditions. Most of the currently known species of thermophilic fungi have been reported from these habitats from a culture-dependent approach.

Soil, the depository of all life and the laboratory where most of the changes that enable life to continue are carried out, is suggested to be the natural habitat of thermophilic fungi, particularly in the upper layers rich in organic matter. The isolation of thermophilic fungi from soil was reported as early as 1939 by Waksman and his coworkers. Soils in tropical countries do not appear to have a higher population of thermophilic fungi than soils in temperate countries, as believed earlier. Contrary to possible expectations, Ellis and Keane (1981) also observed that thermophilic fungi are more frequent in temperate soils than in tropical soils of Australia. Their widespread occurrence could well be due to the dissemination of propagules from self-heating masses of organic material (Maheshwari et al., 1987). Coal spoil tips were found to be another potential habitat of thermophilic fungi by Evans (1971). He isolated 32 thermophilic and thermotolerant fungi from coal spoil tips and observed that the population of thermophilous fungi in well-colonized warm areas was higher than that of fungi in corresponding nonwarm areas. Similarly, forest soils of pine-scrub oak, loblolly pine, pine hardwood, and floodplain hardwood (Ward and Cowley, 1972) and peanut soils (Taber and Pettit, 1975) were found to be potential sources of thermophilic fungi. Tansey and Brock (1978) have reviewed 34 studies reporting the isolation of thermophilic fungi from soil, including studies of both temperate and tropical soils. They also observed that thermophilic fungi are much more common in acid thermal habitats than in neutral to alkaline pH ones.

The occurrence of thermophilic fungi in aquatic sediment of lakes and rivers, as first reported by Tubaki et al. (1974), is mysterious in view of the low temperature (6°C–7°C) and low level of oxygen (average 10 ppm, <1.0 ppm at a depth of 31 m) available at the bottom of the lake. A number of thermophilic fungi survive the stresses, such as increased water pressure, absence of oxygen, and desiccation (Mahajan et al., 1986). Undoubtedly, the thermophilic fungi owe their ubiquity and common occurrence in large measure to this special ability to occupy a temperature niche that most other fungi cannot inhabit. The temperature, humidity, and atmosphere of these environments are favorable attributes for fungal development (Salar and Aneja, 2007). More attempts are needed, however, to provide evidence for active involvement in the substrate from which thermophilic fungi are being reported.

The above account shows that a number of habitats are suitable for the growth of thermophilic fungi, including soil, composts, stored grains, coal mine soils, nesting material of birds, effluents from thermal plants, wood and pulp, and other similar habitats where thermogenic activity occurs. However, exclusive habitats of thermophilic fungi are rare, and these are present along with their mesophilic counterpart, as both enjoy overlapping temperature spectra. It is also known that the minimum temperature for the growth of majority of thermophilic fungi is 30°C or lower which is also the upper or higher limits for many mesophilic fungi. Therefore, one may ask why mesophilic versus thermophilic when the temperature regimes for the two groups are distinctly different. In spite of their wide occurrence, these fungi are not active in certain environments. It is believed that this is because they occur only as propagules and are generally transported by air, mainly from composting systems (Le Goff et al., 2010). These and many more questions remain to be answered.

1.5   ISOLATION AND CULTURE OF THERMOPHILIC FUNGI

Earlier, the microorganisms including thermophilic fungi were isolated using culture-dependent approaches, which were shown to provide inaccurate representations of microbial populations from a particular niche or substrate. In the culture-dependent approach, readily cultured microorganisms often grow quickly, but these may not be the dominant ones within their environment. This is often analogized as growing "weeds" instead of the "desired flowers." In recent times, the methods used to isolate microorganisms from their natural habitats have been revolutionized with the onset of metagenomics and next-generation sequencing technologies. The use of culture-independent 16S rRNA gene library analysis for microbial ecology is a step forward to know the sequences of more than 99% of microorganisms, as the majority of these resist cultivation using traditional approaches. This is particularly important when thermostable enzymes, also called thermozymes, are required for their use in bio-technology industry, mainly because of their high thermal stability (resistance to inactivation at high temperatures) and thermofilia (optimum activity at high temperatures). Morgenstern et al. (2012) studied sequences from 86 fungal genomes and from two out-group genomes, *Arabidopsis thaliana* and *Drosophila melanogaster*, to construct a robust molecular phylogeny of thermophilic fungi. To understand the phylogenetic relationships in the thermophilic-rich orders Sordariales and Eurotiales, they used nucleotide sequences from the nuclear ribosomal small subunit (SSU), the 5.8S gene with ITS1 and ITS2, and the ribosomal large subunit (LSU) to include additional species for analysis. These phylogenetic studies were helpful in clarifying the position of several thermophilic taxa. However, for eukaryotic microorganisms, as well as for thermophilic fungi, the use of these approaches is limited. Nevertheless, they display information about the presence of major microbial populations in a particular environment, yet they provide limited benefit, as these methods present no information regarding the physiology and functions of uncultured microorganisms. Thus, it seems that isolating thermophilic fungi in pure culture will remain a cornerstone of thermophilic fungal research. Until recently, mycologists used traditional culture-dependent methods for the isolation and investigation of the diversity of thermophilic fungi.

In their natural habitats, thermophilic fungi usually grow in complex, mixed populations with many bacterial and fungal species. This presents a problem for mycologists, as a pure culture, a population of cells arising from a single cell, is invariably needed to characterize an individual species. Most of the thermophilic fungi grow in simple media containing carbon and nitrogen sources and a few mineral salts, suggesting that they have very simple nutritional requirements. Simple nitrogen sources, like nitrates of sodium and potassium, asparagine, and yeast extract, support good growth of these fungi (Satyanarayana and Johri, 1984). Pure cultures of thermophilic fungi from different sources or substrates have been isolated in several ways, and a few of the more common techniques are reviewed below.

1.5.1 ISOLATION TECHNIQUES

1.5.1.1 Dilution Plate Technique (Apinis, 1963b)

This is the most common method for isolation of thermophilic fungi. In this method, 10 g of sample is transferred to a 250 mL conical flask containing 100 mL of sterile distilled water. The contents are shaken on a magnetic or mechanical shaker for 15–20 min and then serially diluted to obtain 10^{-1} to 10^{-5} dilutions. From each dilution, 1.0 mL of sample is transferred to a sterile petri dish and cooled melted yeast extract starch agar medium is poured over it. The plate is moved gently on the table in a figure-of-eight motion to effect proper dispersion. Alternatively, the solution can be put on the surface of solidified medium and spread evenly throughout (spread plate). The plates are incubated at 45°C in a humidified biochemical oxygen demand (BOD) incubator for 2–5 days. The resulting colonies are counted and colony-forming units (CFU) are calculated. The accuracy of this technique is low when only one plate is counted. There are several contributing factors, including improper dispersion of spores during dilution, failure to break up spore masses, and mutual inhibition of growth by certain fungi. Greater accuracy is attained by doing several plates (10 or more) at the most desirable dilution.

1.5.1.2 Warcup's Soil Plate Method (Warcup, 1950)

In 1950, Warcup introduced a simple plating technique, whereby soil is distributed throughout a thin layer of nutrient medium. This method was devised because Warcup observed that in the preparation of dilution plates, many fungi are discarded with the residue, whereas the use of soil plates would permit the growth of fungi embedded in humus or attached to mineral particles. In this method, the soil plate is prepared by transferring a small amount of soil sample into a sterilized petri plate. About 15 mL of cooled medium (40°C) is added to the plate and the soil particles are dispersed throughout the medium. Adequate dispersal may be obtained by gently shaking and rotating the plate before the agar solidifies. About 0.005–0.015 g of soil has been found to give a convenient number of colonies on each plate. Czapek–Dox supplemented with 0.5% yeast extract agar has been found to be a satisfactory medium for the growth and sporulation of several soil fungi and is extensively used as an isolation medium. The plates are incubated as above.

1.5.1.3 Humidified Chamber Technique (Buxton and Mellanby, 1964)

This method is employed especially for isolation of thermophiles from the nesting material of birds. Collected bird nesting materials are taken and directly placed in a glass chamber previously arranged with sterile wet filter paper and a sterile glass slide on top of it. The nest materials are directly placed on the sterile glass slide. The internal temperature of the chamber is regularly maintained at 40°C–45°C. Fungus that grows on the nest materials is collected and transferred to sterile yeast extract starch agar plates and checked for its thermophilic nature.

1.5.1.4 Paired Petri Plate Technique (Cooney and Emerson, 1964)

This method is designed to provide moisture and a suitable environment for the isolation and growth of thermophilic fungi. In this method, using paired plates, the upper plate is fixed with sterile filter paper and the paired plates are sealed with cellophane tape to prevent moisture loss. This method gives very good results, allowing the maximum amount of thermophiles to be isolated.

1.5.1.5 Waksman's Direct Inoculation Method (Waksman et al., 1939)

It is convenient to place materials of interest directly on a nutrient agar medium. This technique encourages the rapid spreading of fungi at the expense of other molds. In this method, 20 mL of modified Emerson's (YpSs) medium is poured into a sterile petri plate and allowed to solidify. Small quantities of the samples are sprinkled over the medium in the dishes, and such plates are incubated in an inverted position at 47°C ± 2°C in a high-humidity incubation chamber. After a few days of incubation, mold colonies appear on the surface and can be transferred into pure culture. This method is commonly used in soil studies, requiring only a pinch of soil, evenly dispersed over the surface of the agar.

1.5.1.6 Isolation from Air

The isolation of thermophilic fungi requires an efficient method to remove the organism or its spores from the bioaerosols suspended in air. Simply exposing the agar plates to air and then incubating them at the desired temperature allowed isolation of air mycoflora. However, more precise isolation of fungal spores from the air is performed by using impactor samplers, such as the Andersen air sampler, which is 1.5 m in height. Three types of forced-air sampling instruments are generally used: impactor sampler, impinger sampler, and filteration sampler. Impingement is generally not recommended, as some fungal spores are hydrophobic and may not be retained in the collection fluid. Filter sampling is followed by elution of spores from the filter material. It allows dilution of the sample in locations where high airborne spore concentrations are present and permits extended sampling periods. Exposed or treated agar plates are incubated at 45°C ± 2°C in a high-humidity incubation chamber.

1.5.2 Culture Media for Isolation of Thermophilic Fungi

Thermophilic fungi can be isolated and cultured on a variety of standard mycological media (Table 1.2). The maintenance of adequate moisture in all cultures at a higher incubation temperature is a great problem. The elevated temperature results in the desiccation of the agar in petri plates. Placing the agar plates in humidified containers with sterilized water solves this problem. The media shown in Table 1.2 are the most widely used, even today (Cooney and Emerson, 1964).

TABLE 1.2
Some Common Culture Media to Grow and Maintain Thermophilic Fungi

Frequently Used Culture Media

1. Yeast extract starch agar medium

Yeast extract, Difco-powdered	4.0 g
K_2HPO_4	1.0 g
$MgSO_4 \cdot 7H_2O$	0.5 g
Soluble starch	15.0 g
Agar	20.0 g
Rose bengal	0.0001 g
Water (¼ tap, ¾ distilled)	1000.0 mL
Streptopenicillin	(30 units/mL)

(The pH of the medium is adjusted to 6.2 with 0.1 N HCl/NaOH.)

2. Yeast glucose agar

Yeast extract, Difco-powdered	5.0 g
Glucose	10.0 g
Agar	20.0 g
Tap water	1000.0 mL

3. Czapek–Dox agar medium

$NaNO_3$	3.0 g
$MgSO_4 \cdot 7H_2O$	0.05 g
KCl	0.5 g
$FeSO_4$	0.01 g
Sucrose	30.0 g
Agar-agar	15.0 g
Water (¼ tap, ¾ distilled)	1000.0 mL

4. Dextrose-peptone-yeast extract agar medium

Glucose	1.0 g
Peptone	2.0 g
Yeast extract	0.3 g
KH_2PO_4	0.2 g
$MgSO_4 \cdot 7H_2O$	0.02 g
Soluble starch	15.0 g
Agar-agar	20.0 g
Water (¼ tap, ¾ distilled)	1000.0 mL

5. Modified Czapek–Dox medium

Glucose	2.0 g
L-Asparagine	10.0 g
KH_2PO_4	1.52 g

(Continued)

TABLE 1.2 (CONTINUED)
Some Common Culture Media to Grow and Maintain Thermophilic Fungi

Frequently Used Culture Media

KCl	0.52 g
$MgSO_4 \cdot 7H_2O$	0.52 g
$CuNO_3 \cdot 3H_2O$	Trace
$ZnSO_4 \cdot 7H_2O$	Trace
$FeSO_4 \cdot 7H_2O$	Trace
Agar-agar	20.0 g
Water (¼ tap, ¾ distilled)	1000.0 mL

(The pH of medium is adjusted to 6.2 with 0.1 N HCl/NaOH.)

6. Modified Czapek–Dox medium (Ogundero, 1980)

Malt extract	20.0 g
Yeast extract	2.5 g
$NaNO_3$	0.2 g
KCl	0.05 g
Sodium glycerophosphate	0.05 g
K_2SO_4	0.03 g
$FeSO_4 \cdot 7H_2O$	0.01 g
Agar-agar	15.0 g
Distilled water	1000 mL

(Streptomycin sulfate 40 μg/mL is added to the medium)

Occasionally Used Culture Media

7. Sabouraud dextrose agar medium (Sabouraud, 1892)

Dextrose	40.0 g
Peptone	10.0 g
Agar-agar	20.0 g
Water	1000 mL
pH	5.6

8. Malt extract agar (Reddish, 1919)

Malt extract	30.0 g
Peptone	5.0 g
Agar-agar	15.0 g
Water	1000 mL
pH	5.4

9. Martin's rose bengal agar medium (Martin, 1950)

Glucose	10.0 g
Peptone	5.0 g
KH_2PO_4	1.0 g

(Continued)

TABLE 1.2 (CONTINUED)

Some Common Culture Media to Grow and Maintain Thermophilic Fungi

Occasionally Used Culture Media

$MgSO_4 \cdot 7H_2O$	0.5 g
Rose bengal	33.0 mg
Streptomycin sulfate (10% sol.)	3.0 mL
Agar-agar	15.0 g
Distilled water	1000 mL

Note: For primary isolations, rose bengal (50 mg/L) and streptopenicillin-streptomycin (30 units/mL) are added to all the media to inhibit bacterial growth.

1.6 BIOTECHNOLOGICAL SIGNIFICANCE

In view of their ability to produce thermostable enzymes, thermophilic fungal organisms are gaining importance in biotechnological industries requiring process operation at elevated temperatures. One of the major areas where thermophilic fungal strains have been investigated world over is enzyme technology and protein engineering. It is now well recognized that thermophilic activities are in general associated with protein thermostability. Accordingly, proteins produced by thermophiles tend to be more thermostable than their mesophilic counterparts.

One of the technical problems in industrial fermentation is the maintenance of temperature at optimal levels during the entire cultivation period. The use of thermophilic strains can be an effective solution to this problem. Their biotechnological potential in man's economy has been realized from their ability to efficiently degrade organic matter, acting as biodeteriorants and natural scavengers, to produce extracellular as well as intracellular enzymes, amino acids, antibiotics, phenolic compounds, polysaccharides and sterols, and nutritionally enriched feeds and SCP, and their suitability as agents of bioconversion, for example, their role in the preparation of mushroom compost. Their involvement in genetic manipulations is a much more recent development. Nevertheless, studies that have been done (Cooney and Emerson, 1964; Emerson, 1968; Tansey and Brock, 1978; Mehrotra, 1985; Sharma, 1989; Satyanarayana et al., 1992; Sharma and Johri, 1992; Singh and Satyanarayana, 2014, 2016) indicate that thermophilic fungi appear to be nature-borne biotechnologists. The advantages of using thermostable enzymes for conducting biotechnological processes at high temperatures include a reduced risk of contamination by mesophilic microorganisms, a decrease in the viscosity of the reaction medium, an increase in the bioavailability and solubility of organic compounds, and an increase in the diffusion coefficient of substrates and products, resulting in higher reaction rates (Singh et al., 2016). Cost-effectiveness would be enhanced as a result of longer lifetimes of microorganisms and/or enzymes, particularly for immobilized systems. A number of extracellular enzymes, such as cellulases, xylanases, amylases, lipases, proteases, and pectinases, have been applied in biotechnological processes, such as biobleaching of paper and pulp; production of animal feed; SCP; production of fermentable sugars for obtaining biofuel from cellulosic wastes; fruit juice extraction and clarification; refinement of vegetable fibers; degumming of natural fibers; curing of coffee, cocoa, and tobacco; and wastewater treatment (Singh et al., 2016).

A number of thermophilic fungi have been reported to produce bioactive compounds, such as antibiotics that have antibacterial and antifungal properties. The antibacterial antibiotics include penicillin G, malbranchins A and B, 6-aminopenicillanic acid, sillucin, miehein, and vioxanthin, which are active against both Gram-positive and Gram-negative bacteria. The antifungal antibiotics include myriocin from *Myriococcum albomyces* and thermozymocidin from *Thermoascus aurantiacus* (Mehrotra, 1985; Satyanarayana et al., 1992).

Thermophilic species of *Mucor*, *Talaromyces*, and others have been reported to produce organic acids (lactic and citric acids) and extracellular phenolic compounds. Some thermophilic fungi are known to excrete various amino acids like alanine, glutamic acid, and lysine (Subrahmanyam, 1985).

Lignocellulosic materials are the earth's most abundant renewable organic carbon source and are composed of 65%–72% utilizable polysaccharides (hemicelluloses and cellulose) and 10%–30% lignin. The importance of cellulolytic thermophiles over their mesophilic counterparts is realized due to their higher rate of cellulose breakdown, good sources of proteins, and higher specific growth rates and activity over a wide range of temperatures 20°C–50°C (Seal and Eggins, 1976). *Chaetomium cellulolyticum* and *Sporotrichum pulverulentum* are the most widely used organisms for upgrading animal feed and producing SCP from lignocellulosic wastes (Thomke et al., 1980). Thermophilic *Aspergillus fumigatus*, *Sporotrichum thermophile*, and *Thermoascus aurantiacus* produced increased (twofold) protein contents on sugar beet pulp and molasses solution (Sundman et al., 1981; Grajek, 1988). Ghai et al. (1979) used sag waste from the canning industry as a substrate for the production of SCP using the thermotolerant fungi *Chaetomium cellulolyticum* and *Actinomucor* sp.

The disposal of municipal wastes and refuse has presented a serious problem in big cities, and the odors that characterize the outskirts of such big cities confirm the severity of the problem and the health hazard. Studies carried out on humification and composting processes indicate that thermophilic fungi are quite active and constitute a particularly effective component of the total microbial flora (Mehrotra, 1985). Municipal refuse decomposed by certain thermophilic fungi *Humicola* sp., *Mucor* sp., and *Sporotrichum* sp.—have been found to be richer in N, P, and K (Sen et al., 1979). The process of composting municipal solid waste with thermophilic fungi, that is, *Chaetomium thermophile*, *Humicola lanuginosa*, *Mucor pusillus*, *Thermoascus aurantiacus*, and *Torula thermophila*, has been investigated by Kane and Mullins (1973).

Another major area where thermophilic fungi have been found to play an important role is mushroom composting. *Rhizomucor* spp. and *Aspergillus fumigatus*, which have pH optima near 7 and temperature optima near 40°C, are the typical pioneers of mushroom compost. When self-heating and ammonification start and the pH reaches near 9, the pioneers are replaced by species like *Talaromyces thermophilus* and *Thermomyces lanuginosus*, which have relatively high pH and temperature optima. However, they have a moderate growth rate and do not degrade cellulose.

Torula thermophila (= *Scytalidium thermophilum*) and *Chaetomium thermophile* grow fast and degrade cellulose strongly. Thermophilic fungi arc believed to contribute significantly to the quality of the compost (Ross and Harris, 1983; Straatsma et al., 1989). *Scytalidium thermophilum* has been used to prepare a compost of constant high quality that does not emit ammonia and odor into the environment (Straatsma et al., 1994a, 1994b). The effects of these fungi on the growth of mushroom mycelium and mushroom yield have been described at three distinct levels: (1) they decrease the concentration of ammonia in the compost, which otherwise would counteract the growth of the mushroom mycelium; (2) they immobilize nutrients in a form that apparently is available to the mushroom mycelium; and (3) they may have a growth-promoting effect on the mushroom mycelium, as has been demonstrated for *Scytalidium thermophilum* and for several other thermophilic fungi (Straatsma et al., 1994b). The effectiveness of *Scytalidium thermophilum* in compost preparation for *Agaricus bisporus* has been shown by Straatsma et al. (1994a), who obtained a twofold increase in the yield of mushrooms on inoculated compost compared with that on the pasteurized control. Salar and Aneja (2006) have obtained similar results with *Torula thermophila* and *Malbranchea sulfurea* when used singly and in dual cultures in phase II mushroom compost.

Recent technological advances in genetic manipulation of thermophilic fungi by induced mutants using both chemical and physical mutagens have accelerated the production of enzymes like cellulases by these fungi. In view of their commercial significance, the cloning of genes has been carried out to understand the structure–function relationship of various enzymes of thermophilic fungal strains. In order to enhance cellulase production, Morrison et al. (1987) isolated a mutant of *Talaromyces emersonii* CBS 814•70 named UV7 by ultraviolet (UV) irradiation; this mutant was capable of increased cellulase production during growth on cellulose lactose and glucose-containing media. Moloney et al. (1983) obtained a morphological mutant UCG 42 by treating the strain with nitrosoguanidine. This mutant had an improved filter paper hydrolyzing activity. This is an instance of the applicability of thermophilic fungi in experimental systems for genetic manipulations to achieve the desired objectives.

Fahnrich and Irrgang (1981) obtained a hybrid 7S/7 from the progeny of a cross between two morphological and physiological variants of *Chaetomium cellulolyticum* (i.e., variants 4S and 7S). The hybrid strain 7S/7 was capable of producing more CMCase than either of the parent strains. The possibility of genetic manipulation for achieving genetic recombination following protoplast fusion has been reported by Gupta and Gautam (1995). Protoplasts of a catabolite repression-resistant strain of *Malbranchea sulfurea* and a mutant of it overproducing amylase were isolated and fused via electrofusion. The yield of the hybrids was 5×10^{-5}. One stable hybrid, DGCS 1, was insensitive to glucose repression and produced approximately twice the α-amylase activity as either of its parents. The amount of dinitrosalicylic acid (DNS) in DGCS 1 was also twice that of either parent strain. Similarly, improved xylanolytic activity of *Melanocarpus albomyces* IIS-68 has been reported through mutagenization of the protoplast (Jethro et al., 1993).

REFERENCES

Abdullah, S.K., and Zora, S.E. 1993. *Chaetomium mesopotamicum*, a new thermophilic species from Iraqi soil. *Cryptogamic Botany* 3:387–389.

Allen, P.J., and Emerson, R. 1949. Guayule rubber, microbiological improvement by shrub retting. *Industrial & Engineering Chemistry* 41:346–365.

Ames, L. 1963. *A Monograph of the Chaetomiaceae*. U.S. Army Research and Development Series 2, 1–125. Lehre, Germany: Bibliotheca Mycologica Cramer.

Apinis, A.E. 1963a. Occurrence of thermophilous microfungi in certain alluvial soils near Nottingham. *Nova Hedwigia* 5:57–78.

Apinis, A.E. 1963b. Thermophilic fungi of coastal grasslands. In *Soil Organisms: Proceedings of the Colloquium on Soil Fauna, Soil Microflora and Their Relationship*, ed. J. Doeksen and J. Vander Drift, 427–438. Amsterdam: North Holland.

Apinis, A.E., and Eggins, H.O.W. 1966. *Thermomyces ibadanensis* sp. nov. from oil palm kernel stacks in Nigeria. *Transactions of the British Mycological Society* 49: 629–632.

Apinis, A.E., and Pugh, G.J.F. 1967. Thermophilous fungi of birds' nests. *Mycopathologia et Mycologia Applicata* 33:1–9.

Arx, J.A. von., Figueras, M.J., and Guarro, J. 1988. Sordariaceous ascomycetes without ascospore ejaculation. *Beihefte zur nova hedwigia* 94:1–104.

Awao, T., and Udagawa, S.I. 1983. A new thermophilic species of *Myceliophthora*. *Mycotaxon* 16:436–440.

Basu, M. 1984. *Myceliophthora indica* is a new thermophilic species from India. *Nova Hedwigia* 40:85–90.

Blalock, H.G., Georg, L.K., and Derieux, W.T. 1973. Encephalitis in turkey poults due to *Dactylaria (Diplorhinotrichum) gallopava*: A case report and its experimental reproduction. *Avian Diseases* 17:197–204.

Blochl, E., Rachel, R., Burggraf, S., Hafenbradl, D., Jannasch, H.W., and Stetter, K.O. 1997. *Pyrolobus fumarii*, gen. and sp. nov., represents a novel group of archaea, extending the upper temperature limit for life to 113°C. *Extremophiles* 1:14–21.

Brock, T.D., and Fred, E.B. 1982. *Biology of Microorganisms*. Englewood Cliffs, NJ: Prentice-Hall.

Buxton, P.A., and Mellanby, K. 1934. The measurement and control of humidity. *Bulletin of Entomological Research* 25:171–175.

Carmichael, J.W. 1962. *Chrysosporium* and some other aleurosporic hyphomycetes. *Canadian Journal of Botany* 40:1137–1174.

Chen, K.Y., and Chen, Z.C. 1996. A new species of *Thermoascus* with a *Paecilomyces* anamorph and other thermophilic species from Taiwan. *Mycotaxon* 50:225–240.

Chen, M.-Y., Chen, Z.-C., Chen, K.-Y., and Tsay, S.-S. 2000. Fungal flora of hot springs of Taiwan (1): Wu-Rai. *Taiwania* 45:207–216.

Cooney, D.G., and Emerson, R. 1964. *Thermophilic Fungi: An Account of Their Biology, Activities, and Classification*. San Francisco: W.H. Freeman and Co.

Crisan, E.V. 1964. Isolation and culture of thermophilic fungi. *Contribution from Boyce Thompson Institute Plant Research* 22:291–301.

Eggins, H.O., and Malik, K.A. 1969. The occurrence of thermophilic cellulolytic fungi in a pasture land soil. *Antonie van Leeuwenhoek* 35:178–184.

Ellis, D.H., and Keane, P.J. 1981. Thermophilic fungi isolated from some Australian soils. *Australian Journal of Botany* 29:689–704.

Emerson, R. 1968. Thermophiles. In *The Fungi*, ed. G.C. Ainsworth and A.S. Sussman, 105–128. New York: Academic Press.

Emmons, C.W. 1930. *Coniothyrium terricola* proves to be a species of *Thielavia*. *Bulletin of the Torrey Botanical Club* 57:123–126.

Evans, H.C. 1971. Thermophilous fungi of coal spoil tips. II. Occurrence, distribution and temperature relationships. *Transactions of the British Mycological Society* 57:255–266.

Fahnrich, P., and Irrgang, K. 1981. Cellulase and protein production by *Chaetomium cellulolyticum* strains grown on cellulosic substrates. *Biotechnology Letters* 3:201–206.

Fergus, C.L. 1971. The temperature relationships and thermal resistance of a new thermophile *Papulaspora* from mushroom compost. *Mycologia* 63:426–431.

Gams, W., and Lacey. 1972. *Acremonium thermophilum*. *Transactions of the British Mycological Society* 59:520.

Ghai, S.K., Kahlon, S.S., and Chahal, D.S. 1979. Single cell protein from canning industry waste: Sag waste as substrate for thermotolerant fungi. *Indian Journal of Experimental Biology* 17:789–791.

Gochenaur, S.E. 1975. Distributional patterns of mesophilous and thermophilous microfungi in two Bahamian soils. *Mycopathologia* 57:155–164.

Grajek, W. 1988. Production of protein by thermophilic fungi from sugar-beet pulp in solid-state fermentation. *Biotechnology and Bioengineering* 32:255–260.

Griffon, E., and Maublanc, A. 1911. Deux moisissures thermophiles. *Bulletin of the Society of Mycology, France* 27:68–74.

Guarro, J., Abdullah, S.K., Al-Bader, S.M., Figueras, M.J., and Gene, J. 1996. The genus *Melanocarpus*. *Mycological Research* 100:75–78.

Gupta, A.K., and Gautam, S.P. 1995. Improved production of extracellular α-amylase, by the thermophilic fungus *Malbranchea sulfurea*, following protoplast fusion. *World Journal of Microbiology and Biotechnology* 11:193–195.

Hedger, J.H. 1974. The ecology of thermophilic fungi in Indonesia. In *Biodegradation et humification*, Kilberts *et al.* (eds.), 59–65. France: Sarreguemines.

Hedger, J.N., Samson, R.A., and Basuki, T. 1982. *Scytalidium indonesicum*. *Transactions of the British Mycological Society* 78:365.

Jethro, J., Ganesh, R., Goel, R., and Johri, B.N. 1993. Improvement of xylanase in Melanocarpus albomyces IIS-68 through protoplast fusion and enzyme immobilization. *Journal of Microbiology and Biotechnology* 8:17–28.

Joshi, M.C. 1982. A new species of *Rhizomucor*. *Sydowia* 35:100–103.

Kane, B.E., and Mullins, J.T. 1973. Thermophilic fungi in a municipal waste compost system. *Mycologia* 65:1087–1100.

Kuku, F.O., and Adeniji, M.O. 1976. The effect of moulds on the quality of Nigerian palmkernels. *International Biodeterioration & Biodegradation* 12:37–41.

La Touche, C.J. 1950. On a thermophilic species of *Chaetomium*. *Transactions of the British mycological Society* 33:94–104.

Le Goff, O., Bru-Adan, V., Bacheley, H., Godon, J.J., and Wéry, N. 2010. The microbial signature of aerosols produced during the thermophilic phase of composting. *Journal of Applied Microbiology* 108:325–340.

Lindt, W. 1886. Mitteilungen ber einige neue pathogene Schimmelpilze. *Archiv for Experimentelle Pathologie und Pharmakologie* 21:269–298.

Lodha, B.C. 1978. Generic concepts in some ascomycetes occurring on dung. *In Proceedings of the International Symposium on Taxonomy of Fungi*, Madras, India, 1973, 241–257.

Madigan, M.T., and Orient, A. 1999. Thermophilic and halophilic extremophiles. *Current Opinion in Microbiology* 2:265–269.

Mahajan, M.K., Johri, B.N., and Gupta, R.K. 1986. Influence of desiccation stress in a xerophilic thermophile *Humicola* sp. *Current Science* 55:928–930.

Maheshwari, R., Bharadwaj, G., and Bhat, M.K. 2000. Thermophilic fungi: Their physiology and enzymes. *Microbiology and Molecular Biology Review* 64:461–488.

Maheshwari, R., Kamalam, P.T., and Balasubramanyam, P.V. 1987. The biogeography of thermophilic fungi. *Current Science* 56:151–155.

Malloch, D., and Cain, R.F. 1973. The genus *Thielavia*. *Mycologia* 65:1055–1077.

Martin, J.P. 1950. Use of acid, rose bengal, and streptomycin in the plate method for estimating soil fungi. *Soil Science* 69:215–232.

Mason, E.W. 1933. Annotated account of fungi received at the Imperial Mycological Institute. List II (Fascicle 2). Kew, UK: Imperial Mycological Institute.

Mehrotra, B.S. 1985. Thermophilic fungi—Biological enigma and tools for the biotechnologist and biologist. *Indian Phytopatholoy* 38:211–229.

Meyer, R.D., and Armstrong, D. 1973. Mucormycosis-changing status. *Critical Reviews in Clinical Laboratory Sciences* 4:421–451.

Miehe, H. 1907. *Die Selbsterhitzung des Heus. Eine biologische Studie*. Jena, Germany: Gustav Fischer Verlag.

Miehe, H. 1930a. Die Warmebildung von Reinkulturen im Hinblick auf die Atiologie der Selbsterhitzung pflanzlicher Stoffe. *Archives of Microbiology* 1:78–118.

Miehe, H. 1930b. Uber die Selbsterhitzung des Heues. *Arbeiten der Deutsche Landwirtschafts-Gesellschaft, Berlin* 111:76–91.

Moloney, A.P., Hackett, T.J., Considine, P.J., and Coughlan, M.P. 1983. Isolation of mutants of *Talaromyces emersonii* CBS 814.70 with enhanced cellulase activity. *Enzyme and Microbial Technology* 5:260–264.

Mora, C., Tittensor, D.P., Adl, S., Simpson, A.G.B., and Worm, B. 2011. How many species are there on earth and in the ocean? *PLoS Biology* 9:e1001127.

Morgan-Jones, G. 1974. Notes on Hyphomycetes. V. A new thermophilic species of *Acremonium*. *Canadian Journal of Botany* 52:429–431.

Morgenstern, I., Powlowski, J., Ishmael, N., Darmond, C., Marqueteau, S., Moisan, M., Quenneville, G., and Tsang, A. 2012. A molecular phylogeny of thermophilic fungi. *Fungal Biology* 116:489–502.

Morrison, J., McCarthy, U., and McHale, A.P. 1987. Cellulase production by *Talaromyces emersonii* CBS 814.70 and a mutant UV7 during growth on cellulose, lactose and glucose containing media. *Enzyme and Microbial Technology* 9:422–425.

Mouchacca, J. 1995. Thermophilic fungi in desert soils: A neglected extreme environment. In *Proceedings of the IUBS-IUMS Workshop*, Egham, UK, August 10–13, 1993, ed. D. Allsopp, R.R. Colwell, and D.L. Hawksworth, 265–288. Wallingford, UK: CAB International.

Mouchacca, J. 1997. Thermophilic fungi: Biodiversity and taxonomic status. *Cryptogamie Mycologie* 18:19–69.

Mouchacca, J. 2000. Thermophilic fungi and applied research: A synopsis of name changes and synonymies. *World Journal of Microbiology and Biotechnology* 16:881–888.

Narain, R., Srivastava, R.B., and Mehrotra, B.S. 1983. A new thermophilic species of *Scytalidium* from India. *Zentralblatt für Mikrobiologie* 138:569–572.

Noack, K. 1920. Der Betriebstoffwechsel der thermophilen Pilze. *Jahrbücher für Wissenschaftliche Botanik* 59:593–648.

Nottebrock, H., Scholer, H.J., and Wali, M. 1974. Taxonomy and identification of mucormycosis-causing fungi I. Synonymity of *Absidia ramosa* with *A. corymbifera*. *Sabouraudia* 12:64–74.

Ogundero, V.W. 1980. Lipase activities of thermophilic fungi from mouldy groundnuts in Nigeria. *Mycologia* 72:118–126.

Pan, W.-Z., Huang, X.-W., Wei, K.-B., Zhang, C.-M., Yang, D.-M., Ding, J.-M., and Zhang, K.Q. 2010. Diversity of thermophilic fungi in Tengchong Rehai National Park revealed by ITS nucleotide sequence analyses. *Journal of Microbiology* 48:146–152.

Powell, A.J., Parchert, K.J., Bustamante, J.M., Ricken, J.B., Hutchinson, M.I., and Natvig, D.O. 2012. Thermophilic fungi in an aridland ecosystem. *Mycologia* 104:813–825.

Reddish, A. 1919. *Abstract Bacteriology* 3:6.

Rippon, J.W. 1974. *Medical Mycology*. Philadelphia: W.B. Saunders.

Ross, R.C., and Harris, P.J. 1983. The significance of thermophilic fungi in mushroom compost preparation. *Scientia Horticulturae* 20:61–70.

Sabouraud, R. 1892. *Annales de Dermatologie et de Syphiligraphie* 3:1061.

Salar, R.K., and Aneja, K.R. 2006. Thermophilous fungi from temperate soils of northern India. *Journal of Agricultural Technology* 2:49–58.

Salar, R.K., and Aneja, K.R. 2007. Thermophilic fungi: Taxonomy and biogeography. *Journal of Agricultural Technology* 3:77–107.

Satyanarayana, T., and Johri, B.N. 1984. Thermophilic fungi of paddy straw compost: Their growth, nutrition and temperature relationships. *Journal of the Indian Botanical Society* 63:164–170.

Satyanarayana, T., Johri, B.N., and Klein, J. 1992. Biotechnological potential of thermophilic fungi. In *Handbook of Applied Mycology*, ed. D.K. Arora, R.P. Elander, and K.G. Mukerji, 729–761. New York: Marcel Dekker.

Schipper, M.A.A. 1978. On the genera *Rhizomucor* and *Parasitella*. *Studies in Mycology* 17:53–71.

Seal, K.J., and Eggins, H.O.W. 1976. The upgrading of agricultural wastes by thermophilic fungi. In *Food from Wastes*, ed. G.G. Birch, K.J. Parker, and J.T. Wargan, 58–78. London: Applied Science Publishers.

Seifert, K.A., Samson, R.A., Boekhout, T., and Louis-Seize, G. 1997. *Remersonia*, a new genus for *Stilbella thermophila*, a thermophilic mould from compost. *Canadian Journal of Botany* 75:1158–1165.

Sen, S., Abraham, T.K., and Chakrabarty, S.L. 1979. Biodegradation and utilization of city wastes by thermophilic micro-organisms. *Indian Journal of Experimental Biology* 17:1284–1285.

Sharma, H.A., and Johri, B.N. 1992. The role of thermophilic fungi in agriculture. *Handbook of Applied Mycology* 4:707–728.

Sharma, H.S.S. 1989. Economic importance of thermophilous fungi. *Applied Microbiology and Biotechnology* 31:1–10.

Singh, B., Poças-Fonseca, M.J., Johri B.N., and Satyanarayana, T. 2016. Thermophilic molds: Biology and applications. *Critical Reviews in Microbiology* 42:985–1006.

Singh, B., and Satyanarayana, T. 2014. Thermophilic fungi: Their ecology and biocatalyst. *Kavka* 42:37–51.

Singh, B., and Satyanarayana, T. 2016. Ubiquitous occurrence of thermophilic fungi in various substrates. In *Fungi from Different Substrates*, 201–216. Boca Raton, FL: CRC Press.

Stetter, K.O., Fiala, G., Huber, R., and Segerer, A. 1990. Hyperthermophilic microorganisms. *FEMS Microbiology Reviews* 75:117–124.

Stolk, A.C. 1965. Thermophilic species of *Talaromyces* Benjamin and *Thermoascus* Miehe. *Antonie van Leeuwenhoek* 31:(3)262–276.

Stolk, A.C., and Samson, R.A. 1972. The genus *Talaromyces*—Studies on *Talaromyces* and related genera. II. *Studies in Mycology* 2:1–65.

Straatsma, G., Gerrits, J.P.G., Augustijn, M.P., Op Den Camp, H.J., Vogels, G.D., and Van Griensven, L.J.L.D. 1989. Population dynamics of *Scytalidium thermophilum* in mushroom compost and stimulatory effects on growth rate and yield of *Agaricus bisporus*. *Microbiology* 135:751–759.

Straatsma, G., Olijnsma, T.W., Gerrits, J.P.G., Amsing, J.G., Den Camp, H.J.O., and Van Griensven, L.J.D. 1994a. Inoculation of *Scytalidium thermophilum* in button mushroom compost and its effect on yield. *Applied and Environmental Microbiology* 60:3049–3054.

Straatsma, G., Samson, R.A., Olijnsma, T.W., Den Camp, H.J.O., Gerrits, J.P.G., and Van Griensven, L.J.L.D. 1994b. Ecology of thermophilic fungi in mushroom compost, with emphasis on *Scytalidium thermophilum* and growth stimulation of *Agaricus bisporus* mycelium. *Applied and Environmental Microbiology* 60:454–458.

Subrahmanyam, A. 1985. *Studies on morphology and biochemical activities of some thermophilic fungi.* DSc thesis, Kumaun University, Nainital, India.

Sundman, G., Kirk, T.K., and Chang, H.M. 1981. Fungal decolorization of kraft bleach plant effluent. *Tappi* 64:145–148.

Taber, R.A., and Pettit, R.E. 1975. Occurrence of thermophilic microorganisms in peanuts and peanut soil. *Mycologia* 67:157–161.

Tansey, M.R. 1971. Isolation of thermophilic fungi from self-heated industrial wood chip piles. *Mycologia* 63:537–547.

Tansey, M.R., and Brock, T.D. 1978. Microbial life at high temperature ecological aspects. In *Microbial Life in Extreme Environments*, ed. D. Kushner, 159–216. London: Academic Press.

Tansey, M.R., and Jack, M.A. 1975. *Thielavia australiensis* sp. nov., a new thermophilic fungus from incubator bird (mallee fowl) nesting material. *Canadian Journal of Botany* 53:81–83.

Thakre, R.P., and Johri, B.N. 1976. Occurrence of thermophilic fungi in coal-mine soils of Madhya Pradesh. *Current Science* 45:271–273.

Thomke, S., Rundgren, M., and Eriksson, K.E. 1980. Nutritional evaluation of the white rot fungus *Sporotrichum pulverulentum* as feed stuff to rats, pigs and sheep. *Biotechnology and Bioengineering* 22:2285–2303.

Tsiklinskaya, P. 1899. Sur les mucedinees thermophiles. *Annales de l'Institut Pasteur* 13:500–505.

Tubaki, K., Ito, T., and Matsuda, Y. 1974. Aquatic sediments as a habitat of thermophilic fungi. *Annals of Microbiology* 24:199–207.

Ueda, S., and Udagawa, S.I. 1983. *Thermoascus aegyptiacus*, a new thermophilic ascomycete. *Transactions of the Mycological Society of Japan* 24:135–142.

van Oorschot, C.A.N. 1977. The genus *Myceliophthora. Persoonia* 9:401–408.

van Oorschot, C.A.N., and de Hoog, G.S. 1984. Some hyphomycetes with thallic conidia. *Mycotaxon* 20:129–132.

von Arx, J.A. 1975. *On Thielavia and Some Similar Genera of Ascomycetes.* Studies in Mycology No. 8. Baarn, Netherlands: CBS.

von Arx, J.A. 1981. On *Monilia sitophila* and some families of Ascomycetes. *Sydowia* 34:12–29.

Waksman, S.A., Umbreit, W.W., and Cordon, T.C. 1939. Thermophilic actinomycetes and fungi in soils and in composts. *Soil Science* 47:37–61.

Waldrip, D.W., Padhye, A.A., Ajello, L., and Ajello, M. 1974. Isolation of *Dactylaria gallopava* from broiler-house litter. *Avian Diseases* 18:445–451.

Warcup, J.H. 1950. The soil-plate method for isolation of fungi from soil. *Nature* 166:117–188.

Ward, J.E., and Cowley, G.T. 1972. Thermophilic fungi of some central South Carolina forest soils. *Mycologia* 64:200–205.

2 Origin of Thermophily in Fungi

2.1 INTRODUCTION

As stated in Chapter 1, life originated in a largely thermal and anoxic environment. The ecological adaptation of organisms to adverse environments is largely influenced by both biotic and abiotic factors. Inasmuch as biotic factors are concerned, thermophilic fungi in their natural habitat face competition from their mesophilic counterparts and microfauna comprising potential competitors, antagonists, and predators to complete their life cycle. On the front of abiotic factors, certain organisms have peculiar characteristics that help them survive in adverse environments, such as those with high salt concentrations, extremes of temperature and pH, increased water pressure, and an absence of oxygen; those under desiccation; or those with a combination of all these factors. In response to such stressful environments, some organisms, including thermophilic fungi, have developed adaptations that allow them to grow under these harsh conditions. It is now well recognized that the most influential factor for the growth of microorganisms is the environmental temperature, because it acts directly on the structure and function of biomolecules and on the maintenance of integrity of cellular structures.

It is speculated that geothermally stable environments (habitats having a constant high temperature) have exerted selective pressure on microorganisms, selecting the moderately thermophilic and those that require high temperatures to survive (Gomes et al., 2016). One school of thought (Powell et al., 2012) has suggested that thermophily in the kingdom Fungi arose as an adaptation to transient seasonal changes and high day temperatures, rather than simply an adaptation to specialized new high-temperature niches. This hypothesis is also supported by an earlier study by Tansey and Jack (1976), who studied thermophilic fungi in sun-heated soils. As a conclusion of their study, they provided direct proof that solar heating provides sufficient heat for growth and successful competition of thermophilic fungi in soils. They further reported that thermophilic and thermotolerant fungi were most frequently isolated from sun-heated soil, less so from grass-shaded soil, and still less so from tree-shaded soil. Challenging this hypothesis, however, Maheshwari et al. (2000) and Rajasekaran and Maheshwari (1993) proposed that the presence of thermophilic fungi in soils reflects their dispersal from natural composts. This suggestion was further refuted by Powell et al. (2012) in their comprehensive multiyear study of thermophilic fungi in an arid-grassland ecosystem at the Sevilleta National Wildlife Refuge in central New Mexico, which lack sufficient litter to support natural composting. Powell et al. (2012) further suggested that it is reasonable to speculate that thermophily does not represent an adaptation to environments with constant high temperature, but instead a competitive strategy in environments with fluctuating temperatures. In environments with extreme diurnal shifts in temperature, thermophily could provide a selective advantage under high temperatures, allowing these fungi to compete in complex microbial communities. Another school of thought believes that thermophilic fungi descended from their mesophilic ancestors associated with bird nests, such as those found in an old group of bird, the Megapodiidae (incubator birds), which have probably existed for 50–60 million years (Grzimek, 1984), as a possible natural site of occurrence of thermophilic fungi (Rajasekaran and Maheshwari, 1993). These birds are found in Australia and islands of the southwestern Pacific, incubating their eggs in compost, which they construct from forest litter. In these mounds of composts, microbial activity produces respiratory heat that warms them up to 50°C for several months (Seymour and Bradford, 1992). The occurrence of thermophilic fungi in composts is explained on the basis of favorable growth conditions (prolonged elevated temperature, humid and aerobic conditions, and adequate supply of carbohydrate and nitrogen) for their development. However, it is not safe to consider composts as "natural" sites where thermophilic fungi may have evolved. Nevertheless, it is beyond doubt that thermophilic fungi are distributed worldwide in almost all habitats where thermogenesis occurs for any reason.

Further attempts to increase the thermal tolerance of mesophilic organisms through induced mutations, however, have been relatively unsuccessful. The predicament here is the nature of the adaptive process itself. It is believed that natural selection plays an important role in establishing the capability of the microorganisms for growth and persistence at elevated temperatures. In such a selection process, the organism as a whole must be capable of functioning at high temperature. The distinction between thermophilism and simple thermal stability remains difficult and critical. Thermophilism refers to the stability of the entire functioning organism based on its genetic constitution, while thermal stability encompasses the stability of an individual cellular component or process independent of other components or processes. This chapter focuses on the origin and ecological relationships of thermophilic fungi, and how these fungi have adapted to elevated temperatures in environments in which most of their close relatives fail to survive. Further, the current hypothesis to explain thermophilism in fungi is discussed in relation to various available theories and hypotheses.

2.2   ORIGIN AND ECOLOGICAL RELATIONSHIPS

Thermophilic fungi are virtually ubiquitous and a well-recognized group by way of their physiological growth conditions and taxonomic characterization. They have been reported from natural as well as man-made habitats. The occurrence of self-heating decomposing organic matter all over the globe is believed to be the major natural habitat of thermophilic fungi. High temperature conditions also prevail in sun-heated soils in the tropics; even in a newly built wood chip pile, temperature rapidly increases to ignition. Thermogenesis occurs in these habitats as a result of metabolic activities of the microorganisms. Extensive studies on the isolation and characterization of thermophilic fungi have also been carried out from man-made habitats, such as hay, manure stored peat, stored grains and oil palm kernels, and mushroom composts. In these habitats, thermophiles occur either as resting propagules or as active mycelia, depending on the nutritional status and other favorable environmental conditions. Besides these natural self-heated habitats, thermophilic fungi have also been isolated from man-made thermal habitats, that is, cooling towers, effluents from nuclear power reactors, and ducts employed for thermal insulation (Johri and Satyanarayana, 1986). Commonly occurring pioneer thermophilic fungi, *Mucor pusillus* and *Humicola lanuginosa* (syn. *Thermomyces lanuginosus*), were isolated as chance contaminants from bread and potato by Tsiklinskaya in 1899.

As one may recall, soil is the depository of all life and the laboratory within which most of the changes that enable life to continue are carried out. A small amount of soil contains a variety of microorganisms of diverse form and physiology. From a functional viewpoint, soil may be considered the land surface of the earth that provides the substratum for plant and animal life. Due to their potential as a resource for biological materials, different types of soils have been explored by several workers all over the world for the isolation of thermophilic fungi. Thus, soils remain the major substrate for the isolation of thermophilic fungi. The isolation of thermophilic fungi was reported as early as 1939 by Waksman and his coworkers from soil and the later still continues to provide a good resource for their isolation in laboratory (Apinis, 1963; Eggins and Malik, 1969; Evans, 1971; Eicker, 1972; Ward and Cowley, 1972; Johri and Thakre, 1975; Taber and Pettit, 1975; Tansey and Jack, 1976; Thakre and Johri, 1976; Ellis and Keane, 1981; Sandhu and Singh, 1981; Abdel-Hafez, 1982; Singh and Sandhu, 1986; Maheshwari et al., 1987; Hemida, 1992; Guiraud et al., 1995, Mouchacca, 1995).

Evans (1971) studied coal spoil tips for the occurrence of thermophilic fungi and isolated 32 thermophilic and thermotolerant fungal species representing the genera *Mucor, Rhizopus, Mortierella, Chaetomium, Sphaerospora, Talaromyces, Thielavia, Acrophialophora, Aspergillus, Calcarisporium, Chrysosporium, Geotrichum, Penicillium, Scolecobasidium,* and *Tritirdium.* He observed that the population of thermophilous fungi in well-colonized warm areas was higher than that of thermophilous fungi in corresponding nonwarm areas. Ward and Cowley (1972) explored South Carolina forest soils for thermophilic fungi. The habitat types considered were sand hill pine-scrub oak, loblolly pine, pine hardwood, and floodplain hardwood. They isolated *Aspergillus fumigatus, Aspergillus nidulans, Thermoascus aurantiacus, Thielavia* sp., and an unidentified species. Similarly, Taber and Pettit (1975) isolated thermophilic fungi from peanut soils. Besides *Mucor pusillus,* which was found to be the most common species, the other species isolated were *Humicola lanuginosa, Thermoascus aurantiacus, Malbranchea pulchella, Aspergillus fumigatus,* and *Sporotrichum* sp.

Thermophilic and thermotolerant fungi have also been isolated from the soils of south-central Indiana by Tansey and Jack (1976). They isolated 19 species, found most frequently in sun-heated soil, followed by grass-shaded soil and tree-shaded soil. Thakre and Johri (1976) exploited coal mine soils of Madhya Pradesh, India, for isolating thermophilic fungi. In order to facilitate the recovery of thermophilic fungi, Johri and Thakre (1975) amended soil suspension with glucose, cellulose, ammonium nitrate, ammonium phosphate, and asparagines and incubated the soil samples at

45°C prior to plating. The thermophilic microflora recovered as a result of nutritional enrichment was *Absidia corymbifera, Achaetomium macrosporum, Aspergillus fumigatus, Humicola grisea, Aspergillus nidulans, Penicillium* sp., *Rhizopus microsporus, Rhizopus rhizopodiformis, Sporotrichum* sp., *Thermoascus aurantiacus, Thermomyces lanuginosus, Thielavia minor*, and *Torula thermophila*. Sundaram (1977) recovered 10 species of thermophilic fungi from rice field soils. The golden period of thermophilic fungal research was elegantly reviewed by Tansey and Brock (1978), who made a critical appraisal of 34 studies concerning the isolation of thermophilic fungi from soil, including studies of both temperate and tropical soils.

Ten species of thermophilous fungi were identified from different altitudinal zones of forest soils of Darjeeling in the eastern Himalayas by Sandhu and Singh (1981). They observed that there was a decrease in the prevalence of true thermophiles with increasing altitude. Ellis (1980) isolated four thermophilous fungi from Antarctic and sub-Antarctic soils and presented a possible explanation for the existence of propagules of thermophilous fungi in some Antarctic soils. Ellis and Keane (1981) isolated 13 species of thermophilic fungi from soil samples collected from 100 sites in six climatic regions in Australia. The widest range of species was found in soils of the cool, moist, southeastern climatic zone.

Abdel-Hafez (1982) studied desert soils of Saudi Arabia for thermophilic and thermotolerant fungi. He isolated and cataloged 48 species belonging to 24 genera. Singh and Sandhu (1986) isolated 46 species of thermophilous fungi from Port Blair, India, saline marshy soils collected from eight different sites. The temperature responses of the fungi revealed 14 microthermophiles, 22 thermotolerant species, and 10 true thermophilic species. They cataloged the fungi in order of their ecological importance based on their frequency, relative density, and presence value. Maheshwari et al. (1987) isolated 11 species of thermophilic fungi from soils collected from different places in India. They observed that the thermophilic fungal flora in India is qualitatively similar to that recorded from other countries in temperate or tropical latitudes. They proposed that the widespread occurrence of thermophilic fungi in soils may be a result of the dissemination of propagules from self-heating masses of organic materials (composts) that occur worldwide. Hemida (1992) isolated thermophilic and thermotolerant fungi from cultivated and desert soils of Egypt exposed continuously to cement dust particles. Mouchacca (1995) identified 19 theromophilic and 30 thermotolerant fungi from desert soils of the Middle East (Syria, Iraq, Kuwait, Saudi Arabia, Jordan, and Egypt). He observed a greater abundance of thermophilic fungi in soils under temperate climates. Similarly, Salar and Aneja (2006) explored temperate soils of northern India (Manali, Mussoorie, Nainital, and Shimla) for the presence of thermophilic and thermotolerant fungi, and they isolated 19 species belonging to 14 genera. *Chaetomium senegalense* (ascomycetes) and *Myceliophthora fergusii* (mitosporic ascomycetes) were reported for the first time from India. They observed that the temperature of the soil from which thermophilic fungi were cultured did not correspond with their optimum axenic growth temperature.

The enigmatic growth of these fungi at elevated temperatures remains a subject for which a hypothesis needs to be formulated—not only why these organisms grow at high temperatures, but also why they cannot grow at ordinary temperatures, where the majority of their close relatives thrive.

2.3   FUNGAL ADAPTATIONS TO THERMOPHILY

The growth and development of microorganisms are generally affected by the chemical and physical conditions of their environment. Among these, temperature is the most influential factor affecting the functions of biomolecules and cellular structures. All organisms have an optimum temperature for growth, and different groups of microorganisms have evolved to grow over different temperature ranges. Most microorganisms grow within a narrow range of temperature. Nevertheless, certain microorganisms have evolved to occupy a special niche of extreme environments, allowing for their selection and persistence, which not only resists but also requires elevated temperatures for their survival.

Exposure of nonthermophiles to elevated temperatures normally causes damage to cytoplasmic membranes, ribosome breakdown, irreversible enzyme denaturation, and DNA damage. As a result of adaptation to high temperatures, certain peculiar features are found in thermophilic fungi. Intriguingly, the genome sizes of thermophiles are generally smaller than those of nonthermophiles. The reduction in genome size of thermophilic fungi compared with that of their mesophilic counterparts is one such recent revelation (van Noort et al., 2013). This process may involve loss of genes coding for specific proteins, loss of transposable elements, and reduction in the size of introns and intergenic regions. As is well known, thermophily in prokaryotes is more pronounced than that in eukaryotes. Prokaryotic genomes are compact and contain little intergenic DNA compared with eukaryotes. This genomic feature helps in a short division time for rapid reproduction and also reduces the energy consumption for nucleotide synthesis. On the contrary, duplication of intriguing genes, such as those involved in hyphal melanization and reproductive structure (ascocarp, peridia, ascospores, and conidia) pigmentation, may provide insight into the evolution of thermophily in fungi. Such a phenomenon is also observed in mesophilic fungi to protect their cells from high temperatures, ultraviolet (UV) radiation, and desiccation during unfavorable conditions of atmospheric humidity.

Over the past few years, several studies have been conducted regarding the evolution of thermophily in microorganisms. Interestingly, many of these studies are based on the "omics" approach involving genomics, proteomics, and transcriptomics (Figure 2.1).

However, in eukaryotes, particularly in fungi, such studies are scanty. Recently, the first eukaryotic thermophilic genome, *Chaetomium thermophilum*, gave initial insights into the potential for structural biology (Amlacher et al., 2011). Moreover, with the publication of two more genomes of *Thielavia terrestris* and *Thielavia heterothallica* (Berka et al., 2011), a comparative analysis of their thermophilic nature can be assessed. Nevertheless, in the foregoing account we focus our discussion on the evolution of thermophily in fungi from all perspectives, including traditional approaches and the more recent genomic evolution, as contributing to thermophilism in fungi.

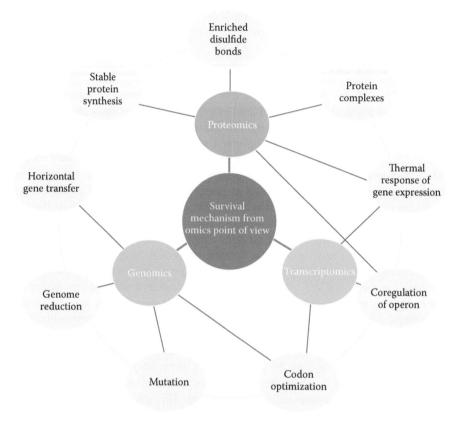

FIGURE 2.1 Web of the survival mechanisms of thermophiles involving omics approaches. (Adapted from Wang et al., 2015.)

2.4 HYPOTHESIS TO EXPLAIN THERMOPHILISM IN FUNGI

A number of theories and hypotheses to describe the origin of thermophily in fungi have been proposed from time to time, but the fundamental question of the origin of thermophilic microorganisms remains. The cause of this curious phenomenon of growth and sustenance of life at elevated temperatures is enigmatic and intriguing. It has been rightly stated that the question is not only why these organisms can grow at high temperatures up to 60°C or more, but also why they cannot grow at ordinary temperatures, 15°C–25°C, where so many of their close relatives thrive (Emerson, 1968). It is likely that the inability of most organisms to grow at high temperatures is often based on the concept of a "prime lethal event"; that is, thermal death is due to the interruption of a single metabolic process or the destruction of a single component essential for survival. Blackman (1905), in his "master reaction" hypothesis, proposed that "where a process is conditioned as to its rapidity by a number of independent factors, the rate of the process is limited by the pace of the slowest factor." Death would occur if a single component was completely destroyed or became nonfunctional. By now, a large number of hypotheses have been offered, some of which are mere speculations, and others are supported by some observations, which have also been contradicted or supported by the findings of others on the same or other organisms under similar conditions. Interestingly, proteins of organisms are essential components of life processes. The diversity of the "recipes" for thermostability straightaway raises two important questions: (1) What are the possible physical mechanisms to increase the thermostability of proteins, and (2) how did evolution use possible physical mechanisms of thermal stabilization to develop strategies of adaptation to high temperatures and other possible demands of the environment? The extensive literature on these aspects has been elegantly reviewed from time to time (Emerson, 1968; Crisan, 1973; Mehrotra, 1985; Satyanarayana et al., 1992; Dix and Webster, 1995; Stetter, 1999; Wang et al., 2015; Singh et al., 2016). In order to remain in their native conformations at elevated temperatures, proteins and nucleic acids require the adjustment of interactions at physiologically relevant temperatures. Several hypotheses for explaining the ability of thermophiles to grow at high temperatures have been proposed from time to time and are discussed below.

2.4.1 PROTEIN THERMOSTABILITY AND STABILIZATION

It is a common belief that life originated in a thermal environment. The role of various factors contributing to protein thermostability has been a subject of intense study over the decades. The most frequently reported mechanisms in protein thermostabilization include increased van der Waals interactions, higher core hydrophobicity, additional networks of hydrogen bonds, enhanced secondary structure propensity, ionic interactions, increased packing density, and decreased length of surface loops (Berezovsky and Shakhnovich, 2005). It has been shown that proteins use various combinations of these mechanisms. However, no general physical mechanism for increased thermostability was found. The involvement of proteins in thermostability largely came from the studies on archaea and bacteria. Analyses of proteins from thermophilic microorganisms have revealed two physical mechanisms of protein thermostabilization.

2.4.1.1 Structure-Based

It has been observed that structurally the proteins of thermophiles are more compact and hydrophobic than their mesophilic counterparts, resulting in increased rigidity and resistance to unfolding. Studies have shown that the adaptation of proteins to high temperature depends on the rigidity–flexibility balance of the molecule. A higher rigidity of the thermophilic protein requires a high temperature to promote the thermal motion, and the flexibility is critical for catalytic activity. Structure-based thermophily in proteins evolved mostly from archaea. Berezovsky and Shakhnovich (2005) have observed that in the thermophilic bacteria *Thermus thermophilus*, hydrolase has a higher total number of van der Waals contacts than mesophilic bacteria. In the secondary structure of thermophilic protein, there are six α-helices, whereas only three α-helices are found in its mesophilic counterpart. Similarly, elements of secondary structures in thermostable hydrolase are quite extended in size, and the density of hydrogen bonds is also higher in a protein from the thermophilic organism. Furthermore, rubredoxin from the thermophilic archaebacteria *Pyrococcus furiosus* has a pronounced predisposition toward enhanced packing density compared with mesophilic proteins, and this is also reflected in the increased number of H-bonds per residue. Similarly, van der Waals interactions and the involvement of more residues in elements of secondary structures contribute to the increased stability of thermophilic 2Fe-2S ferredoxin.

2.4.1.2 Sequence-Based

The sequence-based mechanism of the thermostability of proteins does not show marked structural differences from their mesophilic homologues. Rather, a small number of apparently strong interactions are responsible for the high thermal stability of these proteins. Studies have shown that sequence alignments discriminate proteins from hyperthermostable *Thermotoga maritima* from other thermostable proteins (Berezovsky and Shakhnovich, 2005). They have lower sequence identity than their respective mesophilic proteins and show substantial redistribution or an increased number of charged residues. Some organisms have evolved as mesophilic but later recolonized in hot environments. In such organisms, their already existing proteins should be changed in order to adapt to the new thermal environment, and thus selection of sequence and structure had to occur concurrently. This process results in evolutionary pressure on protein structures to make them more designable. Designability is a property of a protein structure that indicates how many sequences exist that fold into a particular structure at various levels of stability. The sequence-based strategy of thermophilic adaptation is to make sequence substitutions leading to the formation of a restricted set of specific interactions (e.g., ion bridges) without redesign of the whole structure. This strategy is exemplified by *Thermotoga maritima*, which recolonized a hot environment, demonstrating that only 24% of its genes are similar to those of archaea (Nelson et al., 1999). Thus, thermophilic adaptations can be considered from a structure or sequence-centric point of view. However, the choice of particular strategy depends on the evolutionary history of an organism.

2.4.2 Heat Shock Proteins

One of the most highly conserved features of all living organisms is the ability to sense and respond to sudden changes in temperature. Heat shock proteins (HSPs) are defined as those proteins whose synthesis is substantially stimulated by a temperature several degrees in excess of the optimum growth temperature (OGT) (Schlesinger, 1986). When thermal stress is applied to an organism, the most prominent response is the production of a set of novel proteins or an increase in the quantity of certain types of existing proteins. This phenomenon was first discovered in the larvae of fruit fly (*Drosophila buskii*) by Ritossa in 1962. Since then, this phenomenon has been manifested in several other organisms, including thermophilic fungi (Chen and Chen, 2004), and is a universal reaction found in all living organisms. The optimum temperature for the production of HSPs varies in different organisms, being 43°C–47°C in *Escherichia coli* (Neidhardt et al., 1984), 36°C in *Saccharomyces cerevisiae* (McAlister and Finkelstein, 1980), and 40°C–43°C in *Fusarium oxysporum* (Freeman et al., 1989). In general, a temperature of 5°C or more above the OGT of an organism will induce the synthesis of HSPs.

On the basis of their molecular weight, HSPs are classified into several families. In general, three categories are considered: (1) high molecular size, with a molecular weight ranging between 69 and 120 kDa; (2) medium molecular size, with a molecular weight ranging between 39 and 68 kDa; and (3) low molecular size, with a molecular weight below 38 kDa. The synthesis of HSPs in fungi is a rapid process. For example, in *Fusarium oxysporum*, HSP synthesis began after 10 min of heat treatment (Freeman et al., 1989). Chen and Chen (2004) studied HSPs of thermophilic and thermotolerant fungi. They observed that the thermophilic and thermotolerant fungi, when subjected to heat shock (HS) treatments at 40°C, synthesized mostly high- and medium-molecular-weight HSPs, whereas at 50°C they synthesized mostly low-molecular-weight HSPs (Table 2.1). In fungi, HSPs are involved in physiological functions, as in other organisms, conferring thermal resistance by binding to cell organelles. High-molecular-weight HSPs assemble near the nucleus

TABLE 2.1

Molecular Weights of HSPs of Thermophilic and Thermotolerant Fungi Produced as a Result of HS Treatment at 40°C and 50°C

Species	Low-Molecular-Weight HSPs (kDa)		Medium-Molecular-Weight HSPs (kDa)		High-Molecular-Weight HSPs (kDa)	
	40°C	50°C	40°C	50°C	40°C	50°C
Rhizopus microspores var. *chinensis*		35, 28, 22	52, 46, 38			
Rhizopus microsporus var. *rhizopodiformis*		35, 28, 22	52, 46, 38			
Rhizomucor pusillus		35, 25, 22, 20	52, 46, 38	52	97.4	
Rhizomucor miehei	32, 23		62, 52, 46, 38			
Humicola insolens var. *thermoidea*			58, 51, 46, 40		94	
Scytalidium thermophila		28	58, 51, 46, 42		94	
Malbranchea cinnamomea		30, 26	51, 46		94	
Thermoascus taitungiacus		35, 30, 28, 26	51, 42, 40		94, 69	
Thermoascus crustaceus		35, 30, 26	66, 46, 40		94	
Myceliophthora hinnulea			66, 55, 52, 45, 40	52, 45, 40	92, 78	
Myceliophthora thermophila	22	35, 30, 23, 22, 20	52	52	92	
Myceliophthora fergusii	30		50		92	
Talaromyces emersonii	30		53, 46, 38		87, 80	
Aspergillus fumigatus			53, 52, 46, 38		150, 87	

Source: Modified from Chen, K.Y., and Chen, Z.C., *Bot. Bull. Acad. Sin.*, 45, 247–257, 2004.

but are restricted to the cytoplasm (Neuman et al., 1987). In contrast, low-molecular-weight HSPs are bound to the chromatin, as has been observed in *Dictyostelium* (Loomis and Wheeler, 1982). Under HS treatment, HSPs perform the equally important, but quite different, function of molecular chaperones; that is, they are involved in protein stabilization and adaptive modification of cell protein composition, which includes the identification of defective proteins and their partial proteolysis and refolding, as well as control of the folding of newly synthesized polypeptides (Tereshina, 2005).

Another important protein involved in thermotolerance is ubiquitin. It is a small polypeptide consisting of 76 amino acids with a molecular mass of 8 kDA. This protein is involved in universal nonlysosomal ATP-dependent degradation of polypeptides in eukaryotes. Ubiquitin is not required for vegetative growth at an optimal growth temperature, but it is essential for resistance to different kinds of stress, for example, stress induced by high or low temperature, starvation, or amino acid analogues. This protein is synthesized during the sporulation process in fungi and is also found to be essential for the maintenance of spore viability (Plesofsky-Vig, 1996).

2.4.3 Proteome and Genome as Determinants of Thermophilic Adaptation

Recently, the amino acid composition of proteins and the nucleotide composition of genomes of thermophilic organisms have been found to be reflective of their adaptations to thermal environments. In particular, the amino acid composition of Ile, Val, Tyr, Trp, Arg, Glu, and Leu (IVYWREL) has been observed to be related to the OGT in prokaryotes (Zeldovich et al., 2007) and eukaryotes (van Noort et al., 2013). The correlation coefficient (R^2) between the total concentration of seven amino acids in proteomes—IVYWREL—and OGT has been observed to be as high as 0.93. Ponnuswamy et al. (1982), for the first time, made a systematic search for amino acids that are most significant for the thermostability of proteins. They considered a set of 30 proteins and about 65,000 combinations of different amino acids to find the amino acid sets that serve as the best predictors of denaturation temperature. Analyzing the genome sequences of three thermophilic fungi, *Chaetomium thermophilum*, *Thielavia terrestris*, and *Thielavia heterothallica* (Sordariomycetes), van Noort et al. (2013) observed that the total frequency of IVYWREL amino acids in *C. thermophilum* is significantly higher than that in its mesophilic counterpart. Besides, *C. thermophilum* also had high frequencies of isoleucines, tryptophans, and tyrosines. These larger hydrophobic amino acids are likely to play a role in filling the hydrophobic cores of proteins (Hummer et al., 1998). Another important observation with regard to proteins of *C. thermophilum* is the overrepresentation of cysteines, which play a great role in catalytic residues, disulfide bridges, and metal binding, thus helping to contribute to the folding and stability of proteins. A striking difference between prokaryotes and eukaryotes with regard to thermophily is the depletion of glycines in *C. thermophilum*, whereas they are enriched in *Chaetomium globosum* (mesophilic fungus). The exchange of alanines with glycines has been shown to destabilize α-helices, particularly in the center of the helix (Chakrabartty et al., 1991).

It appears that *Chaetomium globosum* has in fact used this strategy to make proteins less thermostable, whereas *C. thermophilum* has evolved in the opposite direction, lowering its glycine content. van Noort et al. (2013) further verified the generalizability of these trends by examining two more thermophilic fungal genomes, *Thermomyces lanuginosus* and *Talaromyces thermophilus* of the subclass Eurotiomycetidae, a different fungal clade that also includes *Aspergillus fumigatus* and *Emericella nidulans*. Compared with their mesophilic neighbors, these species both have a significantly higher total frequency of IVYWREL amino acids. They also show a depletion of glycines and significant enrichment in arginines and alanines, a more or less universal trend in different clades of fungi.

Further, all amino acids from the IVYWREL predictor set and Cys (stabilizing S-S bridges) and Lys (important in the general mechanism of thermostability) are attached to tRNA by class I aminoacyl tRNA synthetases (Eriani et al., 1990). Thus, class I amino acids may constitute a group of amino acids sufficient for the synthesis of thermostable proteins. Moreover, all class I synthetases have the Rossmann fold structure, one of the ancient last universal common ancestor (LUCA) domains (Shakhnovich et al., 2005) with high compactness and therefore stability (England et al., 2003). Therefore, the above observations imply a possible association between thermal adaptation and the evolution of protein biosynthesis.

The thermostability of proteins is, however, only a partial prerequisite of the thermal adaptation of an organism, as its DNA must also remain stable at elevated temperatures. The relationship between OGT and the nucleotide content of genomes of a number of organisms have been explored by several workers (Lambros et al., 2003; Singer and Hickey, 2003; Musto et al., 2005). It is believed that the genomic structure in thermophiles is more stable than that in mesophiles. The contents of guanine (G) and cytodine (C) in the genome are important indicators of DNA stability; therefore, comparisons of genomes of thermophilic and mesophilic fungi would yield conclusive information about the role of GC content in the adaptation of thermophilic fungi at elevated temperatures. The whole genome sequences of thermophilic ascomycetes *Myceliophthora thermophila* and *Thielavia terrestris*, published by Berka et al. (2011), showed that their genome is 38.7 and 36.9 Mb long, respectively. They also observed higher GC contents of the coding regions in the genomes of *Myceliophthora thermophila* and *Thielavia terrestris* compared with their mesophilic counterparts, *Trichoderma reesei* and *Chaetomium globosum*. The G:C pair are more resilient to thermal denaturation and may represent an adaptation of the protein-encoding genes to elevated temperature (Berka et al., 2011). The GC contents of several theromphilic fungi, that is, *Rhizomucor miehei* (43.8%) (Zhou et al., 2014), *Thermomyces lanuginosus* (52.14%) (Mchunu et al., 2013), *Thielavia terrestris* (54.7%) (van Noort et al., 2013), and *Myceliophthora thermophila* (51.4%) (Berka et al., 2011), were found to be much higher than the average value for mesophilic zygomycetes (35.3%). It was therefore hypothesized that a high GC content contributes to the thermostability of the genome and is correlated with the OGT.

On the contrary, Zeldovich et al. (2007) believed that the G + C content of DNA is not correlated with the OGT of organisms. Rather, the only signature of thermal adaptation is the fraction of A + G in coding DNA, which is found to be correlated with OGT to a considerable extent, mainly due to codon patterns of IVYWREL amino acids. However, they have observed a strong and independent correlation between OGT and the frequency with which pairs of A and G nucleotides appear as nearest neighbors in genome sequences. They further observed that sequence correlations in DNA show that ApG dinucleotides are overrepresented in both strands of double-stranded DNA of thermophilic organisms. Some investigators, however, have argued that some microbes have different OGTs but share similar and even lower GC contents; for example, *Caldicellulosiruptor hydrothermalis* contains only 35% GC, with an OGT of 70°C (Blumer-Schuette et al., 2011). Therefore, the GC composition seems to be independent of thermophily, or at least not universal to all thermophiles, including fungi.

2.4.4 REDUCTION IN GENOME SIZE

The genomes of prokaryotes are compact and contain little intergenic DNA compared with eukaryotes. This is an intriguing feature, as it plays an important role in rapid reproduction and smaller generation time, and also reduces the energy consumption for nucleotide synthesis. It has been observed that the genome sizes of thermophiles are usually smaller than those of nonthermophiles, as all species that thrive at temperatures of >60°C have genomes smaller than 4 Mb, whereas species living at temperatures of <45°C have a genome size of larger than 6Mb (van Noort et al., 2013). Thermal fluctuations in the environment induce genomic evolution, which in turn provides thermal tolerance in organisms thus exposed. These evolutionary changes could be achieved through horizontal gene transfer (exchange of DNA among organisms of different species), gene loss, or gene mutations (Averhoff and Muller, 2010). Studies with regard to genome reduction in thermophilic fungi are sporadic, and only limited data are available. van Noort et al. (2013) studied the genomes of thermophilic *Chaetomium thermophilum* (Cth), *Thielavia terrestris* (Tte), and *Thielavia heterothallica* (Tht) and compared them with those of their mesophilic counterparts, *Chaetomium globosum* (Cgl) and *Neurospora crassa* (Ncr). They observed that the genome of the first three species (thermophilic) is smaller than that of the latter two (mesophilic). In agreement with previous studies of prokaryotic thermophiles, they found that the genome size reduction is due mainly to fewer protein-coding genes (Cth 7,267, Tte 9,813, and Tht 9,110 vs. Cgl 11,124 and Ncr 10,620), but also to shorter introns and shorter intergenic regions. As anticipated, the protein lengths and the number of protein family members in thermophiles are reduced compared with those of their homologous counterparts in nonthermophiles, owing to the small size of the genomes of thermophiles (Wang et al., 2015).

Presently, the Genozymes Research Project (http://fungalgenomics.ca/wiki/Fungal _Genomes) is underway to sequence the genomes of thermophilic fungi, including *Acremonium thermophilum* (ATCC 24622), *Chaetomium olivicolor* (CBS 102434), *Dactylomyces thermophilus* (ATCC 26413), *Humicola hyalothermophila* (CBS 454.80), *Myceliophthora hinnulea* (ATCC 52474), *Rhizomucor miehei* (CBS 182.67), *Remersonia thermophila* (ATCC 22073) (syn. *Stilbella thermophila*), *Talaromyces emersonii* (NRRL 3221), *Melanocarpus albomyces* (ATCC 16460),

Thermoascus aegyptiacus (ATCC 56490), and *Thermophymatospora fibuligera* (ATCC 62942). The data from this project will be a landmark in understanding the mechanism of thermophily in fungi, besides an unprecedented contribution of the roles of thermophilic fungi toward their industrial applications.

2.4.5 THERMOTOLERANCE GENES

The possibility of a gene being responsible for thermophily was first suggested by Rahn (1945). It has been suggested that one gene expresses at the lower range of temperature and another at the higher range of temperature. This idea possibly emerged from the observations of Brock (1967) and Bott and Brock (1969), who described that certain bacteria grow at temperatures as high as 95°C and as low as −10°C (Larkin and Stokes, 1968). The reversal of thermophilic strains to mesophilic ones, and vice versa (Fries, 1953), is a point against this contention. A higher number of ribose methylations in the RNA of *Thermomyces lanuginosus* than in any other nonvertebrate has been reported by Nageswara Rao and Cherayil (1979), but its significance in thermophily is not clearly understood. The genome of the ubiquitous laboratory weed *Aspergillus fumigatus* was sequenced in 2005, and nearly one-third of the predicted genes were of unknown function (Nierman et al., 2005; Ronning et al., 2005). *A. fumigatus*, which is able to grow between 12°C and 52°C, has been a subject of study to identify the genes controlling its thermotolerant character (Chang et al., 2004). By employing chemical mutagenesis, Chang et al. isolated a temperature-sensitive mutant (*ts*) of *A. fumigatus*, which was able to grow at 42°C but failed to grow at 48°C, where its wild type could grow well. The genomic library of wild type containing the hygromycin-resistant gene (*hph*) as the selection marker was used to transform the temperature-sensitive mutant. The transformants so selected could grow pretty well at 48°C, and a 35.5 kb thermotolerance gene (*THTA*) complemented the *ts* phenotype. Whether *THTA* is required for thermotolerance was shown by gene disruption, as the strains carrying the disrupted gene failed to grow at 48°C following colony transfer. In *A. fumigatus*, it is most likely that *THTA* is the gene encoding for HSP.

2.4.6 RAPID TURNOVER OF ESSENTIAL METABOLITES

It is a general observation that the growth rates of thermophilic fungi are high, and thus so are the metabolic rates; therefore, a reserve of essential metabolites is always maintained to replace the thermolabile metabolites (Allen, 1950). The mechanism of protein thermostability is not fully understood. It may be due to subtle changes in the amino acid complement or at binding with certain ions: Zn, Ca, or Co. The induction of HS proteins is a possible explanation of protein thermostability. To study the applicability of the rapid-turnover hypothesis in thermophilc fungi, Miller et al. (1974) evaluated and compared the protein breakdown rates in thermophilic (*Penicillium duponti*, *Malbranchea pulchella*, and *Mucor miehei*) and mesophilic (*Penicillium notatum* and *Penicillium chrysogenum*) fungi by examining the breakdown of pulse-labeled proteins. They observed that both types of fungi had a negligible rate of breakdown of bulk proteins in the logarithmic growth phase. However, a protein breakdown rate of 5.2%–6.7% per hour was observed in the stationary growth phase of both types. Moreover, the rate of breakdown of soluble proteins in thermophilic fungi was almost twice than that of mesophilic fungi. They further suggested that an increased turnover rate of soluble proteins in thermophilic fungi is important for their survival and adaptation to elevated temperatures. In an another study, Rajasekaran and Maheshwari (1990) also observed a lower protein breakdown in thermophilic fungi than in mesophilic fungi. However, some specific proteins may have higher turnover rates in thermophilic fungi than in their mesophilic counterparts. In the thermophilic fungus *Thermomyces lanuginosus*, an increase in respiratory rate with an increase in temperature, compared with no such increase in the mesophilic species *Aspergillus niger*, has been reported by Prasad et al. (1979). Endogenous respiration in the spores of *Myrotheciurn verrucaria* was found to be stimulated by heating at 50°C. Koffler (1957), however, found no metabolic apparatus for the rapid biosynthesis in thermophiles.

2.4.7 MACROMOLECULAR THERMOSTABILITY

This hypothesis stems from the fact that thermophiles are capable of producing essential macromolecules, such as enzymes and other proteins, exhibiting an unusual degree of thermostability. A comparative study by De Bertoldi et al. (1973) of mesophilic and thermophilic *Humicola* species has shown that there is no significant difference in their GC contents. Extracellular enzymes from thermophilic flamentous fungi usually have a high glycosylation level, and this has been a factor associated with thermostability (Martins et al., 2013). Thermophilic fungi are known to produce thermostable proteases, lipases, amylases, cellulases, ribonucleases, aminopeptidases, and carboxypeptidases (Zuber, 1967; Somkuti and Babel, 1968; Chapins and Zuber, 1970; Craveri et al., 1974; Johri et al., 1985; Grajek, 1987; Jensen and Olsen, 1992). Thermostable glucose-6-phosphate dehydrogenase from the thermophilic fungus *Penicillium dupontii* has been described by Broad and Shepherd (1971). Wali et al. (1979) reported thermostable malate dehydrogenase from some thermophilic fungi. Many of these fungi do not appear to synthesize heat-stable malate dehydrogenase; some factors *in vivo* appear to impart thermostability to enzymes that are otherwise thermolabile. Salts of sodium, potassium, ammonium, and citrate could protect malate dehydrogenase of *Humicola lanuginosa*, *Mucor pusillus*, and *Chaetomiun thermophile* against thermal inactivation. Proteases of thermophilic fungi have optima at 45°C–55°C (Craveri et al., 1974; Zakirov et al., 1975; Satyanarayana and Johri, 1983) and lipases at 45°C–50°C (Ogundero, 1980). Craveri and Colla (1966) reported the average optimum temperature of thermophilic ribonucleases as 60.5°C. The optimum temperature of α-amylase activity from the thermophilic fungus *Thermomyces lanuginosus* was 60°C (Jensen and Olsen, 1992). The xylanase and cellulases possess optima at 50°C–70°C (Vandamme et al., 1982; Grajek, 1987; Yoshioka et al., 1987; Bernier and Stutzenberger, 1989). These are glycoproteins, and they are associated with various amounts (13%–68%) of carbohydrates (Flannigan and Sellars, 1978; Hayashida and Yoshioka, 1980). The carbohydrate moiety appears to play a key role in stabilizing the conformational structures of such enzymes at high temperatures.

A thermoprotective disaccharide trehalose (α,α-1,1-diglucose, or fungal sugar, mycose) is known to occur in fungi, myxomycetes, some animals, plants, and bacteria. It is the most widespread naturally occurring disaccharide and is primarily localized in the cytoplasm. It is synthesized during the sporulation process in fungi and attains the highest level in resting form, thus its nickname "dormancy sugar." Under HS, glucose levels dramatically increase in cells as a partial inhibition of glycolysis occurs. Trehalose synthesis may protect against the toxic and mutagenic effects of high glucose levels. A 100-fold increase in trehalose concentration is reported in the cells of *Saccharomyces cerevisiae* when it is grown at 40°C (Hottiger et al., 1987). In filamentous fungi, trehalose is a transported form of carbohydrate. Evidence suggests that under oxidative stress, certain derivatives of trehalose impart cell protection by protecting its proteins against oxidative degradation. Tereshina (2005) described the multifarious functions of trehalose, including (1) a reserve carbohydrate used during germination and in the process of storing fungal spores, (2) a membrane protectant under various types of stress (thermal, oxidative, and osmotic stresses and stresses caused by heavy metals, drugs, and metabolic inhibitors), (3) the regulator of the process of glycolysis and of cell concentrations of glucose and ATP, (4) a transported carbohydrate form, and (5) a chemical chaperone involved in protein stabilization and folding. Owing to the characteristic features of its structure and function, it is hypothesized that trehalose is an important macromolecule protecting the subcellular structures during HS, and thus playing a major role in thermophily.

2.4.8 Ultrastructural Thermostability and Pigmentation

In general, there appears to be no specific difference in ultrastructural features, such as organelles, structural modifications, or developmental patterns of thermophiles and mesophiles. Nevertheless, thermophilic fungi cultivated at high temperature (50°C) have been reported to contain dense body vesicles as the predominant storage organelles, containing mainly phospholipids but a few neutral lipid droplets. A different mode of lipid storage in thermophilic fungi is reported to be a remarkable ultrastructural feature compared with that in fungi grown at lower temperatures. In general, thermophilic molds contain more lipids than their mesophilic counterparts (Satyanarayana and Johri, 1992). Similarly, the presence of membrane-bound inclusion bodies in the cytoplasm associated with the plasma membrane of the hyphae of some thermophilic fungi have been reported by Garrison et al. (1975). These are probably lipoidal in nature, as treatment with thiocarbohydrazide enhanced the electron opacity of these bodies. Johri (1983) also observed some electron-dense bodies in the sporangiospores of *Rhizopus rhizopodiformis*. Thermophilic molds exhibited avidity for binding sterols (St) and their derivatives to the mycelium, suggesting a possible role in thermostability (Satyanarayana and Chavant, 1987). Furthermore, bright pigmentation with the deposition of melanin in cell walls of reproductive structures such as ascocarp, ascospores, peridia, and conidia has been observed in some thermophilic fungi (*Talaromyces duponti*, *Thermoascus aurantiacus*, *Malbranchea cinnamomea*, and *Thermomyces lanuginosus*). This pigmentation phenomenon appears to impart protection of the organism under stressful conditions of environmental temperature and atmospheric humidity; a similar phenomenon also occurs in mesophiles when they are grown in unfavorable conditions of atmospheric humidity and temperature. Spectroscopic analysis of these pigments revealed that they are similar to aphins, the polycyclic quinones found in the hemolymph (circulating fluid) of several colored species of aphids. Thus, the possibility of the association of thermostability with ultrastructural elements as organelles appears to be a promising hypothesis.

2.4.9 LIPID SOLUBILIZATION

The lipid solubilization hypothesis suggests that all biological processes would cease if the cell lost its integrity due to solubilization, that is, melting of the membrane lipids at elevated temperatures, eventually resulting in cell death. Satyanarayana and Johri (1992) reported that thermophilic fungal lipids are made up mainly of palmitic, oleic, and linoleic acids, with low levels of lauric, palmitoleic, and stearic acids. Thermophilic fungi thus do not appear to contain unusual fatty acids. The lipid composition of thermophilic fungi therefore plays a critical role in thermophily, as many cellular functions are membrane linked. It has been shown that thermophilic fungi synthesize more saturated lipids than mesophilic species (Meyer and Bloch, 1963; Mumma et al., 1970). By virtue of their higher melting point owing to more saturated lipids, thermophiles are expected to be able to maintain their cellular integrity at elevated temperature compared with mesophiles with their less saturated lipids. Mumma et al. (1970) observed extremely high amounts of phosphatidic acids (PAs) in the composition of phospholipids of *Humicola grisea* var. *thermoidea*, suggesting an extremely important role of these compounds in thermophily. Sumner and Morgan (1969) reported a higher level of lipids in *Rhizomucor miehe* and *Rhizomucor pusillus* at 25°C than at 48°C. Similarly, the lipid content in the mycelium of *Thermomyces lanuginosus* was 75.1% at 37°C and 8.5% at 52°C; the unusually high lipid at 37°C could be due to accumulation of storage lipid at suboptimal growth temperatures (Crisan, 1973).

A comparison of the closely related mesophilic and thermophilic fungi *Chaetomium brasiliense* and *Chaetomium thermophile* var. *thermophile*, respectively, revealed that the fatty acids of the thermophilic fungus were more saturated than those of its mesophilic counterpart (Oberson et al., 1999). It was also observed that under optimal growth temperatures, both fungi synthesized comparable amounts of trehalose, whereas in response to HS, the amount of trehalose increased in both fungi, but the thermophilic fungus produced less trehalose than the mesophile. Nevertheless, these data revealed that there is no direct relationship between trehalose and thermophily. Recently, Yanutsevich et al. (2014) have shown, for the first time, an increase in the proportions of PAs and St, and a decrease in phosphatidylethanolamines (PEs) and phosphatidylcholines (PCs), in response to HS in the thermophilic fungus *Rhizomucor miehei*. However, the degree of unsaturation of the fatty acids in the phospholipids did not decrease. They further detected a high level of trehalose in the cells of fungi during growth under optimum temperatures, but in response to HS, there was a sharp decline in its quantity. Ianutsevich et al. (2016) studied the composition of the soluble cytosol carbohydrates and membrane lipids of the thermophilic fungi *Rhizomucor tauricus* and *Myceliophthora thermophila* at optimum temperature conditions (41°C–43°C) and under HS (51°C–53°C). They observed that under optimum temperatures, the membrane lipid composition was characterized by a high proportion of PAs (20%–35% of the total), which were the main components of the membrane lipids, together with PCs, PEs, and St. In response to HS, the proportion of PAs and St increased, whereas the amount of PCs and PEs decreased. However, they observed no decrease in the degree of fatty acid desaturation in the major phospholipids under HS. The function of PAs under HS seems to be their involvement in the processes of endocytosis and exocytosis due to their ability to form a microdomain and their participation in the formation of membrane curvatures (Kooijman et al., 2003; McMahon and Gallop, 2005). Ianutsevich et al. (2016) have demonstrated that the changes in the membrane lipids under HS in thermophiles and mesophiles were similar, and could be characterized by an increase in the relative content of PAs and St. No specific change in the fatty acid unsaturation induced by HS was observed. Therefore, this HS pattern in thermophilic fungi seems to be conserved. Thus, the lipid solubilization hypothesis suggests that membrane lipids per se are responsible for maintaining the integrity of the cells by protecting their proteins from denaturation at elevated temperatures.

2.5 ACQUIRED THERMOTOLERANCE

Acquired thermotolerance refers to the prolonged survival of organisms at lethal temperatures after they are subjected to sublethal HS. For instance, if a fungal organism growing optimally at 25°C is exposed to a temperature of 37°C (sublethal), it acquires resistance to a lethal temperature of 50°C. In response to HS, mesophilic fungi display synthesis of HSPs, accumulation of trehalose, change of the state of water in the cell compartments, and membrane composition (Piper, 1993). Together, all these changes result in the emergence of a new property—acquired thermotolerance to the lethal temperature. Trent et al. (1994) revealed that similar to thermophilic bacteria and archaea, thermophilic fungi also synthesize HSPs in order to acquire thermotolerance. They demonstrated that conidia of *Thermomyces lanuginosus* germinate at 50°C, however, if heat shocked at 55°C for 1 h prior to exposure to 58°C, exhibited enhanced survival compared with non–heat shocked conidia. They further observed that thermotolerance could be eliminated if protein synthesis is inhibited using cycloheximide during the HS period. After pulse labeling of the proteins during the HS period and their separation on sodium dodecyl sulfate–polyacrylamide gel electrophoresis (SDS-PAGE), they observed the increased synthesis of eight HSPs. Out of these, three small HSPs (molecular weight 31–33 kDa) dominated the HS response. In view of the fact that *Thermomyces lanuginosus* is capable of growth up to 60°C–62°C, the role of HSPs induced at 55°C is not very clear. As explained earlier, proteins of the small HSPs are considered to be important in thermophily.

While studying the HS response of the thermophilic budding yeast *Hansenula polymorpha*, Reinders et al. (1999) observed accumulation of a large amount of trehalose during 2 h if HS at 47°C, thereby increasing the survival rate of wild-type cells toward the challenging HS (40 min at 56.5°C) by more than 1000-fold. In a recent study, Ianutsevich et al. (2016) made an interesting observation with regard to acquired thermotolerance. They observed that in mesophilic fungi trace amounts of trehalose are present at their optimal growth temperature, which sharply increased and was maintained at 6%–8% of dry weight over 6 h of HS (Tereshina et al., 2010). Interestingly, in thermophiles under optimal temperature conditions, high levels of trehalose (8%–18% of dry weight) are always present at all stages of growth, while a marked decrease in trehalose level was observed in response to HS, which may be attributed to the lack of acquired thermotolerance. Thus, their data strongly suggest a relationship between thermophilia and the level of the thermoprotective disaccharide trehalose, and between acquired thermotolerance and an increased trehalose level. They further came to the conclusion that in contrast to mesophilic fungi, thermophilic fungi did not show any "acquired thermotolerance" in response to HS owing to their inability to further increase the synthesis of trehalose. This HS response pattern seems to be conserved in filamentous thermophilic fungi. Thus, thermophilic fungi, unlike mesophilic ones, do not have mechanisms of protection against HS and do not acquire thermotolerance after being exposed to HS. This determines low limits of the fungal thermophily (up to 60°C) (Wharton, 2002).

2.6 HOMEOVISCOUS VERSUS HOMEOPHASIC ADAPTATIONS

All biological systems operate in relatively narrow temperature ranges (tolerance zone) explicit for a particular species. Changes in temperature out of the optimal range may result in abnormalities of almost all biological processes. In response to environmental temperature fluctuation, living systems have evolved various mechanisms of defense for low and high temperatures. The homeoviscous adaptation hypothesis, first postulated by Sinensky in 1974, pertains to the maintenance of a certain membrane viscosity by changing the unsaturation degree of the phospholipid acyl chain. This involves an increase in the proportion of saturated fatty acids in phospholipids at high temperatures, and an increase in proportion of unsaturated fatty acids in phospholipids at lower temperatures. For example, many organisms vary the fatty acid composition of their membrane phospholipids in order to maintain membrane fluidity for the optimal functioning of membrane-localized transporters and enzymes. Wright et al. (1983) observed whether failure to regulate membrane fluidity may be an explanation for the high minimum temperature of growth of thermophilic fungi. To analyze this phenomenon, they shifted *Talaromyces thermophilus* from a high (50°C) to a low (33°C) growth temperature and observed that there was virtually no change in the degree of unsaturation of fatty acids. This may be due to metabolic limitation of the fungus, most probably because of a nonfunctional fatty acid desaturase, which restricted the ability of the fungus to convert oleate to linoleate at lower temperatures. Contrary to this, Rajasekaran and Maheshwari (1990) reported that in *Thermomyces lanuginosus* the concentration of linoleic acid at 30°C was twice that at 50°C. They observed that the degree of unsaturation of phospholipid fatty acid was 0.88 in mycelia grown at 50°C, whereas it was 1.0 in the temperature-shifted cultures (from 50°C to 30°C) and 1.06 in cultures grown at 30°C constantly. However, no decrease in the degree of fatty acid desaturation in the major phospholipids in response to HS in the thermophilic *Rhizomucor tauricus* and *Myceliophthora thermophila* was observed (Ianutsevich et al., 2016). Therefore, the conception of the inability of thermophilic fungi to adjust membrane fluidity does not seem to be a reason for their high minimum temperature of growth.

The homeophasic hypothesis ascribes the greatest importance to maintaining a certain balance between the "bilayer" and "nonbilayer" lipids. According to this hypothesis, put forward by Hazel (1995), adaptation to thermal effects is due to the maintenance of a certain ratio between the stabilizing lipids, which are cylindrical in shape (PCs) and form the bilayer, and the destabilizing ones, which are conical in shape (PEs and PAs) and form nonbilayer structures. However, both homeoviscous and homeophasic hypotheses do not make a difference between the thermal effects within the tolerance zone and under HS. Therefore, comparison of the responses to these two types of effects revealed fundamental differences. But how changes in protein state and content correlate with those of lipids and sugars, and what types of contributing links exist between these types of adaptation remain obscure.

In summarizing the origin of thermophily in the kingdom Fungi, the adaptation of thermophiles to high temperatures is a combination of different strategies, including genetic selection and functional acclimatization. Cellular components, such as lipids, nucleic acids, and proteins, are usually thermolabile. Genomics studies of thermophiles have revealed that the evolution of thermal tolerance depends on the levels of heritable variations, for example, reduction in genome size, horizontal DNA transfer, and gene mutation. The available information suggests that the analyses of gene expression in the transcriptome and proteome in thermophiles show that the global transcriptional regulation of certain operons containing key functional genes is an efficient response to thermal stress, supported by the thermostable and efficient protein synthetic machinery. Further, the functional system is vital for thermophilic adaptation, in particular for those proteins functioning under thermal stress. Analyses of the proteomes of thermophilic fungi have shown that most thermophiles contain gene clusters coding for proteins with tolerance to high temperatures, as a number of HSP families have been reported. Similarly, homeoviscous and homeophasic adaptation of molecules of thermophilic fungi allows them to carry out their biological processes at elevated temperatures. However, contradictory reports concerning the occurrence of the acquired thermotolerance phenomenon in thermophilic fungi put the later hypothesis in a state of disarray.

REFERENCES

Abdel-Hafez, S.I.I. 1982. Thermophilic and thermotolerant fungi in the desert soils of Saudi Arabia. *Mycopathologia* 80:15–20.

Allen, M.B. 1950. The dynamic nature of thermophily. *Journal of General Physiology* 33: 205–214.

Amlacher, S., Sarges, P., Flemming, D., van Noort, V., Kunze, R., Devos, D.P., Arumugam, M., Bork, P., and Hurt, E. 2011. Insight into structure and assembly of the nuclear pore complex by utilizing the genome of a eukaryotic thermophile. *Cell* 146:277–289.

Apinis, A.E. 1963. Occurrence of thermophilous microfungi in certain alluvial soils near Nottingham. *Nova Hedwigia* 5:57–78.

Averhoff, B., and Muller, V. 2010. Exploring research frontiers in microbiology: Recent advances in halophilic and thermophilic extremophiles. *Research in Microbiology* 161:506–514.

Berezovsky, I.N., and Shakhnovich, E.I. 2005. Physics and evolution of thermophilic adaptation. *Proceedings of the National Academy of Sciences of the United States of America* 102:12742–12747.

Berka, R.M., Grigoriev, I.V., Otillar, R., Salamov, A., Grimwood, J., Reid, I., Ishmael, N. et al. 2011. Comparative genomic analysis of the thermophilic biomass-degrading fungi *Myceliophthora thermophila* and *Thielavia terrestris*. *Nature Biotechnology* 29:922–927.

Bernier, R., and F. Stutzenberger. 1989. β-Glucosidase biosynthesis in *Thermomonospora curvata*. *MIRCEN Journal* 5:15–25.

Blackman, F.F. 1905. Optima and limiting factors. *Annals of Botany (London)* 19:281–295.

Blumer-Schuette, S.E., Ozdemir, I., Mistry, D., Lucas, S., Lapidus, A., Cheng, J.F., Goodwin, L.A. et al. 2011. Complete genome sequences for the anaerobic, extremely thermophilic plant biomass-degrading bacteria *Caldicellulosiruptor hydrothermalis*, *Caldicellulosiruptor kristjanssonii*, *Caldicellulosiruptor kronotskyensis*, *Caldicellulosiruptor owensensis*, and *Caldicellulosiruptor lactoaceticus*. *Journal of Bacteriology* 193:1483–1484.

Bott, T.L., and T.D. Brock. 1969. Bacterial growth rates above 90°C in Yellowstone hot springs. *Science* 164:1411–1412.

Broad, T.E., and Shepherd, M.G. 1971. Purification and properties of glucose-6-phosphate dehydrogenase from the thermophilic fungus *Penicillium dupontii*. *Biochimica et Biophysica Acta* 198:407–414.

Brock, T.D. 1967. Life at high temperatures. *Science* 158:1012–1019.

Chakrabartty, A., Schellman, J.A., and Baldwin, R.L. 1991. Large differences in the helix propensities of alanine and glycine. *Nature* 351:586–588.

Chang, Y.C., Tsai, H.F., Karos, M., and Kwon-Chung, K.J. 2004. THTA, a thermotolerance gene of *Aspergillus fumigatus*. *Fungal Genetics and Biology* 41:888–896.

Chapins, R., and Zuber, H. 1970. Thermophilic aminopeptidases: API from *Talaromyces duponi*. *Methods in Enzymology* 19:552–556.

Chen, K.Y., and Chen, Z.C. 2004. Heat shock proteins of thermophilic and thermotolerant fungi from Taiwan. *Botanical Bulletin of Academia Sinica* 45:247–257.

Craveri, R., and Colla, C. 1966. Ribonuclease activity of mesophilic and thermophilic molds. *Annals of Microbiology* 16:97–99.

Craveri, R., Manachini, P.L., and Aragozzini, F. 1974. Thermostable and alkaline lipolytic and proteolytic enzymes produced by a thermophilic mold. *Mycopathologia et Mycologia Applicata* 54:193–204.

Crisan, E.V. 1973. Current concepts of thermophilism and the thermophilic fungi. *Mycologia* 65:1171–1198.

De Bertoldi, M., Lepidi, A.A., and Nuti, M.P. 1973. Significance of DNA base composition in classification of *Humicola* and related genera. *Transactions of the British Mycological Society* 60:77–85.

Dix, N.J., and Webster, J. 1995. *Fungal Ecology*. London: Chapman & Hall.

Eggins, H.O., and Malik, K.A. 1969. The occurrence of thermophilic cellulolytic fungi in a pasture land soil. *Antonie van Leeuwenhoek* 35:178–184.

Eicker, A. 1972. Occurrence and isolation of South African thermophilic fungi. *South African Journal of Science* 68:150–155.

Ellis, D.H. 1980. Thermophilous fungi isolated from some Antarctic and sub-Antarctic soils. *Mycologia* 72:1033–1036.

Ellis, D.H., and Keane, P.J. 1981. Thermophilic fungi isolated from some Australian soils. *Australian Journal of Botany* 29:689–704.

Emerson, R. 1968. Thermophiles. In *The Fungi*, ed. G.C. Ainsworth and A.S. Sussman, 105–128. New York: Academic Press.

England, J.L., Shakhnovich, B.E., and Shakhnovich, E.I. 2003. Natural selection of more designable folds: A mechanism for thermophilic adaptation. *Proceedings of the National Academy of Sciences of the United States of America* 100:8727–8731.

Eriani, G., Delarue, M., Poch, O., Gangloff, J., and Moras, D. 1990. Partition of tRNA synthetases into two classes based on mutually exclusive sets of sequence motifs. *Nature* 347:203–206.

Evans, H.C. 1971. Thermophilous fungi of coal spoil tips. II. Occurrence, distribution and temperature relationships. *Transactions of the British Mycological Society* 57: 255–266.

Flannigan, B., and Sellars, P.N. 1972. Activities of thermophilous fungi from barley kernels against arabinoxylan and carboxymethyl cellulose. *Transactions of the British Mycological Society* 58:338–341.

Flannigan, B., and Sellars, P.N. 1978. Production of xylanolytic enzymes by Aspergillus fumigatus. *Transactions of the British Mycological Society* 71:353–358.

Freeman, S., Ginzburg, C., and Katan, J. 1989. Heat shock protein synthesis in propagules of *Fusarium oxysporum* f. sp. *niveum. Phytopathology* 79:1054–1058.

Fries, L. 1953. Factors promoting growth of *Coprinus fimetarius* (L.) under high temperature conditions. *Physiologia Plantarum* 6:551–563.

Garrison, R.G., Boyd, K.S., and Lane, J.W. 1975. Ultrastructural studies on *Thermomyces lanuginosa* and certain other closely related thermophilic fungi. *Mycologia* 67:961–971.

Gomes, E., de Souza, A.R., Orjuela, G.L., Da Silva, R., de Oliveira, T.B., and Rodrigues, A. 2016. Applications and benefits of thermophilic microorganisms and their enzymes for industrial biotechnology. In *Gene Expression Systems in Fungi: Advancements and Applications, Fungal Biology*, ed. M. Schmoll and C. Dattenböck, 459–492.

Grajek, W. 1987. Comparative studies on the production of cellulases by thermophilic fungi in submerged and solid-state fermentation. *Applied Microbiology and Biotechnology* 26:126–129.

Grzimek, B. 1984. *Animal Life Encyclopedia*, vol. 7, Birds I, 432–443. New York: Van Nostrand Reinhold.

Guiraud, P., Steiman, R., Seigle-Murandi, F., and Sage, L. 1995. Mycoflora of soil around the Dead Sea II: Deuteromycetes (except *Aspergillus* and *Penicillium*). *Systematic and Applied Microbiology* 18:318–322.

Hayashida, S., and Yoshioka, H. 1980. Thermostable cellulases from *Humicola insolens* YH-8. *Agricultural and Biological Chemistry* 44:1721–1728.

Hazel, J.R. 1995. Thermal adaptation in biological membranes: Is homeoviscous adaptation the explanation? *Annual Review of Physiology* 57:19–42.

Hemida, S.K. 1992. Thermophilic and thermotolerant fungi isolated from cultivated and desert soils, exposed continuously to cement dust particles in Egypt. *Zentralblatt für Mikrobiologie* 147:277–281.

Hottiger, T., Schmutz, P., and Wiemken, A. 1987. Heat-induced accumulation and futile cycling of trehalose in *Saccharomyces cerevisiae. Journal of Bacteriology* 169:5518–5522.

Hummer, G., Garde, S., Garcia, A.E., Paulaitis, M.E., and Pratt, L.R. 1998. The pressure dependence of hydrophobic interactions is consistent with the observed pressure denaturation of proteins. *Proceedings of the National Academy of Sciences of the United States of America* 95:1552–1555.

Ianutsevich, E.A., Olga, D.A., Groza, N.V., Kotlova, E.R., and Tereshina, V.M. 2016. Heat shock response of thermophilic fungi: Membrane lipids and soluble carbohydrates under elevated temperatures. *Microbiology* 162:989–999.

Jensen, B., and Olsen, J. 1992. Physicochemical properties of a purified alpha-amylase from the thermophilic fungus *Thermomyces lanuginosus. Enzyme and Microbial Technology* 14:112–116.

Johri, B.N. 1983. Fine structure in freeze fracture spore of a thermophilic *Rhizopus. Tropical Plant Science Research* 1:39–41.

Johri, B.N., Jain, S., and Chauhan, S. 1985. Enzymes from thermophilic fungi: Proteases and lipases. *Proceedings of the Indian Academy of Sciences (Plant Science)* 94:175–196.

Johri, B.N., and Satyanarayana, T. 1986. Thermophilic moulds: Perspectives in basic and applied research. *Indian Review of Life Sciences* 6:75–100.

Johri, B.N., and Thakre, R.P. 1975. Soil amendments and enrichment media in the ecology of thermophilic fungi. *Proceedings of the Indian National Science Academy* 41:564–570.

Koffler, H. 1957. Protoplasmic differences between mesophiles and thermophiles. *Bacteriological Reviews* 21:227–270.

Kooijman, E.E., Chupin, V., de Kruijff, B., and Burger, K.N. 2003. Modulation of membrane curvature by phosphatidic acid and lysophosphatidic acid. *Traffic* 4:162–174.

Lambros, R.J., Mortimer, J.R., and Forsdyke, D.R. 2003. Optimum growth temperature and the base composition of open reading frames in prokaryotes. *Extremophiles* 7:443–450.

Larkin, J.M., and Stokes, J.L. 1968. Growth of psychrophilic microorganisms at subzero temperatures. *Canadian Journal of Microbiology* 14:97–101.

Loomis, W.F., and Wheeler, S.A. 1982. Chromatin-associated heat shock proteins of *Dictyostelium*. *Developmental Biology* 90:412–418.

Maheshwari, R., Bharadwaj, G., and Bhat, M.K. 2000. Thermophilic fungi: Their physiology and enzymes. *Microbiology and Molecular Biology Reviews* 64:461–488.

Maheshwari, R., Kamalam, P.T., and Balasubramanyam, P.V. 1987. The biogeography of thermophilic fungi. *Current Science* 56:151–154.

Martins, E.S., Leite, R.S., and da Silva, R. 2013. Purification and properties of polygalacturonase produced by thermophilic fungus *Thermoascus aurantiacus* CBMAI-756 on solid-state fermentation. *Enzyme Research* 2013, 438645. doi: 10.1155/2013/438645.

McAlister, L., and Finkelstein, D.B. 1980. Heat shock protein and thermal resistance in yeast. *Biochemical and Biophysical Research Communications* 93:819–824.

Mchunu, N.P., Permaul, K., Rahman, A.Y.A., Saiot, J.A., Singh, S., and Alam, M. 2013. Xylanase superproducer: Genome sequence of a compost-loving thermophilic fungus, *Thermomyces lanuginosus* strain SSBP. *Genome Announcement* 1:e00388-13.

McMahon, H.T., and Gallop, J.L. 2005. Membrane curvature and mechanisms of dynamic cell membrane remodelling. *Nature* 438:590–596.

Mehrotra, B.S. 1985. Thermophilic fungi—Biological enigma and tools for the biotechnologist and biologist. *Indian Phytopathology* 38:211–229.

Meyer, F., and Bloch, K. 1963. Effect of temperature on the enzymatic synthesis of unsaturated fatty acids in *Torulopsis utillis*. *Biochimica et Biophysica Acta* 77:671–673.

Miller, H.M., Sullivan, P.A., and Shepherd, M.G. 1974. Intracellular protein breakdown in thermophilic and mesophilic fungi. *Biochemical Journal* 144:209–214.

Mouchacca J., 1995. Thermophilic fungi in desert soils, a neglected extreme environment. In *Microbial Diversity and Ecosystem Function*, ed. D. Allsopp et al., 265–288. London: CAB International and Royal Holloway, University of London.

Mumma, R.O., Fergus, C.L., and Sekura, R.D. 1970. The lipids of thermophilic fungi: Lipid composition comparisons between thermophilic and mesophilic fungi. *Lipids* 5:100–103.

Musto, H., Naya, H., Zavala, A., Romero, H., Alvarez-Valin, F., and Bernardi, G. 2005. The correlation between genomic G_C and optimal growth temperature of prokaryotes is robust: A reply to Marashi and Ghalanbor. *Biochemical and Biophysical Research Communication* 330:357–360.

Nageswara Rao, J.S., and Cherayil, J.D. 1979. Minor nucleotides in the ribosomal RNA of *Thermomyces langinosus*. *Current Science* 48:1–5.

Neidhardt, F.C., VanBogelen, R.A., and Vaughn, V. 1984. The genetics and regulation of heat-shock proteins. *Annual Review of Genetics* 18:295–329.

Nelson, K.E., Clayton, R.A., Gill, S.R., Gwinn, M.L., Dodson, R.J., Haft, D.H., Hickey, E.K. et al. 1999. Evidence for lateral gene transfer between Archaea and Bacteria from genome sequence of *Thermotoga maritima*. *Nature* 399:323–329.

Neuman, D., Nieden, U., Manteuffel, R., Walter, G., Scharf, K.D., and Nover, L. 1987. Intracellular localization of heat-shock proteins in tomato cell cultures. *European Journal of Cell Biology* 43:71–81.

Nierman, W.C., Pain, A., Anderson, M.J., Wortman, J.R., Kim, H.S., Arroyo, J., Berriman, M. et al. 2005. Genomic sequence of the pathogenic and allergenic filamentous fungus *Aspergillus fumigatus*. *Nature* 438:1151–1156.

Oberson, J., Rawyler, A., Brandle, R., and Canevascini, G. 1999. Analysis of the heat-shock response displayed by two *Chaetomium* species originating from different thermal environments. *Fungal Genetics and Biology* 26:178–189.

Ogundero, V.W. 1980. Lipase activities of thermophilic fungi from mouldy groundnuts in Nigeria. *Mycologia* 72:118–126.

Piper, P.W. 1993. Molecular events associated with acquisition of heat tolerance by the yeast *Saccharomyces cerevisiae*. *FEMS Microbiology Reviews* 11:339–356.

Plesofsky-Vig, N. 1996. The heat-shock protein and stress response. In *The Mycota. III. Biochemistry and Molecular Biology*, ed. R. Brambl and G.A. Marzluf, 171–190. Berlin: Springer.

Ponnuswamy, P., Muthusamy, R., and Manavalan, P. 1982. Amino acid composition and thermal stability of globular proteins. *International Journal of Biological Macromolecules* 4:186–190.

Powell, A.J., Parchert, K.J., Bustamante, J.M., Ricken, J.B., Hutchinson, M.I., and Natvig, D.O. 2012. Thermophilic fungi in an aridland ecosystem. *Mycologia* 104:813–825.

Prasad, A.R.S., Kurup, C.R., and Maheshwari, R. 1979. Effect of temperature on respiration of a mesophilic and a thermophilic fungus. *Plant Physiology* 64:347–348.

Rahn, O. 1945. Physical methods of sterilization of microörganisms. *Bacteriological Reviews* 9:1–47.

Rajasekaran, A.K., and Maheshwari, R. 1990. Growth kinetics and intracellular protein breakdown in mesophilic and thermophilic fungi. *Indian Journal of Experimental Biology* 28:46–51.

Rajasekaran, A.K., and Maheshwari, R. 1993. Thermophilic fungi: An assessment of their potential for growth in soil. *Journal of Bioscience* 18:345–354.

Reinders, A., Romano, I., Wiemken, A., and De Virgilio, C. 1999. The thermophilic yeast *Hansenula polymorpha* does not require trehalose synthesis for growth at high temperatures but does for normal acquisition of thermotolerance. *Journal of Bacteriology* 181:4665–4668.

Ronning, C.M., Fedorova, N.D., Bowyer, P., Coulson, R., Goldman, G., Kim, H.S., Turner, G. et al. 2005. Genomics of *Aspergillus fumigatus*. *Revista Iberoamericana de Micologia* 22:223–228.

Salar, T.K., and Aneja, K.R. 2006. Thermophilous fungi from temperate soils of northern India. *Journal of Agricultural Technology* 2:49–58.

Sandhu, D.K., and Singh, S. 1981. Distribution of thermophilous microfungi in forest soils of Darjeeling (Eastern Himalayas). *Mycopathologia* 74:79–85.

Satyanarayana, T., and Chavant, L. 1987. Bioconversion and binding of sterols by thermophilic moulds. *Folia Microbiologica* 32:353–359.

Satyanarayana, T., and Johri, B.N. 1983. Extracellular protease production of thermophilic fungi of decomposing paddy straw. *Tropical Plant Science Research* 1:137–140.

Satyanarayana, T., and Johri, B.N. 1992. Lipids of thermophilic moulds. *Indian Journal of Microbiology* 32:1–14.

Satyanarayana, T., Johri, B.N., and Klein, J. 1992. Biotechnological potential of thermophilic fungi. In *Handbook of Applied Mycology*, ed D.K. Arora, R.P. Elander, and K.G. Mukerji, 729–761. New York: Marcel Dekker.

Schlesinger, M.J. 1986. Heat shock proteins: The search for functions. *Journal of Cell Biology* 103:321–325.

Seymour, R.S., and Bradford, D.F. 1992. Temperature regulation in the incubation mounds of the Australian brush-turkey. *Condor* 94:134–150.

Shakhnovich, B.E., Deeds, E., Delisi, C., and Shakhnovich, E. 2005. Protein structure and evolutionary history determine sequence space topology. *Genome Research* 15: 385–392.

Sinensky, M. 1974. Homeoviscous adaptation—A homeostatic process that regulates the viscosity of membrane lipids in *Escherichia coli*. *Proceedings of the National Academy of Sciences of the United States of America* 71:522–525.

Singer, G.A., and Hickey, D.A. 2003. Thermophilic prokaryotes have characteristic patterns of codon usage, amino acid composition and nucleotide content. *Gene* 317:39–47.

Singh, B., Poças-Fonseca, M.J., Johri B.N., and Satyanarayana, T. 2016. Thermophilic molds: Biology and applications. *Critical Reviews in Microbiology* 42:985–1006.

Somkuti, G.A., and Babel, F.J. 1968. Lipase activity of *Mucor pusillus*. *Applied Microbiology* 16:617–619.

Stetter, K.O. 1999. Extremophiles and their adaptation to hot environments. *FEBS Letters* 452:22–25.

Sumner, J.L., and Morgan, E.D. 1969. The fatty acid composition of sporangiospores and vegetative mycelium of temperature adapted fungi in the order Mucorales. *Journal of General Microbiology* 59:215–221.

Sundaram, B.M. 1977. Isolation of thermophilic fungi from soils. *Current Science* 46: 106–107.

Taber, R.A., and Pettit, R.E. 1975. Occurrence of thermophilic microorganisms in peanuts and peanut soil. *Mycologia* 67:157–161.

Tansey, M.R., and Brock, T.D. 1978. Microbial life at high temperature ecological aspects. In *Microbial Life in Extreme Environments*, ed. D. Kushner, 159–216. London: Academic Press.

Tansey, M.R., and Jack, M.A. 1976. Thermophilic fungi in sun heated soils. *Mycologia* 68:1061–1075.

Tereshina, V.M. 2005. Thermotolerance in fungi: The role of heat shock proteins and trehalose. *Microbiology* 74:247–257.

Tereshina, V.M., Memorskay, A.S., Kotlova, E.R., and Feofilova, E.P. 2010. Membrane lipid and cytosol carbohydrate composition in *Aspergillus niger* under heat shock. *Microbiology* 79:40–46.

Thakre, R.P., and Johri, B.N. 1976. Occurrence of thermophilic fungi in coal mine soils of Madhya Pardesh. *Current Science* 45:271–273.

Trent, J.D., Gabrielsen, M., Jensen, B., Neuhard, J., and Olsen, J. 1994. Acquired thermotolerance and heat shock proteins in thermophiles from the three phylogenetic domains. *Journal of Bacteriology* 176:6148–6152.

Vandamme, E.J., Logghe, J.M., and Geeraerts, H.A. 1982. Cellulase activity of a thermophilic *Aspergillus fumigatus* (fresenius) strain. *Journal of Chemical Technology and Biotechnology* 32:968–974.

van Noort, V., Bradatsch, B., Arumugam, M., Amlacher, S., Bange, G., Creevey, C., Falk, S., Mende, D.R., Sinning, I., Hurt, E., and Bork, P. 2013. Consistent mutational paths predict eukaryotic thermostability. *BMC Evolutionary Biology* 13:7.

Wali, A.S., Mattoo, A.K., and Modi, V.V. 1979. Comparative temperature-stability properties of malate dehydrogenases from some thermophilic fungi. *International Journal of Peptide and Protein Research* 14:99–106.

Wang, Q., Cen, Z., and Zhao, J. 2015. The survival mechanisms of thermophiles at high temperatures: An angle of omics. *Physiology* 30:97–106.

Ward, J.E., and Cowley, G.T. 1972. Thermophilic fungi of some central South Carolina forest soils. *Mycologia* 64:200–205.

Wharton, D.A. 2002. The hot club. In *Life at the Limits: Organisms in Extreme Environments*, 129–149. Cambridge: Cambridge University Press.

Wright, C., Kafkewitz, D., and Somberg E.W. 1983. Eucaryote thermophily: Role of lipids in the growth of *Talaromyces thermophilus*. *Journal of Bacteriology* 156:493–497.

Yanutsevich, E.A., Memorskaya, A.S., Groza, N.V., Kochkina, G.A., and Tereshina, V.M. 2014. Heat shock response in the thermophilic fungus *Rhizomucor miehei*. *Microbiology* 83:498–504.

Yoshioka, H., Chavnich, S., Nilubol, N., and Hayashida, S. 1987. Production and characterization of thermostable xylanase from *Talaromyces byssochlamydoides* YH-50. *Agricultural and Biological Chemistry* 45:579–586.

Zakirov, M.Z., Shchrlokova, S.S., and Karavaeva, N.N. 1975. *Torula thermophila* UzPT-1 thermophilic fungus producing proteolytic enzymes. *Applied Biochemistry and Microbiology* 2:608–612.

Zeldovich, K.B., Berezovsky, I.N., and Shakhnovich, E.I. 2007. Protein and DNA sequence determinants of thermophilic adaptation. *PLoS Computational Biology* 3:e5.

Zhou, P., Zhang, G., Chen, S., Jiang, Z., Tang, Y., Henrissat, B., Yan, Q., Yang, S., Chen, C.F., Zhang, B., and Du, Z. 2014. Genome sequence and transcriptome analyses of the thermophilic zygomycete fungus *Rhizomucor miehei*. *BMC Genomics* 15:294.

Zuber, H. 1967. Sequence analysis of peptides with citrus carboxypeptidase and with a thermophilic carboxypeptidase from *Talaromyces dupontii*. In *Abstracts: 7th International Congress of Biochemistry, Tokyo*, 541–542. Vol. 3. Tokyo: Science Council of Japan.

3 Physiology of Thermophilic Fungi

3.1 INTRODUCTION

Fungi comprise a diverse group of heterotrophic eukaryotic microorganisms that occupy a variety of habitats. The majority of fungi are mesophiles growing in the temperature range between 10°C and 40°C, but a small group of fungi are capable of growth above 50°C and are categorized as thermophilic or thermotolerant forms. All those fungi growing at or above 40°C were termed as thermophilic by Crisan (1964). The latter are the only eukaryotic organisms that grow in temperatures up to 62°C (Tansey and Brock, 1972). The majority of fungal species are composed of filamentous hyphae and are often referred to as molds, whereas yeasts, which are also included in the fungal kingdom, are unicellular fungi. All fungi have a nonphotosynthetic mode of nutrition and are classified as chemoorganoheterotrophs (sometimes called chemo-organotrophs or chemoheterotrophs) on the basis of their nutritional category.

Filamentous fungi originate from either fragments of hyphae, which may be septate or aseptae (coenocytic), or disseminated spores that germinate under suitable environmental conditions. The fungal hyphae can grow rapidly, up to several micrometers per minute in length, traversing the substrate on which they grow. Depending on the species, the individual fungal hyphae are 1–15 μm in diameter. The major components of their cell walls comprise the polysaccharides (80%–90%), some lipids, and protein constituents. The growth rates of thermophilic fungi are comparatively much higher than those of mesophilic or psychrophilic fungi. Because of their fast growth rates, thermophilic fungi can be harnessed for their potential in the production of industrial metabolites and biochemicals used for human welfare.

Physiological studies of thermophilic fungi in relation to their environment are scant. The more we understand their life and physiology, the better we can utilize them in industries and medicine. Whatever little information is available on the metabolism of thermophilic fungi has helped us to improve the industrial production of various substances, such as enzymes, proteins, vitamins, and antibiotics. As described in Chapter 2, attempts have been made to study their unique biochemical and molecular characteristics in order to understand the phenomenon of thermophily. Efforts have been made to investigate their nutritional requirements, growth and metabolism, temperature relationships, pigmentation, protein induction, and enzymes, compared with those of their mesophilic counterparts, to find any significant differences that might help to explain their ability to grow at relatively high temperatures, at which most of their close relatives perish.

3.2 NUTRITIONAL REQUIREMENTS OF THERMOPHILIC FUNGI

All organisms are composed of a variety of elements, known as macroelements or macronutrients (carbon, oxygen, nitrogen, hydrogen, sulfur, and phosphorus). These elements are essential components of organic molecules, such as proteins (C, H, O, N, and S), lipids (C, H, O, and P), carbohydrates (C, H, and O) and nucleic acids (C, H, O, N, and P). The biosynthesis of cellular structures necessary for growth, reproduction, and maintenance requires a continuous supply of these basic nutrients and an energy source. Nutritional studies play a very important role in understanding the close relationship and physiological behavior of two species of unrelated fungi. Until the 1980s, thermophilic fungi were thought to have complex or unusual nutritional requirements (Maheshwari et al., 2000). With limited information available on the nutritional behavior of thermophilic fungi, it is emphasized that they can grow on simple media containing a carbon and nitrogen source and mineral salts, suggesting that they do not have any special nutritional requirements for their growth. Moreover, they are mostly autotrophic for vitamins synthesizing them *de novo*. It has been suggested that nitrates of sodium and potassium are better sources of nitrogen than ammonium nitrate and sulfate, whereas asparagine permits a moderate growth response (Satyanarayana and Johri, 1984). The remarks of Miller et al. (1974), that "no defined medium could be produced in which thermophilic fungi would grow," seem unrealistic, at least for the moment, as thermophilic fungi have been grown on a variety of culture media (Table 1.1). However, the requirement of certain special nutrients or growth factors for optimal growth of thermophilic fungi is reported by several workers. For example, Rosenberg (1975) reported that approximately 50% of the 21 thermophilic and thermotolerant fungi studied required 0.01% yeast extract for growth on a solid medium. Asundi et al. (1974) and Wali et al. (1978) studied the growth pattern of thermophilic fungi in a liquid medium containing glucose and ammonium sulfate and reported that supplementation of succinic acid, a tricarboxylic acid (TCA) intermediate, stimulated growth mainly due to its buffering action rather than its nutritional role. In contrast, Gupta and Maheshwari (1985) suggested that the requirement of succinate for the growth of *Thermomyces lanuginosa* could be eliminated if the pH of the medium was properly controlled. They further demonstrated the anaplerotic function of CO_2 fixation in the synthesis of C_4 acids by analyzing the distribution of ^{14}C in proteins and nucleic acids in the mycelium exposed to $NaH^{14}CO_3$. Although CO_2 is not considered a nutritional requirement for fungi, the growth of *Thermomyces lanuginosus* was severely affected if the gas phase in the culture flask was devoid of CO_2 (Gupta and Maheshwari, 1985). Similarly, the concentration of CO_2 can be as high as 10%–15% inside the composts; therefore, it is likely that its assimilation plays an important nutritional role in the development of thermophilic fungi, which are the main colonizers of such habitats (Straatsma et al., 1989, 1994).

Many thermophilic fungi grow favorably on starch, cellulose, hemicellulose, lignin, and pectin (Deploey, 1976; Basu, 1980; Satyanarayana and Johri, 1984), utilizing these polysaccharides as a carbon source. Subrahmanyam (1977) reported that *Thermoascus auraniacus* showed luxuriant growth on dulcitol, mannitol, oxalic acid, and citric acid, whereas *Scytalidium thermophilum* exhibited poor growth on oxalic and citric acids, with no growth on dulcitol. In another study, thermophilic fungi failed to grow on formaldehyde, except *Torula thermophila*, *Sporotrichum thermophile*, *Thermomucor indicae-seudaticae*, and *Thermoascus aurantiacus*, which grew on methanol and formate (C_1 compounds) as sources of carbon and energy (Chauhan et al., 1985). The utilization of amino acids as a nitrogen source was investigated by Meyer (1970). He reported that the thermophilic fungi *Thermomyces lanuginosus* and *Penicillium duponti* grow on single amino acids as a nitrogen source without any added growth factor. Subrahmanyam (1980) observed poor growth and sporulation of *Thermoascus aurantiacus* in the presence of magnesium nitrate, glycine, peptone, thiourea, L-serine, DL-phenylalanine, DL-leucine, DL-alanine, and ammonium acetate. Boonsaeng et al. (1974) reported repression of succinate dehydrogenase by glucose and its de-repression in the presence of succinate in several thermophilic fungi. Mixed carbon sources, such as glucose and sucrose, are effectively utilized by *Thermomyces lanuginosus* and *Penicillium duponti* at 50°C; however, the rate of utilization was lower when the sugars were supplied singly to the culture media (Maheshwari and Balasubramanyam, 1988). Concurrent utilization of sucrose in the presence of glucose was suggested by them because of (1) invertase insensitivity to catabolite repression by glucose and (2) repression of the activity of the glucose uptake system by glucose as well as sucrose, as both sugars could be utilized concomitantly at 30°C at almost equal rates.

3.3   GROWTH AND METABOLISM OF THERMOPHILIC FUNGI

Growth refers to an increase in mass of the whole or part of a living organism by synthesis of macromolecules, whereas metabolism pertains to the sum total of all biochemical reactions occurring in living cells. Growth processes of filamentous fungi, including thermophilic ones, are complex compared with those of unicellular fungi, for example, yeast. Filamentous fungi are composed of aerial hyphae, which need to be nourished through the mycelium in contact with the medium. This involves transport of nutrients over considerable distances, particularly in sporangiphores and aerial fruiting bodies, which are also routinely produced in thermophilic fungi. It is evident that thermophilic fungi grow at a faster rate than mesophilic or psychrophilic fungi. Therefore, a reserve of essential nutrients is to be maintained in the medium for their continuous supply, particularly in industrial fermentation. In mushroom composts, thermophilic fungi form the major components of the total microflora. Of the 21 species of thermophilic fungi isolated from mushroom compost, *Scytalidium thermophilum*, *Humicola lanuginosa*, *Thermoascus aurantiacus*, *Chaetomium thermophile*, *Absidia corymbifera*, and *Talaromyces emersonii* are fast growing and exhibit high growth rates (Straatsma et al., 1994). The pioneering studies of Noack (1920) on the physiology of thermophilic fungi revealed that the economic yield (grams of fungal biomass produced per gram of sugar consumed) of the thermophilic *Thermoascus aurantiacus* grown in minimal medium at 45°C was similar to that of the mesophilic *Aspergillus niger* grown at 25°C. Thus, the overall metabolism in both types of fungi must be quite similar, as on an average, both utilized 55% of sugars for the synthesis of fungal biomass and 45% for metabolism.

Thermophilic fungi are well adapted for polysaccharide utilization, as they can produce an array of extracellular enzymes to hydrolyze cellulose and hemicellulose, chief components of organic biomass. Bhat and Maheshwari (1987) reported that the growth rate of *Sporotrichum thermophile* on cellulose (paper) was identical to that on glucose. However, *Chaetomium thermophile* and *Humicola insolens* grew better on xylan than on simple sugars (Chang, 1967). A systematic investigation would clearly express growth rates of these fungi under strictly controlled conditions. Prasad and Maheshwari (1978a) observed that the growth rate of *Thermomyces lanuginosus* was more than double in shake flasks than in static cultures, clearly indicating an immensely important role of dissolved oxygen and nutrient distribution in the liquid medium. Similarly, *Chaetomium cellulolyticum* grew rapidly in Czapek's nutrient broth containing 15% glucose as the carbon source, with a specific growth rate of 0.14 per hour and a generation (doubling) time of 5 h (Chahal and Hawksworth, 1976). In furtherance to their work, Chahal and Wang (1978) showed that the rate of protein synthesis was 0.09 per hour at 1% cellulose concentration, and 0.3 per hour when an additional 1% cellulose was added immediately after the complete assimilation of the first batch of substrate.

To compare the growth and metabolism of thermophilic fungi, two parameters are generally used. The first is the exponential growth rate (μ), which is determined from the exponential part of the semilogarithmic growth curve as follows:

$$\mu = [2.303(\log x_2 - \log x_1)/t_2 - t_1]h^{-1}$$

where x_2 and x_1 are dry weight (in mg) at times t_2 and t_1, respectively.

The second parameter is the molar growth yield, which is determined as the yield (dry weight) of mycelium per mole of glucose utilized. Interestingly, the exponential growth rate and molar growth yield of mesophilic and thermophilic fungi were found to be quite similar.

Three distinct patterns of growth (mycelial biomass production) of thermophilic fungi isolated from coal mine soils have been observed (Johri, 1980):

1. In the first pattern, the lag phase is fairly long and gradually merges with the log phase of growth. As a result, mycelia biomass production is slow and autolysis does not occur until the 15th day of incubation under surface cultivation in *Absidia corymbifera* and *Torula thermophila*.
2. The second pattern was found in *Humicola grisea* and *Thielavia minor*, where extended lag and log phases were observed, with a sharp rise in mycelia production between 9 and 12 days of incubation.
3. In the third pattern, the lag and log phases combined did not exceed 3 days, and a decline in mycelial growth commenced after the sixth day of incubation, resulting in a sharp rise in dry mycelia weight, followed by a significant decline, as observed in *Rhizopus rhizopodiformis*, *Thermomyces lanuginosus*, *Acremonium alabamensis*, and *Thermoascus aurantiacus*.

The length of the lag phase for the growth of microorganisms at different temperatures is affected by the temperature of incubation of inocula. A shorter lag phase is expected at growth temperatures nearer to those of inocula and is especially significant for organisms adapted to extreme environments (Farrell and Rose, 1967). Using radial growth on solid media, Kumar and Aneja (1999b) investigated the growth rate of 15 thermophilic and thermotolerant fungi, which varied from 0.30 to 1.90 mm/h at their optimum temperature. On the basis of their growth rate, these fungi were categorized as slow, moderate, and fast growing.

3.4 EFFECTS OF ENVIRONMENTAL FACTORS ON GROWTH

In the foregoing account, we have seen that thermophilic fungi are able to respond to variations in nutrient levels. They are also greatly affected by the chemical and physical nature of their surroundings. An understanding of the environmental influences will further help in the development of better control strategies for their growth and ecological distribution. The adaptation of thermophilic fungi to inhospitable thermal environments that would kill most other organisms is truly remarkable. In this section, we briefly review the effects of the most important environmental factors on the growth of this very important group of fungi. Major emphasis is given to temperature, hydrogen ion concentration (pH), oxygen requirements, water activity, solutes, hydrostatic pressure, and radiation.

3.4.1 EFFECT OF TEMPERATURE

Microorganisms are prone to external temperatures because they are unable to regulate their internal temperature. An important factor influencing the effect of temperature on the growth and metabolism of microorganisms is that the rate of biochemical reactions increases as the temperature rises. It is believed that this rate roughly doubles for every 10°C rise in temperature, enabling the cells to grow faster. However, there are maximum limits beyond which some temperature-sensitive molecules (proteins, nucleic acids, and lipids) are denatured and hence become nonfunctional. Similarly, there is also a minimum temperature for growth, below which the rate of reactions slows down, as the lipid bilayer is not fluid enough to function properly. Therefore, when microorganisms are above or below their optimum temperature, their cellular structure and function are severely affected.

Thermophilic fungi are characterized by an amazingly high optimum temperature for their growth and metabolism. Their temperature dependence with distinct cardinal temperatures—minimum, optimum, and maximum, as presented in Table 3.1—makes them interesting organisms for biotechnological explorations. These fungi have widespread distribution in both tropical and temperate regions of the world (Table 4.3). Tendler et al. (1967) commented that "the ubiquitous distribution of organisms whose minimal temperature for growth exceeds the temperatures obtainable in the natural environment from whence they were isolated, still stands a 'perfect crime' story in the library of biological mysteries." To test whether obligate thermophily is an artifact of the nutritional environment, they cultivated thermophilic isolates of *Humicola*, *Thermoascus*, and *Aspergillus* in a nutritionally rich broth containing glucose, mannitol, starch, casamino acids, yeast extract, and peptone. At 22°C, these isolates produced good growth after 10 days of incubation but failed to grow below

TABLE 3.1
Cardinal Temperatures of Some Thermophilic Fungi

Sr. No.	Organisms	Cardinal Temperatures (°C)			References
		Minimum	Optimum	Maximum	
1.	*Rhizomucor miehei*	24–25	35–45	55–57	Cooney and Emerson, 1964
2.	*Rhizomucor nainitalensis*	25	48	–	Joshi, 1982
3.	*Rhizomucor pusillus*	22–27	35–55	55–60	Cooney and Emerson, 1964; Evans, 1971; Kuster and Locci, 1964
4.	*Rhizopus microsporus*	12	40	50	Evans, 1971
5.	*Rhizopus rhizopodiformis*	–	45	–	Thakre and Johri, 1976
6.	*Canariomyces thermophila*	–	45	–	von Arx et al., 1988
7.	*Chaetomidium pingtungium*	30	48	–	Chen and Chen, 1996
8.	*Chaetomium senegalensis*	–	45	–	Aneja and Kumar, 1994
9.	*Chaetomium thermophile* var. *coprophile*	25–28	45–55	58–60	Cooney and Emerson, 1964; Evans, 1971
10.	*Chaetomium thermophile* var. *dissitum*	25–28	45–50	58–60	Cooney and Emerson, 1964; Evans, 1971
11.	*Coonemeria aegyptiaca*	25	40	55	Mouchacca, 1997
12.	*Coonemeria verrucosa*	20	37	55	Evans, 1971; Stolk, 1965; Apinis, 1967
13.	*Melanocarpus albomyces*	25–25	37–45	55–57	Cooney and Emerson, 1964; Evans, 1971
14.	*Rasamsonia byssochlamydoides*	<35	40–45	–	Stolk and Samson, 1972; Awao and Otsuka, 1974; Houbraken et al., 2012
15.	*Rasamsonia emersonii*	25–30	40–45	55–60	Cooney and Emerson, 1964; Evans, 1971; Stolk, 1965
16.	*Talaromyces thermophilus*	25–30	45–50	57–60	Cooney and Emerson, 1964; Evans, 1971; Stolk, 1965
17.	*Thermoascus aurantiacus*	20–35	40–46	55–62	Cooney and Emerson, 1964; Evans, 1971; Stolk, 1965; Apinis, 1967
18.	*Melanocarpus thermophilus*	20	45	55	Fergus and Sinden, 1969; Hedger and Hudson, 1970

(Continued)

TABLE 3.1 (CONTINUED)
Cardinal Temperatures of Some Thermophilic Fungi

Sr. No.	Organisms	Cardinal Temperatures (°C)			References
		Minimum	Optimum	Maximum	
19.	*Acremonium thermophilum*	20	25–40	47	Gams and Lacey, 1972
20	*Arthrinium pterospermum*	<24	40	50	Cooney and Emerson, 1964; Bunce, 1961
21.	*Malbranchea connamomea*	–	45	–	Cooney and Emerson, 1964
22.	*Myceliophthora fergussi*	25	45	55	Kumar and Aneja, 1999b
23.	*Myceliophthora hinnulea*	20	40–45	50	Awao and Udagawa, 1983
24.	*Myceliophthora thermophila*	18–24	40–50	55	Cooney and Emerson, 1964; Evans, 1971; Hedger and Hudson, 1970
25.	*Rasamsonia composticola*	–	45–50	–	Su and Cai, 2013
26.	*Papulospora thermophila*	–	45	–	Fergus, 1971
27.	*Scytalidium indonesicum*	–	45	–	Hedger et al., 1982
28.	*Scytalidium thermophilum*	23	35–45	58	Cooney and Emerson, 1964; Craveri and Colla, 1966
29.	*Remersonia thermophila*	–	45	–	Seifert et al., 1997
30.	*Thermomyces ibadanensis*	31–35	42–47	60–61	Apinis and Eggins, 1966; Eggins and Coursey, 1964
31.	*Thermomyces lanuginosus*	28–30	45–55	60	Cooney and Emerson, 1964

30°C in a sucrose-salt medium that was devoid of complex supplements. They suggested that the complex nutrient broth contained factors that the organisms could not synthesize at the lower temperature. This suggestion was further corroborated by Wright et al. (1983), who reported that the growth of *Talaromyces thermophilus* at the suboptimal temperature of 33°C benefited by supplementation of culture medium with 5 µg/mL ergosterol.

It has been suggested that eukaryotes fail to construct stable and functional organellar membranes at temperatures above 60°C, a point that is also considered as the upper temperature limit of thermophilic fungi. The optimum temperature of thermophilic fungi, as determined by various workers (Crisan, 1973; Rosenberg, 1975; Kumar and Aneja, 1999a), varies from 35°C to 55°C. Thermotolerant fungi, such as *Aspergillus fumigatus*, *Rhizopus microsporus*, and *Emericella nidulans*, grow at or below 20°C, while thermophilic forms such as *Thermomyces lanuginosus*, *Thermoascus auran-tiacus*, and *Talaromyces emersoni* show little or no growth at 20°C–30°C (Tansey, 1973; Rosenberg, 1975). In determining the temperature requirements of thermophilic fungi, it is pertinent to make a clear-cut distinction between growth behavior and survival. The growth temperature range for thermophilic fungi usually spans 30°. While vegetative growth takes place in a much wider temperature range, asexual and sexual reproduction are affected in thermophilic fungi in a comparatively narrow temperature range. For example, *Rhizomucor miehei* produces zygospores at 35°C but not at 50°C. Similarly, *Chaetomium thermophile* and *T. aurantiacus* produce asco-acarp at 45°C–50°C, beyond which they form sterile ascocarp or vegetative mycelia (Tansey and Brock, 1972). Satyanarayana and Johri (1984) observed that the majority of thermophilic fungi survive exposure to 60°C for 1–48 h. Two thermophilic fungi, *T. aurantiacus* and *Thermomyces lanuginosus*, survived exposure to 72°C for 2–10 min. In *T. aurantiacus*, the lowest temperature required for germination of ascospores was 10°C–12°C higher than that for the mycelial growth (Deploey, 1995). In particular, spores of thermophilic fungi are more heat resistant than the vegetative mycelia, as these structures have thick spore coats and low water activity. Thus, it is more or less established that temperature maxima of true thermophilic fungi fall within 50°C–60°C, optima within 35°C–55°C, and minima within 20°C–35°C. It is further suggested that there is no single temperature that is optimal for the growth of a par-ticular group of fungi. The minimum and maximum temperature relationships of most thermophilic fungi demarcate them into three major groups (Johri et al., 1999):

1. Species that grow at <20°C but show a maximum up to 50°C
2. Species that fail to grow at <20°C and at or above 50°C
3. Species that do not grow at <30°C but show a maximum of 60°C or slightly above

Little information is available on thermophilic fungi with regard to tolerance to subminimal temperatures; as such, fungi have been frequently reported from tem-perate habitats (Salar and Aneja, 2006). Noack (1920) subjected a *Thermoascus aurantiacus* culture to a low temperature of 31°C (4°C below its minimum temper-ature for growth) and observed that its respiration declined only marginally, but if the culture was kept at 21°C for 24 h, it stopped respiring. Subsequent heating of the culture medium to 46°C practically showed no respiration. This phenomenon needs to be explored further in order to establish the viability of thermophilic fungi under refrigeration or their storage at subminimal temperatures.

3.4.2 EFFECT OF pH

Similar to temperature, every microorganism has a pH range and pH optima over which it grows. pH is a measure of the relative acidity of a solution and is defined as the negative logarithm of hydrogen ion concentration and is expressed in molarity.

$$\text{pH} = -\log[\text{H}^+] = \log\left(\frac{1}{[\text{H}^+]}\right)$$

In general, most fungi tend to grow in the acidic range (pH 4–6). The pH of the surrounding medium exerts a profound influence on the transport of nutrients, nutrient solubilities, enzymatic reactions, and availability of specific metallic ions that may, at a specific pH, form insoluble complexes. Metals like magnesium, iron, calcium, and zinc are available to the fungus at low pH values, becoming insoluble at higher pH values. However, the pH preferences of individual fungi vary. All microorganisms respond to external pH changes using mechanisms that maintain a neutral cytoplasmic pH. Growth in fungi is usually stopped on the acidic side at pH 3 and on the alkaline side at pH 8–9. Drastic variations in cytoplasmic pH can harm microorganisms by disrupting the plasma membrane or inhibiting the activity of enzymes and membrane transport proteins.

Thermophilic fungi are tolerant to a broad range of pH values, spanning between 4.0 and 8.0 (Table 3.2). Rosenberg (1975) studied the temperature and pH optima of more than 21 thermophilic and thermotolerant fungi isolated from various habitats. It was reported that the pH optima (pH 6.5) of *Rhizomucor pusillus* and *Rhizomucor miehei* corresponded to the pH optima of hay at baling from where these species were isolated. The pH optima of *Thermomyces lanuginosus*, *Malbranchaea cinnamomea*,

---

TABLE 3.2
Optimum pH of Some Thermophilic Fungi

Sr. No.	Organism	Optimum pH	References
1.	*Rhizomucor miehei*	5.5	Rosenberg, 1975; Schipper, 1978
2.	*Rhizomucor pusillus*	6.1	Rosenberg, 1975; Schipper, 1978
3.	*Chaetomium thermophile* var. *coprophile*	6.8	Rosenberg, 1975
4.	*Chaetomium thermophile* var. *dissitum*	5.5	Rosenberg, 1975
5.	*Rasamsonia emersonii*	3.4–5.4	Rosenberg, 1975; Houbraken et al., 2011
6.	*Talaromyces thermophilus*	7.2–8.1	Rosenberg, 1975
7.	*Thermoascus aurantiacus*	6.8	Rosenberg, 1975
8.	*Melanocarpus albomyces*	7.6	Rosenberg, 1975
9.	*Corynascus thermophilus*	6.8	Rosenberg, 1975
10.	*Arthrinium pterospermum*	5.4–5.9	Rosenberg, 1975
11.	*Malbranchea connamomea*	7.2	Rosenberg, 1975; van Oorschot and de Hoog, 1984
12.	*Myceliophthora thermophila*	6.1–6.8	Rosenberg, 1975; van Oorschot, 1977
13.	*Scytalidium thermophilum*	7.3–7.6	Rosenberg, 1975
14.	*Remersonia thermophila*	7.9	Rosenberg, 1975; Seifert et al., 1997
15.	*Thermomyces lanuginosus*	6.8–7.3	Rosenberg, 1975; Straatsma and Samson, 1993

and *Talaromyces thermophilus* was found to be near neutral (pH 7). On the other hand, *Talaromyces emersoni* and *Allescheria terrestris* have been reported to grow in the acidic environment of sugarcane bagasse (pH 3.4–6.0). However, Rosenberg (1975) clearly observed that no correlation was found between temperature optimum and pH optimum among the thermophilic and thermotolerant fungi investigated.

3.4.3 Effect of Oxygen

Barring a few, almost all multicellular organisms require atmospheric oxygen for growth and metabolism; that is, they are obligate aerobes. Oxygen is required as the terminal electron acceptor, critical to the functioning of the electron transport chain (ETC) in aerobic respiration. Studies on the requirement of oxygen in thermophilic fungi are limited. Most fungi require at least 0.2% oxygen for trace growth, 0.7%–1.05% for moderate growth, and 1.0% for sporulation. The first detailed study was made by Noack (1920) during his pioneering investigations of the physiology of *Thermoascus aurantiacus*. He made an interesting observation that although the rate of respiration was decreased because of reduced oxygen supply, the respiratory quotient (CO_2/O_2) remained practically the same, even at very low concentrations of oxygen. Moreover, the vegetative mycelium of *Thermoascus aurantiacus* could withstand anaerobic environments continuously for 8 days without the loss of viability. Although thermophilic fungi are essentially aerobic (Kane and Mullins, 1973), Henssen's (1957) study of anaerobiosis of *Humicola insolens* is contentious; he reported better growth of the fungus under anaerobic than aerobic conditions. Earlier, Cooney and Emerson (1964) also observed that the perfect (ascocarpic) stage of *Talaromyces dupontii* is dependent on the initiation of anaerobic conditions. Aerobic cultures usually formed the imperfect (conidial) *Penicillium dupontii* stage only. This phenomenon is particularly of genetic interest in fungi producing only sterile mycelia or where sexual stages have not been discovered so far. On the contrary, Deploey and Fergus (1975), while studying the growth and sporulation behavior of thermophilic fungi, observed a complete inhibition of growth in the presence of 100% nitrogen; fungi sporulated only in the presence of oxygen, suggesting that oxygen availability is vital for growth. Prasad and Maheshwari (1978b) also correlated the cessation of growth in *Thermomyces lanuginosus* with a depleting oxygen concentration in static cultures. Similarly, Prasad et al. (1979) compared the metabolic rates of mesophilic and thermophilic fungi and reported the average Q O_2 (μL of O_2 uptake/mg of dry weight/h) of *Aspergillus niger* and *Thermomyces lanuginosus* to be 53.1 and 42.4, respectively, suggesting that thermophilic fungi that are adapted for growth at elevated temperatures did not show much higher metabolic rates than mesophilic fungus.

3.4.4 Effect of Solutes and Water Activity

The activities of microorganisms are greatly affected by changes in the osmotic concentration of their surroundings. A selectively permeable plasma membrane separates microorganisms from their environment; if microorganisms with rigid cell walls are placed in hypotonic solution (lower osmotic concentration than cellular contents), water will enter the cell, making the cell become turgid, while those without walls swell and ultimately burst. Alternatively, if it is placed in hypertonic solution (higher osmotic concentration than cellular contents), water will flow out of the cell. Thus, in order to be metabolically active, microorganisms need to make a balance between inflow and outflow of the dissolved salts of the external environment.

The degree of availability of water to a microorganism is denoted by the term water activity (a_w). The water activity of a solution is the ratio of the vapor pressure of its water to the vapor pressure of pure water at identical conditions of temperature and pressure. It is expressed as

$$a_w = \frac{P_{soln}}{P_{water}}$$

Like all other organisms, thermophilic and thermotolerant fungi are faced with alterations in the water activity, to which they must adjust, to be able to grow. A change in temperature proportionately affects the vapor pressure over the substance and water, and therefore the value of a_w does not alter drastically (Johri et al., 1999). Water activity has a profound influence on the rate of release of energy and heat in thermophiles, which affects their activities. Mahajan et al. (1986) reported a xerophilic strain of thermophilic *Humicola* from the sand dunes of Thar Desert that was able to grow in a medium containing 50% sucrose. The strain showed an increase in the level of proline and sterol contents, depicting their role in desiccation stress.

3.4.5 HYDROSTATIC PRESSURE

Organisms living on land and water surfaces are never subjected to a pressure greater than 1 atm (1.013 bar), and higher pressure generally inhibits microbial growth. Earlier, microbial life was thought to be restricted to the thin surface layer of the planet having abundant organic matter as its nutrient source; however, the discovery of deep-sea hydrothermal vents in the Galapagos Rift in 1977 opened up new vistas in deep-sea biology where the hydrostatic pressure is 350 bar or more. Such a high hydrostatic pressure affects membrane fluidity and membrane-associated functions. In spite of that, several microorganisms with a distinct metabolism, particularly prokaryotes, are found in the deep sea. Presently, more than 300 deep-sea hydrothermal vents are known throughout the world (Desbruyères et al., 2006). These hydrothermal vents are the most fascinating environments for the presence of barotolerant microorganisms. Some of the microorganisms are truly piezophilic (barophilic), growing more rapidly at high pressure. A piezophile is an organism that grows maximally at a pressure greater than 1 atm but less than about 590 atm. Microbial communities explored from deep-sea hydrothermal vents mostly belong to the Bacteria and Archaea domains (Mehta and Satyanarayana, 2013).

The exploration of fungi (both filamentous and yeasts) from deep-sea hydrothermal vents remains neglected, and only scanty reports are available. Culture-independent techniques by Bass et al. (2007) reported the presence of sequences affiliated with *Debaromyces hansenii* and novel sequences close to those of *Malassezia furfur* in hydrothermal vents. Using polymerase chain reaction (PCR)–mediated internal transcribed spacer (ITS) regions of an rRNA gene clone, Nagano et al. (2010) reported the diversity of fungal communities (belonging to Ascomycetes) in 10 different deep-sea sediments. Barotolerant organisms are able to adapt to high pressure by changing their membrane lipids and further increasing the amount of unsaturated fatty acids in their membrane lipids as pressure increases.

3.4.6 EFFECT OF LIGHT

Light of the visible spectrum (Figure 3.1) affects various fungal processes, including mycelial growth, sporulation, and spore germination. These affects may be morphogenetic, influencing structural growth, or nonmorphogenetic, in which light influences the rate or direction of movement, the growth of a structure, or the synthesis of a compound. Both vegetative and reproductive structures respond to light. Further, ultraviolet (UV) light causes mutations, which may be lethal or benign. UV light with a wavelength in the range of 260–265 nm is the most mutagenic. It is generally observed that UV light of shorter wavelengths is more lethal than that of the longer ones, which are more effective as mutagenics and cause nonlethal mutations. The effect of light on the reproduction in fungi is conspicuous. Colonies exposed to diurnal periodicity exhibit alternating zones of sporulating and nonsporulating hyphae. Kumar (1996) observed alternating zones of a sporulating and nonsporulating pattern in thermophilic *Chaetomium thermophile* var. *coprophile* growing on YpSs medium in petri dishes. The zones of spores are usually stimulated by light but may actually form during the subsequent dark periods. Furthermore, light also influences the formation of pigments in certain fungi; for example, in *Neurospora sitophila*, the formation of conidia and protoperithecia are photoresponses.

3.4.7 RELATIVE HUMIDITY

All fungi require a high level of relative humidity in order to grow and sporulate. The association of moisture and fungal growth is well known to cause deterioration of stored agricultural produce. In general, fungi require 95%–100% relative humidity for their optimum growth, whereas a relative humidity below 80%–85% is inhibitory for

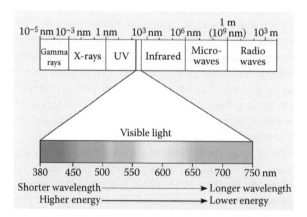

FIGURE 3.1 Electromagnetic spectrum showing the spectrum of light visible to the human eye (visible spectrum).

the growth of most fungi. The growth of thermophilic fungi requires maintenance of a sufficient moisture level in the culture media or composts, as at their optimum temperature (45°C), water tends to evaporate quickly. When thermophilic fungi are grown on agar media, the petri plates need to be incubated in humidified chambers, as in a conventional biochemical oxygen demand (BOD) incubator the agar medium becomes charred, affecting the growth of the fungus adversely. In a natural environment, all fungi have to face a testing time in want of favorable humid conditions.

3.5 COMPLEX CARBON SOURCES AND ADAPTATIONS FOR MIXED SUBSTRATE UTILIZATION

Nutrients available to microorganisms growing in nature are often of a mixed kind. It is therefore not surprising that thermophilic fungi are well adapted for mixed substrate utilization, a property that gives an edge to these organisms over their mesophilic counterparts to thrive in adverse environments. In natural environments, thermophilic fungi are found in compost or soil having organic debris to support their requirements of carbon and nitrogen. As the temperature rises in a compost heap, mesophilic microflora is replaced by thermophilic fungi in succession (Hedger and Hudson, 1974). Because the initially available soluble carbon sources (sugars, organic acids, and amino acids) would have been depleted by the growth of pioneer mesophiles, the carbon sources available to the thermophilic fungi are mainly complex polysaccharides of cellulose and hemicellulose. With an array of extracellular cellulolytic enzymes, most thermophilic fungi are able to utilize these complex polysaccharides as a carbon source for their growth. The noncellulolytic thermophilic fungal species, for example, *Thermomyces lanuginosus* (Chang, 1967; Hedger and Hudson, 1974), *Talaromyces dupontii, Malbranchea pulchella* var. *sulfurea, Mucor pusillus* (Chang, 1967), and *Melanocarpus albomyces* (Maheshwari and Kamalam, 1985), which are frequently isolated from composts, are able to grow commensally with their cellulolytic partner, as the latter releases sugars during hydrolysis of cellulose and hemicellulose. The noncellulolytic fungi are also reported to grow faster in mixed cultures with cellulolytic partners. For example, Hedger and Hudson (1974) reported that the noncellulolytic *Thermomyces lanuginosus* grew profusely with a cellulolytic fungus, *Chaetomium thermophile*, in a mixed culture. Xylan, which is present external to cellulose in the plant cell wall, is also an important carbon source. In fact, some fungi, for example, *Chaetomium thermophile* and *Humicola insolens*, preferentially utilize xylan (Chang, 1967) over simple sugars.

A mixture of sugars is released during composting of plant material by the secretary enzymes produced by thermophilic fungi. In order to know if thermophilic fungi utilize one sugar at a time or a mixture of sugars simultaneously, Maheshwari and Balasubramanyam (1988) grew *Thermomyces lanuginosus* and *Penicillium dupontii* in a medium containing a mixture of glucose and sucrose as the carbon source. It was observed that both fungi concurrently utilized glucose and sucrose at 50°C and 30°C, although sucrose was utilized faster than glucose. Glucose is utilized preferentially, as it is more easily catabolized and yields more energy molecules than other carbon sources, which may be more difficult to degrade or yield less energy. This phenomenon is known as diauxic growth. However, when *Thermomyces lanuginosus* was grown on sugars provided singly and in mixtures in the medium, Maheshwari and Balasubramanyam observed that the rate of utilization of individual sugars in mixtures was lowered, as both sugars reciprocally influenced their utilization. They further reported that the simultaneous utilization of sucrose in the presence of glucose is due to the insensitivity of invertase to catabolite repression by glucose, and the activity of the glucose uptake system was repressed by both glucose and sucrose.

3.6   NUTRIENT TRANSPORT

Nutrients from the environment must be transported across the cell membrane inside the cell, often carried out by specific carrier proteins in the plasma membrane via specific active mechanisms, as membranes are selectively permeable. One of the deterrents in the growth of thermophilic fungi is the reduction in the rate of nutrient uptake at ordinary temperatures. Although little information is available on the mechanisms of nutrient uptake in thermophilic fungi, some studies give insight into the occurrence of a proton gradient–driven symport (Palanivelu et al., 1984; Maheshwari et al., 2000). It has been observed that transporter proteins of thermophilic fungi are transformed into a rigid conformation, affecting the binding and subsequent release of ions and nutrients (Maheshwari and Kamalam, 2005).

As stated in the previous section, there is increasing evidence on the concurrent utilization of mixed substrates, for example glucose and sucrose, at low concentrations. Palanivelu et al. (1984) reported that in *Thermomyces lanuginosus*, specific sucrose transport activity and invertase activity in mycelia exposed to β-fructofuranosidases (sucrose or raffinose) appear to occur simultaneously, as well as decline concurrently upon the exhaustion of sucrose in the culture media. They further reported that the transport of sucrose is H+ coupled and is inhibited by ionophores that dissipate the proton gradient. Proton gradients may also be used to establish a sodium ion gradient across the cell membrane, driving the uptake of nutrients, for example, sugars and amino acids. Similarly, Maheshwari and Balasubramanyam (1988) reported that at 50°C, *Thermomyces lanuginosus*, utilized sucrose preferentially compared with glucose and followed Michaelis–Menten kinetics, with an apparent K_m value of 250 μm. However, at 30°C, both sucrose and glucose were utilized at nearly comparable rates. They contended that simultaneous utilization of the two sugars might be possible owing to (1) the insensitivity of invertase to catabolite repression by glucose, (2) the lowered activity and affinity of the glucose transport system after sucrose was added to the growth medium, and (3) the substrate-level feedback inhibition of glucose itself. Therefore, concurrent utilization of nutrients by thermophilic fungi appears to be an important adaptive strategy in view of their opportunistic growth in thermogenic habitats with low nutrient availability.

3.7   PROTEIN BREAKDOWN AND TURNOVER

Survival of thermophilic microorganisms at elevated temperatures is explained by protein breakdown and their rapid resynthesis. The rapid-turnover theory, initially proposed by Allen (1953) and supported by Bubela and Holdsworth (1966), explains that the thermophilic microorganisms survive at high temperatures because they have the ability to replace heat-damaged intracellular materials rapidly. Crisan (1973) explained that the rapid-resynthesis hypothesis proposes that growth of microorganisms at high temperatures is not due to the presence of unusually thermostable metabolites, but to a particularly active metabolism that replaces thermolabile metabolites at a rate equal to or greater than the rate at which they are being destroyed. The applicability of the rapid-resynthesis hypothesis, also popularly known as the dynamic hypothesis of thermophily, was tested by Miller et al. (1974). They compared the rate of protein breakdown in growing and nongrowing cultures of thermophilic (*Penicillium dupontii, Malbranchaea pulchella* var. *sulfurea*, and *Mucor miehei*) and mesophilic (*Penicillium notatum, Penicillium chrysogenum*, and *Aspergillus oryzae*) fungi by monitoring the breakdown of pulse-labeled protein. It was observed that in growing cells of both types of fungi, the protein breakdown was negligible. However, in the nongrowing cells of both types, the protein breakdown rate varied from 5.2% to 6.7% per hour. They further observed a marked difference in the rate of breakdown of soluble proteins, which in thermophilic fungi was almost twice that in mesophilic fungi. The authors' observation that, in thermophilic fungi, the rate of soluble protein breakdown is higher than the rate of total protein breakdown suggests that in order to survive at elevated temperatures, these fungi have developed the capacity to rapidly resynthesize the labile soluble proteins. As the insoluble proteins are associated with other cellular components, this might protect them from degradation. On the contrary, the lower rate of breakdown of soluble proteins in mesophilic fungi further suggests that these proteins are relatively stable at the temperature of their synthesis.

It is presumed that the energy spent in increased protein turnover in thermophilic fungi would affect their mycelial yield compared with their mesophilic counterpart. However, Maheshwari et al. (2000) reported no significant difference in growth yield of thermophilic and mesophilic fungi examined at their respective temperature optima, substantiating the hypothesis of the rapid turnover of proteins. Earlier, Rajasekaran and Maheshwari (1990) radioactively labeled cellular proteins of thermophilic and mesophilic fungi by adding radioactive amino acids at two different times to measure radioactivity of the whole cells as an index of radioactivity of total proteins, rather than measuring the extractable protein by employing ultrasonic disruption of mycelia. The results revealed a definite protein turnover in the growing cells of both mesophilic and thermophilic species; however, the rate of protein breakdown varied depending on the number of thermophilic fungi in the species being examined. Although they only measured protein breakdown, similar results would have been expected had they measured protein turnover. Nevertheless, certain enzymes, for example, invertase in thermophilic fungi, have a high turnover rate.

3.8   VIRULENCE

Toxigenic studies on thermophilic and thermotolerant fungi are very scanty. As the growth temperature of most thermophilic fungi corresponds to that of warm-blooded animals, the probability that thermophilic fungi are mycotoxin producers is of significance to human and animal health. Thus, the presence of thermophilic fungi in grain, food, and feed may pose a hitherto unrecognized hazard to man.

These substrates, when they become moist and subsequently overheat, may provide thermophilic fungi with a competitive growth advantage over other microorganisms. Thus, the importance of fungi as agents of decay in stored grains is of growing concern in the field of mycotoxicology, as are their toxigenic effects in man's food supply. The ability to grow at 37°C is a characteristic of all human pathogens and has long been suspected to play a great role in the pathogenesis of aspergillosis. *Aspergillus fumigatus*, a thermotolerant fungus, is an opportunistic fungal pathogen that has become the leading infectious mold in immunocompromised patients (Bhabhra and Askew, 2005). These authors further reviewed the relationship between thermotolerance and virulence in pathogenic fungi, emphasizing the link to ribosome biogenesis in *Aspergillus fumigatus*. Future research should be conducted on the development of novel antifungals that disrupt the thermotolerant ribosome assembly to determine their potential in the clinical management of fungal infections.

Growth at elevated temperature is known to be an important virulence factor in several fungal pathogens. Kaerger et al. (2015), in a recent study, investigated several thermotolerant species of *Rhizopus* as causative agents of mucormycoses and observed that they are more virulent than the nonclinically mesophilic species. The virulent clad included *R. microsporus*, *R. arrhizus*, *R. caespitosus*, *R. homothallicus*, and *R. schipperae*, which optimally grow at 37°C (human body temperature) or higher. While thermotolerance is a prerequisite for a pathogen to cause infection in warm-blooded animals, additional virulence factors, such as pathways that facilitate adaptation to the host environment, for example, to elevated temperatures, unfavorable pH values, unbalanced osmotic conditions, and limited nutrients (Cooney and Klein, 2008), have been found to be involved in the process of infection of fungal pathogens. Further, a positive correlation in spore size and virulence in *Mucor circinelloides* has been reported (Li et al., 2011). Similarly, Lee et al. (2013) observed that in *Mucor circinelloides* the hyphal stage is more virulent than the yeast stage, advocating the role of morphogenesis as an important factor linked to virulence in zygomycetes. Earlier, Davis et al. (1975) investigated 23 thermophilic and thermotolerant fungal species that appear to produce mycotoxins. The authors reported that 13 isolates were highly toxic to brine shrimp, chicken embryos, and rats. Of these, *Aspergillus fumigatus* and *Chaetomium thermophile* var. *coprophile* were found to be highly toxigenic. Gregory and Lacey (1963) examined moldy hay and found that *Thermomyces lanuginosus* is implicated in bovine abortion and "farmer's lung" disease. The genus *Rhizopus* presents the largest impact on mankind, being important in agriculture and industry, and is in addition the main causal agent of mucormycoses, followed by *Lichtheimia* and *Mucor*. All three genera account for 70%–80% of the reported infections, predominantly as rhinocerebral, pulmonary, or disseminated manifestations, and are associated with high mortality rates (Gomes et al., 2011).

REFERENCES

Allen, M.B. 1953. The thermophilic aerobic spore forming bacteria. *Bacteriological Reviews* 17:125–173.

Aneja, K.R., and Kumar, R. 1994. *Chaetomium senegalense*—A new record from India. *Proceedings of the National Academy of Sciences, India: Section B* 64:229–230.

Apinis, A.E. 1967. *Dactylomyces* and *Thermoascus*. *Transactions of the British Mycological Society* 50:573–582.

Apinis, A.E., and Eggins, H.O.W. 1966. *Thermomyces ibadanensis* sp. nov. from oil palm kernel stacks in Nigeria. *Transactions of the British Mycological Society* 49:629–632.

Asundi, S., Mattoo, A.K., and Modi, V.V. 1974. Studies on a thermophilic *Talaromyces* sp.: Mitochondrial fatty acids and requirement for a C_4 compound. *Indian Journal of Microbiology* 14:75–80.

Awao, T., and Otsuka, S. 1974. Notes on thermophilic fungi in Japan. *Transactions of the Mycological Society of Japan* 15:7–22.

Awao, T., and Udagawa, S.I. 1983. A new thermophilic species of *Myceliophthora*. *Mycotaxon* 16:436–440.

Bass, D., Howe, A., Brown, N., Barton, H., Demidova, M., Michelle, H., Li, L., Sanders, H., Watkinson, S.C., Willcock, S., and Richard, T.A. 2007. Yeast forms dominate fungal diversity in the deep oceans. *Proceedings of the Royal Society B* 274. doi: 10.1098/rspb .2007.1067.

Basu, P.K. 1980. Production of chlamydospores of *Phytophthora megasperma* and their possible role in primary infection and survival in soil. *Canadian Journal of Plant Pathology* 2:70–75.

Bhabhra, R., and Askew, D.S. 2005. Thermotolerance and virulence of *Aspergillus fumigatus*: Role of the fungal nucleolus. *Medical Mycology* 43:87–93.

Bhat, K.M., and Maheshwari, R. 1987. Sporotrichum thermophile growth, cellulose degradation, and cellulase activity. *Applied and Environmental Microbiology* 53:2175–2182.

Boonsaeng, V., Sullivan, P.A., and Shepherd, M.G. 1974. Succinate dehydrogenase of *Mucor rouxii* and *Penicillium duponti*. *Canadian Journal of Biochemistry* 52:751–761.

Bubela, B., and Holdsworth, E.S. 1966. Amino acid uptake, protein and nucleic acid synthesis and turnover in *Bacillus stearothermophilus*. *Biochimica et Biophysica Acta* 123:364–375.

Bunce, M.E. 1961. *Humicola stellatus* sp. nov., a thermophilic mould from hay. *Transactions of the British Mycological Society* 44:372–376.

Chahal, D.S., and Hawksworth, D.L. 1976. *Chaetomium cellolyticum*, a new thermotolerant and cellulolytic chaetomium. *Mycologia* 68:600–610.

Chahal, D.S., and Wang, W.I.C. 1978. *Chaetomium cellulolyticum* growth behaviours on cellulose and protein production. *Mycologia* 70:160–170.

Chang, Y. 1967. The fungi of wheat straw compost II. Biochemical and physiological studies. *Transactions of the British Mycology Society* 50:667–677.

Chauhan, S., Prakash, A., Satyanarayana, T., and Johri, B.N. 1985. Utilization of C-1 compounds by thermophilic fungi. *National Academy Science Letters Part B* 8:167–169.

Chen, K.Y., and Chen, Z.C. 1996. A new species of *Thermoascus* with a *Paecilomyces* anamorph and other thermophilic species from Taiwan. *Mycotaxon* 50:225–240.

Cooney, D.G., and Emerson, R. 1964. *Thermophilic Fungi: An Account of Their Biology, Activities, and Classification*. San Francisco: W.H. Freeman and Co.

Cooney, N.M., and Klein, B.S. 2008. Fungal adaptation to the mammalian host: It is a new world, after all. *Current Opinion in Microbiology* 11:511–516.

Craveri, R., and Colla, C. 1966. Ribonuclease activity of mesophilic and thermophilic molds. *Annals of Microbiology* 16:97–99.

Crisan, E.V. 1964. Isolation and culture of thermophilic fungi. *Contributions from Boyce Thompson Institute* 22:291–301.

Crisan, E.V. 1973. Current concepts of thermophilism and the thermophilic fungi. *Mycologia* 65:1171–1198.

Davis, N.D., Wagener, R.E., Morgan-Jones, G., and Diener, U.L. 1975. Toxigenic thermophilic and thermotolerant fungi. *Applied and Environmental Microbiology* 29: 455–457.

Deploey, J.J. 1976. Carbohydrate nutrition of *Mucor miehei* and *M. pusillus*. *Mycologia* 68:190–194.

Deploey, J.J. 1995. Some factors affecting the germination of *Thermoascus aurantiacus* ascospores. *Mycologia* 87:362–365.

Deploey, J.J., and Fergus C.L. 1975. Growth and sporulation of thermophilic fungi and actinomycetes in O2-N2 atmosphere. *Mycologia* 67:780–797.

Desbruyères, D., Segonzac, M., and Bright, M., eds. 2006. *Handbook of Deep-Sea Hydrothermal Vent Fauna*. 2nd ed. Linz: State Museum of Upper Austria.

Eggins, H.O.W., and Coursey, D.G. 1964. Thermophilic fungi associated with Nigerian oil palm produce. *Nature* 203:1081–1084.

Evans, H.C. 1971. Thermophilous fungi of coal spoil tips. I. Taxonomy. *Transactions of the British Mycological Society* 57:241–254.

Farrell, J., and Rose, A. 1967. Temperature effects on microorganisms. *Annual Review of Microbiology* 21:101–120.

Fergus, C.L. 1971. The temperature relationships and thermal resistance of a new thermophile *Papulaspora* from mushroom compost. *Mycologia* 63:426–431.

Fergus, C.L., and Sinden, J.W. 1969. A new thermophilic fungus from mushroom compost: *Thielavia thermophila* spec. nov. *Canadian Journal of Botany* 47:1635–1637.

Gams, W., and Lacey, M.E. 1972. *Acremonium thermophilum*. *Transactions of the British Mycological Society* 59:520.

Gomes, M.Z., Lewis, R.E., and Kontoyiannis, D.P. 2011. Mucormycosis caused by unusual mucormycetes, non-*Rhizopus*, -*Mucor*, and -*Lichtheimia* species. *Clinical Microbiology Reviews* 24:411–445.

Gregory, P.H., and Lacey, M.E. 1963. Mycological examination of dust from mouldy hay associated with farmer's lung disease. *Journal of General Microbiology* 30:75–88.

Gupta, S.D., and Maheshwari, R. 1985. Is organic acid required for nutrition of thermophilic fungi? *Archives of Microbiology* 141:164–169.

Hedger, J.N., and Hudson, H.J. 1970. *Thielavia thermophila* and *Sporotrichum thermophile*. *Transactions of the British Mycological Society* 54:497–500.

Hedger, J.N., and Hudson, H.J. 1974. Nutritional studies of *Thermomyces lanuginosus* from wheat straw compost. *Transactions of the British Mycological Society* 62:129–143.

Hedger, J.N., Samson, R.A., and Basuki, T. 1982. *Scytalidium indonesicum*. *Transactions of the British Mycological Society* 78:365.

Henssen, A. 1957. Über die Bedeutung der thermophilen Mikroorganismen für die Zersetzung des Stallmistes. *Archives of Microbiology* 27:63–81.

Houbraken, J., Frisvad, J.C., and Samson, R.A. 2011. Fleming's penicillin producing strain is not *Penicillium chrysogenum* but *P. rubens*. *IMA Fungus* 2:87–95.

Houbraken, J., Spierenburg, H., and Frisvad, J.C. 2012. *Rasamsonia*, a new genus comprising thermotolerant and thermophilic *Talaromyces* and *Geosmithia* species. *Antonie van Leeuwenhoek* 101:403–421.

Johri, B. 1980. Biology of thermophilous fungi. In *Recent Advances in the Biology of Micro-Organisms*, ed. K.S. Bilgrami and K.M. Vyas, 265–278. Dehradun, India: Bishen Singh Mahendra Pal Singh.

Johri, B.N., Satyanarayana, T., and Olsen, J. 1999. *Thermophilic Moulds in Biotechnology*. Dordrecht, Netherlands: Kluwer Academic Publishers.

Joshi, M.C. 1982. A new species of *Rhizomucor*. *Sydowia* 35:100–103.

Kaerger, K., Schwartze, V.U., Dolatabadi, S., Nyilasi, I., Kovács, S.A., Binder, U., Papp, T., Sd, H., Jacobsen, I.D., and Voigt, K. 2015. Adaptation to thermotolerance in *Rhizopus* coincides with virulence as revealed by avian and invertebrate infection models, phylogeny, physiological and metabolic flexibility. *Virulence* 6:395–403.

Kane, B.E., and Mullins, J.T. 1973. Thermophilic fungi in a municipal waste compost system. *Mycologia* 65:1087–1100.

Kumar, R. 1996. *Taxophysiological studies on thermophilous fungi from north Indian soils.* PhD thesis, Kurukshetra University, Kurukshetra, India.

Kumar, R., and Aneja, K.R. 1999a. Biotechnological applications of thermophilic fungi in mushroom compost preparation. In *From Ethnomycology to Fungal Biotechnology*, ed. J. Singh and K.R. Aneja, 115–126. New York: Plenum Publishers.

Kumar, R., and Aneja, K.R. 1999b. Influence of incubation temperature on growth rates of fifteen thermophilous fungi. *Journal of Mycopathological Research* 37(1):5–8.

Kuster, E., and Locci, R. 1964. Studies on peat and peat microorganisms II. Occurrence of thermophilic fungi in peat. *Archiv für Mikrobiologie.* 48:319–324.

Lee, S.C., Li, A., Calo, S., and Heitman, J. 2013. Calcineurin plays key roles in the dimorphic transition and virulence of the human pathogenic zygomycete *Mucor circinelloides.* *PLoS Pathogens* 9:e1003625.

Li, C.H., Cervantes, M., Springer, D.J., Boekhout, T., Ruiz-Vazquez, R.M., Torres-Martinez, S.R., Heitman, J., and Lee, S.C. 2011. Sporangiospore size dimorphism is linked to virulence of *Mucor circinelloides.* *PLoS Pathogens* 7:e1002086.

Mahajan, M.K., Johri, B.N., and Gupta, R.K. 1986. Influence of desiccation stress in a xerophilic thermophile *Humicola* sp. *Current Science* 55:928–930.

Maheshwari, R., and Balasubramanyam, P.V. 1988. Simultaneous utilization of glucose and sucrose by thermophilic fungi. *Journal of Bacteriology* 170:3274–3280.

Maheshwari, R., Bharadwaj, G., and Bhat, M.K. 2000. Thermophilic fungi: Their physiology and enzymes. *Microbiology and Molecular Biology Reviews* 64:461–488.

Maheshwari, R., and Kamalam, P.T. 1985. Isolation and culture of a thermophilic fungus, *Melanocarpus albomyces,* and factors influencing the production and activity of xylanase. *Journal of General Microbiology* 131:3017–3027.

Mehta, D., and Satyanarayana, T. 2013. Dimerization mediates thermoadaptation, substrate affinity and transglycosylation in a highly thermostable maltogenic amylase of *Geobacillus thermoleovorans.* *PloS One* 8:1–13.

Meyer, G.W. 1970. Amino acid utilization by thermophilic fungi. *Bulletin of the Torrey Botanical Club* 97:227–229.

Miller, H.M., Sullivan, P.A., and Shepherd, M.G. 1974. Intracellular protein breakdown in thermophilic and mesophilic fungi. *Biochemistry Journal* 144:209–214.

Mouchacca, J. 1997. Thermophilic fungi: Biodiversity and taxonomic status. *Cryptogamie Mycologie* 18:19–69.

Nagano, Y., Stuart Elborn, J., Cherie Millar, B., Walker, J.M., and Go, C.E. 2010. Comparison of techniques to examine the diversity of fungi in adult patients with cystic fibrosis. *Medical Mycology* 48:166–176.

Noack, K. 1920. Der Betriebsstoffwechsel der thermophilen Pilze. *Jahrbücher für Wissenschaftliche Botanik* 59:593–648.

Palanivelu, P., Balasubramanyam, P.V., and Maheshwari, R. 1984. Coinduction of sucrose transport and invertase activities in a thermophilic fungus *Thermomyces lanuginosus.* *Archives of Microbiology* 139:44–47.

Prasad, A.R.S., and Maheshwari, R. 1978a. Growth of and trehalase activity in the thermophilic fungus *Thermomyces lanuginosus. Proceedings of the Indian Academy of Sciences* 87B:231–241.

Prasad, A.R.S., and Maheshwari, R. 1978b. Temperature response of trehalase from a mesophilic (*Neurospora crassa*) and a thermophilic (*Thermomyces lanuginosus*) fungus. *Archives of Microbiology* 122:275–280.

Prasad, A.R.S., Kurup, C.R., and Maheshwari, R. 1979. Effect of temperature on respiration of a mesophilic and a thermophilic fungus. *Plant Physiology* 64:347–348.

Rajasekaran, A.K., and Maheshwari, R. 1990. Growth kinetics and intracellular protein breakdown in mesophilic and thermophilic fungi. *Indian Journal of Experimental Biology* 28:46–51.

Rosenberg, S.L. 1975. Temperature and pH optima for 21 species of thermophilic and thermotolerant fungi. *Canadian Journal of Microbiology* 21:1535–1540.

Salar, R.K., and Aneja, K.R. 2006. Thermophilous fungi from temperate soils of northern India. *Journal of Agricultural Technology* 2:49–58.

Satyanarayana, T., and Johri, B.N. 1984. Nutritional studies and temperature relationships of thermophilic fungi of paddy straw compost. *Journal of the Indian Botanical Society* 63:165–170.

Schipper, M.A.A. 1978. On the genera *Rhizomucor* and *Parasitella*. *Studies in Mycology* 17:53–71.

Seifert, K.A., Samson, R.A., Boekhout, T., and Louis-Seize, G. 1997. *Remersonia*, a new genus for *Stilbella thermophila*, a thermophilic mould from compost. *Canadian Journal of Botany* 75:1158–1165.

Stolk, A.C. 1965. Thermophilic species of *Talaromyces* Benjamin and *Thermoascus* Miehe. *Antonie van Leeuwenhoek* 31:262–276.

Stolk, A.C., and Samson, R.A. 1972. The genus *Talaromyces*. Studies on *Talaromyces* and related genera. II. *Studies in Mycology* 2:1–65.

Straatsma, G., and Samson, R.A. 1993. Taxonomy of *Scytalidium thermophilum*, an important thermophilic fungus in mushroom compost. *Mycological Research* 97:321–328.

Straatsma, G., Samson, R.A., Olijnsma, T.W., Den Camp, H.J.O., Gerrits, J.P.G., and Van Griensven, L.J.L.D. 1994. Ecology of thermophilic fungi in mushroom compost, with emphasis on *Scytalidium thermophilum* and growth stimulation of *Agaricus bisporus* mycelium. *Applied and Environmental Microbiology* 60:454–458.

Su, Y.Y., and Cai, L. 2013. Rasamsonia composticola, a new thermophilic species isolated from compost in Yunnan, China. *Mycological Progress* 12:213–221.

Subrahmanyam, A. 1977. Nutritional requirements of *Torula thermophila* Cooney & Emerson at two different temperatures. *Nova Hedwigia* 19:85–89.

Subrahmanyam, A. 1980. A new thermophilic variety of *Humicola grisea* var. *indica*. *Current Science* 49:30–31.

Tansey, M.R. 1973. Isolation of thermophilic fungi from alligator nesting material. *Mycologia* 65:594–601.

Tansey, M.R., and Brock, T.D. 1972. The upper temperature limits for eukaryotic organisms. *Proceedings of the National Academy of Sciences of the United States of America* 69:2426–2428.

Tendler, M.D., Korman, S., and Nishimoto, M. 1967. Effects of temperature and nutrition on macromolecule production by thermophilic Eumycophyta. *Bulletin of the Torrey Botanical Club* 94:175–181.

Thakre, R.P., and Johri, B.N. 1976. Occurrence of thermophilic fungi in coal-mine soils of Madhya Pradesh. *Current Science* 45:271–273.

van Oorschot, C.A.N. 1977. The genus *Myceliophthora*. *Persoonia* 9:404–408.

van Oorschot, C.A.N., and de Hoog, G.S. 1984. Some hyphomycetes with thallic conidia. *Mycotaxon* 20:129–132.

von Arx, J.A., Figueras, M.J., and Guarro, J. 1988. Sordariaceous ascomycetes without ascospore ejaculation. *Beihefte zur Nova Hedwigia* 94:1–104.

Wali, A.S., Mattoo, A.K., and Modi, V.V. 1978. Stimulation of growth and glucose catabolite enzymes by succinate in some thermophilic fungi. *Archives of Microbiology* 118:49–53.

Wright, C., Kafkewitz, D., and Somberg, E.W. 1983. Eucaryote thermophily: Role of lipids in the growth of *Talaromyces thermophilus*. *Journal of Bacteriology* 156:493–497.

4 Habitat Diversity

4.1 INTRODUCTION

All ecosystems are unique owing to their distinctive biotic and abiotic factors. Some ecosystems display extremes of characteristics, such as high salt concentrations and extremes of temperature and pH, chemical content, and/or pressure. In response to such adverse environments, some organisms have developed adaptations that allow them to inhabit and survive under such harsh conditions, which are considered harmful to most living organisms. This existence is due to certain genetic and/or physiological adaptations (Cooney and Emerson, 1964; Dix and Webster, 1995; Aguilar, 1996; Stetter, 1999). Microbial species that live in such environments are broadly classified as extremophiles. Of the three domains of life, most of the thermophilic species that have been described belong to Archaea or Bacteria (Barns et al., 1996). The Eukarya domain is less temperature tolerant than the other groups. The upper temperature limit for some fungi among eukaryotes has been found to be 62°C (Tansey and Brock, 1978). Depending on their cardinal growth temperature, these heat-tolerant fungi can be classified as thermotolerant or thermophilic. Although thermophilic fungi were discovered as early as 1887, they were clearly defined by Cooney and Emerson (1964) in their book *Thermophilic Fungi: An Account of Their Biology, Activities, and Classification*. They considered all those fungi to be thermophilic if they grow at or above 50°C and fail to grow at or below 20°C, and thermotolerant if they have growth maxima up to 50°C and minima well below 20°C.

Generally, there is an inverse relationship between biological diversity and the amount of adaptation required to survive in a specific habitat (Whittaker, 1975). Thermophilic fungi are ubiquitous in nature and have been obtained from an array of natural and man-made habitats. These fungi are commonly found in soils and in habitats wherever organic matter heats up for any reason. In these habitats, thermophiles may occur either as resting propagules or as active mycelia, depending on the availability of nutrients and favorable environmental conditions. Furthermore, it is now well established that ecological diversity in extreme environmental habitats does not result in wide-ranging competition, and therefore few species occur in these habitats. Thus, until now approximately 75 species of truly thermophilic fungi have been reported by various workers all over the world and account for barely 0.1% of the total fungal species known to exist.

In the past few decades, thermophilic fungi have been isolated from a large variety of habitats, both natural and man-made. Over the years, the majority of thermophilic fungi have been isolated from manure, composts, stored grains, bird feathers and nests, industrial coal mine soils, beach sands, nuclear reactor effluents, Dead Sea valley soils, and desert soils. This chapter deals with the diversity of thermophilic fungi found in a variety of habitats suitable for their growth. Although many thermophilic fungi have been cultured *in vitro* and are being exploited in a variety of biotechnological applications, an unanticipated number of thermophilic fungi within thermal environments are marked by direct microscopic observation, 16S rDNA amplification, and other culture-independent techniques, indicating that there are many more thermophilic fungi that are yet to be cultivated in the laboratory.

4.2 NATURAL HABITATS

Natural habitats for the occurrence of thermophilic fungi are distributed all over the world. These habitats are characterized by the prevalence of relatively high temperatures throughout the year, or at least for a longer period during the year, and may be terrestrial or marine in origin. The majority of the known thermophilic fungi have been obtained from these habitats using a culture-dependent approach. Some of the innate habitats that harbor a variety of thermophilic fungi are discussed in this section.

4.2.1 Soil

Soil, as is well known, is the depository of all life and the laboratory within which are carried out most of the changes that enable life to continue. A small amount of soil contains a variety of microorganisms of diverse form and physiology. From a functional viewpoint, soil may be considered the land surface of the earth that provides the substratum for plant and animal life. Due to their potential resource for biological materials (e.g., soil fungi), different types of soils have been explored by several workers all over the world. Thermophilic fungi are common in soils; however, variation in the abundance of individual species depends on the type of soil, depth, seasonality, amount of organic materials, and isolation procedures followed. Several workers have suggested that some of the thermophilic fungi may have their natural habitat in soil, particularly in the upper layers rich in organic matter and in the plant debris on the soil surface (Apinis, 1963; Eggins and Malik, 1969; Evans, 1971; Eicker 1972; Ward and Cowley, 1972; Johri and Thakre, 1975; Taber and Pettit, 1975; Tansey and Jack, 1976; Thakre and Johri, 1976; Ellis and Keane, 1981; Sandhu and Singh, 1981; Abdel-Hafez, 1982; Singh and Sandhu, 1986; Maheshwari et al., 1987; Hemida, 1992; Guiraud et al., 1995; Mouchacca, 1995). Based on competitive growth in mixed cultures, Rajasekaran and Maheshwari (1993) measured respiratory rates of thermophilic and mesophilic fungi in order to forecast the potential of thermophilic fungi to grow in soils. They observed that the respiratory rate of thermophilic fungi was markedly responsive to changes in temperature, but that of mesophilic fungi was relatively independent of such changes, suggesting that in a thermally fluctuating environment, thermophilic fungi may be at a physiological disadvantage compared with mesophilic fungi. Their results suggested that although widespread, thermophilic fungi are ordinarily not an active component of soil microflora. Their presence in soil most likely may be the result of the aerial dissemination of propagules from composting plant material.

The isolation of thermophiles from soils was reported as early as 1939 by Waksman and his coworkers, and the soil still continues to provide a good resource for their isolation in the laboratory. Ward and Cowley (1972) explored South Carolina forest soils for thermophilic fungi. The habitat types considered were sand hill pine-scrub oak, loblolly pine, pine hardwood, and floodplain hardwood. They isolated five species: *Aspergillus fumigatus, Aspergillus nidulans, Thermoascus aurantiacus, Thielavia* sp., and an unidentified species. Taber and Pettit (1975) isolated thermophilic fungi from peanut soils. Besides *Mucor pussilus*, which was found to be the most common species, the other species isolated were *Humicola lanuginosa, Thermoascus aurantiacus, Malbranchea pulchella, Aspergillus fumigatus*, and *Sporotrichum* sp.

Thermophilic and thermotolerant fungi were isolated from the sun-heated soils of south-central Indiana by Tansey and Jack (1976). They isolated 18 species (*Myriococcum albomyces, Aspergillus fumigatus, Talaromyces thermophilus, Humicola lanuginosa, Allescheria terristris, Malbranchea pulchella* var. *sulfurea, Thielavia heterothallica, Mucor pusillus, Chaetomium thermophile* var. *dissitum, Thielavia minor, Thermoascus aurantiacus, Mucor miehei, Torula thermophila, Humicola stellata, Acrophialophora nainana, Thielavia sepedonium, Dactylomyces thermophilus*, and *Talaromyces emersonii*) most frequently from sun-heated soil, less so from grass-shaded soil, and still less so from tree-shaded soil. In order to facilitate the appearance of thermophilic fungi, Johri and Thakre (1975) amended soils with glucose, cellulose, ammonium nitrate, ammonium phosphate, and asparagine and incubated the soil samples at 45°C prior to plating. The thermophilic mycoflora consisted of *Absidia corymbifera, Achaetomium macrosporum, Aspergillus fumigatus, Humicoa grisea, Aspergillus nidulans, Penicillium* sp., *Rhizopus microsporus, Rhizopus rhizopodiformis, Sporotrichum* sp., *Thermoascus aurantiacus, Thermomyces lanuginosus, Thielavia minor*, and *Torula thermophila*. Sundaram (1977) isolated and identified 10 species of thermophilic fungi from rice field soils. Tansey and Brock (1978) have reviewed 34 studies, reporting the isolation of thermophilic fungi from soil, including studies of both temperate and tropical soils.

Ten species of thermophilous fungi were identified from different altitudinal zones of forest soils of Darjeeling in the eastern Himalayas (Sandhu and Singh, 1981). They observed that there was a decrease in the prevalence of true thermophiles with increasing altitude. Ellis (1980) isolated four thermophilous fungi from Antarctic and sub-Antarctic soils and presented a possible explanation for the existence of propagules of thermophilous fungi in some Antarctic soils. Similarly, Ellis and Keane (1981) isolated 13 species of thermophilic fungi from soil samples collected at 100 sites in six climatic regions in Australia. The widest range of species was found in soils of the cool, moist, southeastern climatic zone.

Singh and Sandhu (1986) isolated 46 species of thermophilous fungi from Port Blair, India, saline marshy soils collected from eight different sites. The temperature responses of the fungi revealed 14 microthermophilic, 22 thermotolerant, and 10 true thermophilic species. They catalogued the fungi in order of ecological importance based on their frequency, relative density, and presence value. Maheshwari et al. (1987) isolated 11 species of thermophilic fungi from soils collected from different places in India. They observed that the thermophilic fungal flora in India is qualitatively similar to that recorded from other countries in temperate or tropical latitudes. They proposed that the widespread occurrence of thermophilic fungi in soils may be a result of the dissemination of propagules from self-heating masses of organic materials (composts) that occur worldwide. Abdullah and Al-Bader (1990) isolated 35 thermophilic and thermotolerant fungi from 200 soil samples collected from different parts of Iraq. Out of these, six species were truly thermophilic and the rest were thermotolerant. *Aspergillus terreus*, *Aspergillus fumigatus*, and *Aspergillus niger* were present with frequencies of occurrence of 70%, 68%, and 60%, respectively. Their study revealed that thermophilic and thermotolerant fungi are widely represented in the mycoflora of Iraqi soils.

Contrary to possible expectations, however, thermophiles are more frequent in temperate soils than in tropical soils (Ellis and Keane, 1981). Similarly, there is no evidence that thermophiles are more common than mesophiles in tropical soils (Dix and Webster, 1995). Thus, Salar and Aneja (2006) investigated temperate soils of the Himalayan region from north India for the presence of thermophilic fungi in order to determine the diversity of culturable thermophilic and/or thermotolerant fungal species in temperate soils of north India, to determine the optimal *in vitro* growth conditions for these species, and to determine the importance value index (IVI) for each fungal species (Table 4.1). We observed that soil samples of Shimla, Nainital, and Manali were acidic, whereas the soil of Mussoorie was in the alkaline range. The relationship between the number of fungal species and altitude was mainly due to the high moisture content and high organic carbon in the soil samples (Table 4.1). We cultured 10 true thermophilic fungi and 6 thermotolerant organisms from the four study sites. Two species, *Chaetomium senegalense* belonging to Ascomycetes and *Myceliophthora fergusii* belonging to Deuteromycetes, were reported for the first time from India (Aneja and Kumar, 1994; Sigler et al., 1998). In my study on thermophilous fungi of north Indian soils, the fungal colony-forming units (CFUs) were greater in soils of cooler climate than in those of tropical climates (Kumar, 1996). This could be due to high moisture contents in soils of temperate regions. On the contrary, in tropical soils, low moisture contents during the summer may be responsible for physiological stress and desiccation of the spores of these fungi. The species observed in this study are worldwide in distribution, and there appears to be no thermophilous fungal flora characteristic of the region. Although the study revealed a variety of thermophilous fungi (Table 4.2), they may not represent the whole spectrum of fungal diversity in soils. This may be due to the inhibition of the development of propagules of slow-growing fungi by the competitive and faster-growing fungi, or some species are not represented because their population numbers

TABLE 4.1

Physical and Biological Parameters of Temperate Soil Samples Examined for Thermophilic and Thermotolerant Fungi

		Parameters			
Sites	Altitude (m)	Moisture Content (%)	pH	CFU/g (dry wt.)	Number of Species Isolated
Manali	1871	17.2	6.3	491	7
Mussoorie	2042	29.3	7.3	271	11
Nainital	1938	17.6	6.7	43	7
Shimla	2202	26.3	5.8	1456	12

TABLE 4.2

Distribution of Various Thermophilic and Thermotolerant Fungi in Soils of North India

Sr. No.	Species	Soil Sites							
		1	2	3	4	5	6	7	8
1.	*Absidia corymbifera*	+	−	−	−	−	−	−	−
2.	*Absidia spinose*	+	−	−	−	−	−	−	−
3.	*Acrophialophora fusispora*	+	−	−	−	−	−	−	−
4.	*Acrophialophora* sp.	+	−	−	−	−	−	−	−
5.	*Aspergillus flavus*	+	−	−	−	−	−	−	−
6.	*Aspergillus fumigatus*	+	+	+	−	+	+	+	+
7.	*Aspergillus nidulans*	+	−	−	−	−	−	−	−
8.	*Aspergillus niger*	+	−	−	−	−	−	−	−
9.	*Aspergillus terreus*	+	−	−	−	−	−	−	−
10.	*Aspergillus* sp.	−	+	+	+	−	−	+	−
11.	*Chaetomium thermophile* var. *coprophile*	+	−	−	−	+	+	−	−
12.	*Chaetomium thermophile* var. *dissitum*	+	−	−	−	−	−	−	−
13.	*Chaetomium senegalense*	−	+	−	−	−	−	−	−
14.	*Chaetomium luteum*	−	−	−	−	−	−	+	−
15.	*Chrysosporium tropicum*	−	−	−	−	−	+	−	−
16.	*Chrysosporium* sp.	−	−	−	−	−	−	+	−
17.	*Emericella nidulans*	−	−	+	+	−	−	−	−
18.	*Emericella rugulosa*	−	−	+	−	−	−	+	−
19.	*Humicola grisea*	+	+	−	−	+	−	−	−
20.	*Humicola insolens*	+	+	+	+	+	−	+	+
21.	*Malbranchea sulfurea*	+	−	−	+	−	−	−	−
22.	*Melanocarpus albomyces*	−	−	−	−	−	−	−	+
23.	*Penicillium chrysogenum*	+	+	+	−	−	−	−	−
24.	*Rhizomucor pusillus*	+	−	+	−	−	−	+	+
25.	*Rhizopus microcarpus*	+	−	+	+	−	−	+	−
26.	*Thermoascus aurantiacus*	+	−	−	−	+	+	+	−
27.	*Thermomyces lanuginosus*	+	+	−	−	+	−	+	+
28.	*Torula thermophile*	+	+	−	+	+	+	−	+
29.	*Stilbella thermophile*	−	−	−	−	−	−	+	−
30.	Sterile form I	−	−	+	−	−	−	−	−
31.	Sterile form II	−	−	+	−	−	−	−	−

Note: 1 = Kurukshetra; 2 = Manali; 3 = Mussoorie; 4 = Dehra Dun; 5 = Nainital; 6 = Jim Corbett National park; 7 = Shimla; 8 = Jaipur; + = present; − = absent.

are too low. There was a definite influence of climatic seasonality on species diversity, with temperate climates (Shimla, Manali, and Mussoorie) being conducive for maximum diversity. From this study, it follows that thermophilous fungi were as common in soils in the temperate regions as in the tropical regions. Further, the rare species are probably less successful thermophiles in terms of adaptation to north Indian environments and in competition with other species (Singh and Sandhu, 1986).

A variety of underexplored typical soils (desert, geothermal, coal mine, and Dead Sea valley) as promising biotopes for the occurrence of thermophilic and thermotolerant fungi have been investigated by several workers, as described below.

4.2.1.1 Desert Soils

Desert soils are characterized by a barren area of land where little precipitation occurs, and consequently, conditions for plant and animal life are hostile. Deserts can be classified by the amount of precipitation that falls, by the temperature that prevails, by the causes of desertification, or by their geographical location. About one-third of the land surface of the world is arid or semi-arid. Several workers have isolated thermophilic fungi from desert soils that heat up to 50°C or more, with the implied conclusion that desert soil is a natural habitat of thermophilic fungi. However, since liquid water is essential for growth, it is unlikely that the presence of spores of thermophilic fungi in dry soils is a consequence of their growth *in situ*. Abdel-Hafez (1982) studied desert soils of Saudi Arabia for thermophilic and thermotolerant fungi and isolated and catalogued 48 species belonging to 24 genera. The most frequent species were *Aspergillus fumigatus*, *Aspergillus terreus*, *Humicola grisea* var. *thermoidae*, and *Chaetomium thermophile* var. *copropile* on glucose; *A. fumigatus*, *C. thermophile* var. *copropile*, *A. terreus*, *Aspergillus nidulans*, and *C. thermophile* var. *dissitum* on cellulose; and *A. fumigatus* and *A. terreus* on 40% sucrose—Czapek's agar plates. It was further reported by the author that 16 species and 4 varieties were particularly thermophilic: *A. fumigatus*, *Humicola grisea* var. *thermoidae*, *Humicola insolens*, *Humicola lanuginosa*, *C. thermophile* var. *copropile*, *C. thermophile* var. *dissitum*, *Chaetomium virginicum*, *Mucor pusillus*, *Stilbella thermophila*, *Sprotrichum pulverulentum*, *Talaromyces thermophilus*, *Talaromyces emersoni*, *Talaromyces aurantiacus*, *Torula thermophila*, *Malbranchia pulchella* var. *sulfurea*, *Myriococcus albomyces*, *Allescheria terrestris*, *Chrysosporium pruinosum*, *Thielavia thermophila*, and *Papulaspora thermophila*.

Hemida (1992) isolated thermophilic and thermotolerant fungi from cultivated and desert soils of Egypt, exposed continuously to cement dust particles. The author recovered 10 genera and 16 species, in addition to two varieties of *Aspergillus flavus* and *Malbranchia pulchella* from all soil samples, on three types of media at 45°C. In this investigation, the most dominant species was *Aspergillus fumigatus*. Six truly thermophilic fungi. (*Chaetomium thermophile*, *Malbranchea pulchella* var. *sulfurea*, *Rhizomucor pusillus*, *Myriococcum albomyces*, *Talaromyces thermophilus*, and *Torula thermophila*) were obtained. Mouchacca (1995) reviewed the isolation of several thermophilic and thermotolerant fungi (*Myceliophthora thermophila*, *Scytalidium thermophilum*, *Thermomyces lanuginosus*, *Malbranchea cinnamomea*, *Chaetomium thermophile*, *Talaromyces thermophilus*, *Melanocarpus albomyces*,

Thermoascus aurantiacus, Rhizomucor pusillus, Myceliophthora fergusii, Myriococcum thermophilum, Thielavia terrestris, Rhizomucor miehei, Talaromyces byssochlamydoides, and *Talaromyces emersonii*) from desert soils of the Middle East (Syria, Iraq, Kuwait, Saudi Arabia, Jordan, and Egypt). He observed a greater abundance of thermophilic fungi in soils under temperate climates.

4.2.1.2 Coal Mine Soils

Johri and Thakre (1975) and Thakre and Johri (1976) exploited coal mine soils for isolating thermophilic fungi. They collected soil samples from the vicinity of three collieries: Chandameta, Parasia, and Newton Chikly in the Chhindwara District of Madhya Pradesh, India. Besides heating of the sampling site due to self-combustion, the average temperature (21.1°C) and average precipitation (64 cm) of the district anticipated the occurrence of thermophilic fungi in this unique habitat. A total of 14 fungi were isolated, which included *Absidia corymbifera, Rhizopus microspores, Rhizopus rhizopodiformis, Achaetomium macrosporum, Emericella nidulans, Thermoascus aurantiacus, Thielavia minor, Aspergillus fumigatus, Humicola grisea, Penicillium* sp. 1, *Penicillium* sp. 2, *Sporotrichum* sp., *Thermomyces lanuginosus,* and *Torula thermophila. Aspergillus fumigatus* dominated in all the samples investigated.

4.2.1.3 Geothermal Soils

Geothermal activity is perhaps responsible for the creation of the most abundant high-temperature environments. Geothermal soils represent unique and comparatively unexplored model systems to address ecological questions using soil microbial communities, as harsh conditions in these soils exert strong filters on most organisms. These soils are often a direct product of geothermal waters and their aqueous chemical constituents (Figure 4.1). Conditions unique to geothermal soils include

FIGURE 4.1 Grand Prismatic Spring at Midway Geyser Basin showing hydrothermal features, which are magnificent evidence of earth's volcanic activity.

high soil temperature that increases with depth, periodic inundation of chemically diverse waters, belowground influence from steam and geothermal water, and aqueous saturation of chemical elements that act as soil parent material and drivers of soil–chemical processes (Meadow, 2012).

Geothermal soils have received little attention for the isolation and characterization of thermophilic and thermotolerant fungi. Redman et al. (1999) isolated two thermophilic (*Dactylaria constrictum* var. *gallopava* and *Acremonium alabamense*) and six thermotolerant (*Absidia cylindrospora*, *Aspergillus fumigatus*, *Aspergillus niger*, *Penicillium* sp. 1, *Penicillium* sp. 2, and *Penicillium* sp. 3) fungal species from geothermal soils near Amphitheater Springs in Yellowstone National Park. As expected, these soils were characterized by high temperature (up to 70°C), high heavy metal content, low pH values (down to pH 2.7), sparse vegetation, and limited organic carbon. They observed that for all the fungal species studied, the temperature of the soil from which the organisms were cultured corresponded with their optimum axenic growth temperature. They suggested that many other fungal species could be present in the geothermal soils investigated, but they were not detected by traditional culturing methods Further studies performed with molecular biology–based techniques, such as polymerase chain reaction (PCR), followed by denaturing gradient gel electrophoresis, should reveal whether culturable fungal species are, indeed, the predominant fungal species in these soils and whether these soils harbor uncultured fungi, which can present a more precise evaluation of the fungal diversity of this unique ecosystem.

4.2.1.4 Dead Sea Valley Soil

The Dead Sea is a salt lake bordered by Jordan to the east and Israel and Palestine to the west (Figure 4.2). It is known as dead sea as its surface, and the shores are approximately 300 m below the mean sea level, the earth's lowest elevation on land. The Dead Sea area is an interesting ecological niche characterized by high salinity (mainly Na, Mg, Ca, and chlorides, and also KOH and bromides) and sparse vegetation, and its oxygen level is one of the highest in the world, making this site a great pecularity (Guiraud et al., 1995). Steiman et al. (1995) studied mycoflora of soil around the Dead Sea and obtained 256 isolates representing 90 species of 23 genera of Ascomycetes (including *Aspergillus* and *Penicillium*) and Zygomycetes, in addition to some unidentified Ascomycetes and Basidiomycetes. Some of the species showed thermotolerance, but none were strict thermophiles. In their subsequent study of the Dead Sea valley, Guiraud et al. (1995) isolated 106 species belonging to 51 genera of Deuteromycetes. They reported one thermophilic fungus, *Scytalidium thermophilum*, from these samples. However, there does not seem to be a fungus flora characteristic of these soils. In fact, only a few strict halophilic or thermophilic fungi have been described in this study, and the authors concluded in favor of the adaptation of cosmopolitan soil fungi, which may be considered thermotolerant or osmotolerant species.

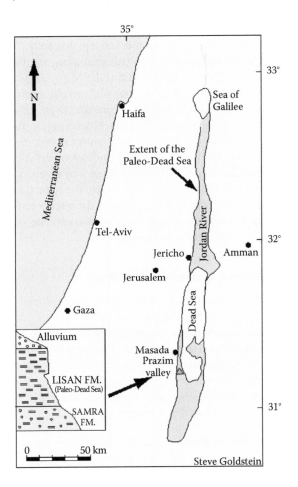

FIGURE 4.2   Dead Sea valley; it is approximately 300 m below the mean sea level and earth's lowest elevation on land.

4.2.2   BEACH SAND

Beaches represent the unconsolidated sediment that lies at the junction between water (oceans, lakes, and rivers) and land and are usually composed of sand, mud, or pebbles. Microorganisms are significant components of beach sand. Recently, special attention has been paid to the care and cleaning of the coasts, as incidences of infections associated with beach sand and coastal water have steadily increased over the past several decades (Velonakis et al., 2014). Filamentous fungi are significant components of the sandy beaches and play an important role as primary decomposers of organic matter in the beach ecosystem. Exclusive studies on the isolation of thermophilic fungi from beach sand are still lacking. Recently, Yee et al. (2016) isolated filamentous fungi from beach sand collected along the Batu Ferringhi and Teluk Bahang beach areas, Penang Island, Peninsular Malaysia. None of the species were strict thermophiles; however, some species of *Aspergillus* and *Penicillium* were thermotolerant, and also showed the highest occurrence. In an earlier study (Zakaria et al., 2011), *Penicillium chrysogenum*, a thermotolerant fungus, was recovered from the sandy soil samples. Similarly, Fatma (2003) reported that *Penicillium* was the most frequent species isolated from sandy soil samples of the Alexandria beach in Egypt. *Penicillium* was also one of the most common fungal genera isolated from the Ipanema beach in Brazil (Sarquis and Oliveira, 1996). Based on the literature survey, it appears that sandy beach soils, which are part of the marine ecosystem, contain a reservoir of a variety of filamentous microfungi that contribute significantly to the ecological functioning of the beach ecosystem.

4.2.3 NESTING MATERIAL OF BIRDS AND ANIMALS

Nests of birds and animals are mainly composed of plant material and made by the incubating birds or other animals. Several megapode species construct enormous *mound* nests made of soil, branches, sticks, twigs, and leaves, and lay their eggs within the decomposing mass. Heat is generated in these nests mainly by the respiration of compost microorganisms. Thus, the nests are an ideal habitat for the development of thermophilic microflora. Several workers have attempted isolation of thermophilic and thermotolerant fungi from the nesting material of birds and animals. Some of these works are discussed in this section.

4.2.3.1 Bird Nests and Feathers

Owing to the incubating behavior of birds, their nests are potential habitats for the occurrence of thermophilic fungi. Bird nests are usually a collection of a large mass of plant material, with admixtures of animal material (feathers and hairs in the nest lining, the fluff of chicks, remnants of prey, and pellets of fledgling raptors and droppings). The moisture content and pH values of the nesting material are among those factors that determine the selection of fungal communities in bird nests (Hubalek, 1967). Apinis and Pugh (1967) isolated 27 species of thermophilous fungi from plant debris in the nests of 12 passerine bird species in Nottinghamshire. In total, they surveyed 54 nests for the presence of thermophilous fungi in which certain species, such as *Aspergillus fumigatus*, *Chaetomium thermophile*, *Coprinus delicatulus*, *Humicola insolens*, *Thermoidium sulphureum*, and *Thermomyces languinosus*, were present at a high frequency. They observed that the species populations of thermophilous nest fungi are similar to those obtained from plant debris on the soil surface and in the grassland vegetation. Further, the authors divided all 27 thermophilous fungi into two groups (1) saprophytes with no known harmful relationships to birds and other animals, such as *Allescheria terrestris*, *Botryotrichum* sp., *Chaetomium thermophile*, *Coprinus delicatulus*, *Humicola insolens*, *Penicillium duponti*, *Sporotrichum thermophile*, *Stilbella thermophila*, *Thermoidium sulphureum*, *Thielavia sepedonium*, and *Torula thermophila*, and (2) saprophytes on the plant debris of the nests but with known potential pathogenicity to birds, other animals, and humans, for example, *Absidia ramosa, Aspergillus fumigatus, Aspergillus terreus, Dendrostilbella boydii, Emericella nidulans, Endomyces lactis, Mucor pusillus, Paecilomyces varioti, Rhizopus arrhizus, Rhizopus cohnii,* and *Thermomyces lanuginosus.*

Seasonal variation in the mycoflora of nesting material of birds was reported by Satyanarayana et al. (1977). They recovered nine thermophilic, two thermotolerant, and eight mesophilic fungi from nesting material of five bird species: crow (*Corvus splendens*), house sparrow (*Passer domesticus*), pipit (*Anthus rufulus*), bee-eater (*Merops superviliesus*), and crow pheasant (*Centropus sinensis*). Ogundero (1979) obtained 10 species of thermophilic and thermotolerant fungi from poultry droppings in Nigeria, of which 6 (*Mucor pusillus, Talaromyces dupontii, Talaromyces emersonii, Thermoascus aurantiacus, Thermomyces lanuginosus,* and *Torula thermophila*) were true thermophiles and the other 4 (*Acremonium albamensis, Aspergillus fumigatus, Byssochlamys verrucosa,* and *Rhizopus microspores*) were thermotolerant. *Aspergillus fumigatus, Mucor pusillus,* and *Thermoascus aurantiacus* are known human pathogens. The presence of these pathogens in poultry droppings is an indication of the potential health risks posed to poultry by infrequent removal of their droppings as they accumulate below the cages (Ogundero, 1979). Rajavaram et al. (2010) isolated 46 species of thermophilic fungi from different substrates in Andhra Pradesh, India, including three species from the nesting material of birds. Korniłłowicz-Kowalska and Kitowski (2013) examined the species diversity of thermophilic fungi in 38 nests of nine species of wetland birds in Poland. They isolated 18 species of thermophilic fungi belonging to 12 genera. They found a good correlation between the weight of the nest and the number and diversity of thermophilic fungi recovered from these nests. Also, the diversity of the thermophilic biota was positively correlated with the individual mass of the bird. In the nests of large birds (mute swan) and medium-sized birds (grey heron, marsh harrier, coot, and great crested grebe), the diversity of thermophilic fungi was significantly higher than that in the nests of smaller birds (gulls and terns), which remains in close relation with the size of the zone warmed by the birds.

4.2.3.2   Alligator Nesting Material

As pointed out earlier, thermophilic fungi are most commonly found in natural habitats, including the nesting material of higher animals. In the case of alligator nesting material, considerable doubt exists as to the role of heat of microbial metabolisms for the incubation of their eggs. Neill (1971), an expert of alligator behavior, however, observed that "in spite of the frequent assertion that the alligator's eggs are 'incubated' by the warmth of decaying nest debris, I believe that this heat of fermentation is often a hazard to the eggs, and that the female's choice of a cool nesting site reflects an imperative need to minimize the ever present danger that the eggs will become overheated. … The alligator embryo should not be significantly more heat-tolerant than the adult; and the temperature in the nest is very close to a body temperature that would be fatal to an adult." He did, however, support the role of heat of fermentation in the incubation of alligator eggs in the sense that this heat counteracts the drop in temperature during night in the swamps and marshes.

The study by Neill prompted Tansey (1973) to undertake studies on the isolation of thermophilic fungi from alligator nesting material of the American alligator (*Alligator mississipiensis*). He isolated five thermophilic (*Chaetomium thermophile* var. *coprophile*, *Chaetomium thermophile* var. *dissitum*, *Humicola lanuginose*, *Talaromyces thermophilus*, and *Thermoascus aurantiacus*) and two thermotolerant (*Aspergillus fumigatus* and *Burgoa-Papulaspora* sp.) fungi from the decomposing nesting material. The author observed that alligator nesting material provides a habitat suitable for the thermophilic fungi, where the heat required for the growth of these fungi is not provided by the body heat of the animal. Based on the data obtained, he further suggested that habitats suitable for the origin and evolutionary survival of obligately thermophilic fungi have existed for long periods of time.

4.2.4 COAL SPOIL TIPS

Coal spoil tip refers to a pile built with the accumulated spoil (material removed when digging a foundation, tunnel, or other large excavation, including mining) of coal. Coal spoil tips are a suitable habitat for the establishment of thermophilic microflora. Evans (1971) studied coal spoil tips for the occurrence of thermophilous fungi and observed that due to greater plant turnover, extreme heat, and a lack of suitable soil crumbs, thermophilic microorganisms predominate in this unique habitat. He also reported that with the gradual accumulation of humus, the temperature decreases, favoring the development of thermotolerant and mesophilic forms. The survival of obligate thermophilic fungi appeared dependent on the production of resting propagule structures that could remain viable at low temperatures during the winter months. He isolated 32 thermophilic and thermotolerant fungal species that represented the genera *Mucor*, *Rhizopus*, *Mortierella*, *Chaetomium*, *Sphaerospora*, *Talaromyces*, *Thielavia*, *Acrophyalophora*, *Aspergillus*, *Calcarisporium*, *Chrysosporium*, *Geotrichum*, *Penicillium*, *Scolecobasidium*, and *Tritirdium*. He observed that the population of thermophilous fungi in well-colonized warm areas was higher than that of corresponding nonwarm areas. The author further opined that the population of thermophiles in coal spoil tips is very similar to that reported from other habitats, like grasslands, but distinctly inferior to the population density reported from composts, manure, and sewage.

4.2.5 HOT SPRINGS

A hot spring is a type of thermal spring generated by the emergence of geothermally heated groundwater from the earth's crust. The water emanating from a hot spring in a nonvolcanic area is heated either geothermally, that is, heat from the earth's mantle, or by coming into contact with hot rocks. On the other hand, in active volcanic zones water gets heated by coming into contact with magma (molten rock). The high temperature gradient near magma may cause water to be heated enough that it starts boiling or becomes superheated. The resulting superheated water builds up steam pressure that erupts in the form of a jet above the earth's surface, known as a geyser. An innumerable number of geothermal hot springs occur in many locations all over the crust of the earth. Hot springs are found on all continents and in many countries around the world. Countries that are renowned for their hot springs include Honduras, Canada, Chile, Hungary, Iceland, Taiwan, Japan, Israel, Fiji, New Zealand, and the United States. However, there are interesting and unique hot springs in many other places of the world, including India. The water of these springs contains a large number of dissolved minerals supporting the growth of microbiota. Thermal hot springs represent extreme niches that have maintained some degree of pristine quality, and their biotechnological potential has largely remained unexplored. In the past few decades, several attempts have been made to isolate and characterize thermophilic microflora from thermal springs in different parts of the world (Hedger, 1975; Chen et al., 2000; Reysenbach et al., 2000; Skirnisdottir et al., 2000; Pan et al., 2010; Sharma et al., 2013).

Hot springs harbor microorganisms from diverse groups, including bacteria, archaea, algae, and fungi, with the majority of them being bacteria and archaea. Now it is well accepted that only 1% of the extant microbial species have been cultured (Ward et al., 1998) using culture-dependent techniques that grow on conventional isolation media. However, culture-independent approaches employing various molecular techniques may provide a true assessment of the microbial diversity present in these hot springs. With regard to the isolation of thermophilic fungi from hot springs, relatively few studies have been conducted. Hedger (1974) observed that although hot springs of Indonesia had high populations of thermophilic fungi, their total microbial flora was restricted to a few species, indicating that not all thermophilic fungi were able to grow under the conditions prevailing in hot volcanic springs. Chen et al. (2000) isolated five species of thermophilic fungi from hot springs of Wu-Rai of northern Taiwan, which were identified as *Aspergillus fumigatus*, *Thermomyces lanuginosus* (syn. *Humicola lanuginosa*), *Humicola insolens*, *Penicillium dupontii*, and *Rhizoctonia* sp. All these isolates can grow at a temperature ranging from 55°C to 64°C. An investigation of the mycoflora in Yangmingshan National Park, northern Taiwan, from August 1999 to June 2000, particularly of thermophilic and thermotolerant fungi inhabiting sulfurous hot spring soils, resulted in the identification of 12 taxa (Chen et al., 2003). They reported four thermophilic fungal species from the Hsiaoyukeng sulfurous hot spring area: *Chrysosporium* sp., *Papulaspora thermophila*, *Scytalidium thermophilium*, and *Sporotrichum* sp.

In another study, Pan et al. (2010) isolated a number of thermophilic fungi from geothermal sites with near-neutral and alkalescent thermal springs in Tengchong Rehai National Park, China. By employing internal transcribed spacer (ITS) sequencing combined with morphological analysis, they identified these fungi to the species level. In total, 102 strains were isolated and identified as *Rhizomucor miehei*, *Chaetomium* sp., *Talaromyces thermophilus*, *Talaromyces byssochlamydoides*, *Thermoascus aurantiacus* Miehe var. *levisporus*, *Thermomyces lanuginosus*, *Scytalidium thermophilum*, *Malbranchea flava*, *Myceliophthora* sp. 1, *Myceliophthora* sp. 2, *Myceliophthora* sp. 3, and *Coprinopsis* sp. Two species, *Thermomyces lanuginosus* and *Scytalidium thermophilum*, were found to be dominant, representing 34.78% and 28.26% of the sample, respectively. Most of these species thrived at alkaline growth conditions. Sharma et al. (2013) isolated 101 thermophilic microbial strains from the thermal hot springs of Himachal Pradesh, India. However, they could isolate and identify only one fungal isolate as *Myceliophthora thermophila* SH1 using

ITS sequencing. Kambura et al. (2016) investigated the diversity of fungal communities in the hot springs of Lake Magadi and Little Magadi, Kenya. They used amplicons of the ITS region on total community DNA using Illumina sequencing to explore the fungal community composition within the hot springs and obtained a total of 334,394 sequence reads, from which 151 operational taxonomic units (OTUs) were realized at 3% genetic distance. Taxonomic analysis further showed that 80.33% of the OTUs belonged to the phylum Ascomycota, and 11.48% to Basidiomycota, and the rest consisted of Chytridiomycota, Glomeromycota, and early diverging fungal lineages. Their results revealed representatives of thermophilic and alkaliphilic fungi within the hot springs of Lake Magadi and Little Magadi.

4.3 MAN-MADE HABITATS

A man-made habitat refers to an environment created by humans that contains all vital elements to support the growth of a living species. Man-made thermal habitats conducive for the development and growth of thermophilic fungi include cooling towers, effluents from nuclear power reactors, ducts employed for thermal insulation, compost piles, wood chip piles, stored grains, stored peat, and retting guayule (Johri et al., 1999). In these habitats, heat generation is due to self-heating or solar heating elevating the temperature up to 70°C. The man-made habitats are relatively low-temperature habitats compared with natural thermal habitats, and best suited for the isolation of thermophilic microorganisms. Thermophilic fungi have been isolated from a multitude of man-made habitats, some of which are discussed below.

4.3.1 HAY

Ever since the dawn of agriculture and the domestication of animals, humans have depended on the storage of food and fodder for use in time of need. Hay pertains to grass, legumes, or other herbaceous plants that have been cut, dried, and stored in the form of bales for later use as animal fodder, particularly for grazing animals. The hay bales are stacked one over another, forming haystacks. Moist haystacks generate heat due to microbial fermentation. Little information is available on thermophilic fungal flora from hay. Miehe (1907) described the presence of thermophilic fungi from a self-heated damp stack of hay. These include *Rhizomucor pusillus*, *Thermomyces lanuginosus*, *Thermoascus aurantiacus*, and *Malbranchea pulchella* var. *sulfurea*. To examine the role of thermophilic fungi in the self-heating of agricultural residues, Miehe inoculated damp hay and other plant materials kept inside thermal flasks with pure cultures of individual fungi. It was observed that sterilized hay (control) did not generate heat, whereas that inoculated with the fungus did, and the peak temperature achieved by the material correlated with the maximum temperature of growth of the fungus used. Further, by controlled experiments, Miehe demonstrated that the naturally occurring microorganisms in damp haystacks or other plant materials caused its heating. Thermophilic fungi have also been reported from stacked peat (Isachenko and Malchevskaya, 1936; Mishustin, 1950; Kuster and Locci, 1964), where the problems of self-heating and combustion are similar. Further, Gregory and Lacey (1963) observed that different types of microflora can develop in hay depending on the water content in the bale or stack. He demonstrated that hay having a 16% moisture content heated only moderately, while that having 25% moisture heated to 45°C, allowing the development of thermotolerant molds. But, the temperature in wet bales with 45% moisture soared to 60°C–65°C, and a range of thermophilic fungi, including *Aspergillus fumigatus*, *Mucor pusillus*, and *Humicola lanuginose*, and some actinomycetes were recovered from it.

4.3.2 WOOD CHIP PILES

Wood chips are medium-sized solid peels of wood made by cutting or chipping larger pieces of freshly harvested wood, and are generally used for producing wood pulp and as a raw material for technical wood processing. These wood chips are stored in piles, and thus remain wet and may hold more than 50% of their weight as water (Brown et al., 1994). The available moisture initiates bacterial growth, thereby promoting fermentation. The respiring parenchymatous cells in wood and microbial metabolism raise the temperature in the piles. The presence of thermophilic and thermotolerant microbiota is dominant in wood chip piles (Tansey, 1971; Flannigan and Sagoo, 1977; Miller et al., 1982) as the temperature reaches up to 50°C–60°C in the interior of the wood chip piles. Occasionally, the temperature in wood chip piles can rise even to ignition within a few weeks of storage. Therefore, it constitutes a potential habitat for the growth and development of thermophilic fungi, which generally cause deterioration in the quality of wood chips.

A detailed study of the occurrence and growth of thermophilic fungi in self-heated industrial wood chip piles was pioneered by Tansey (1971). He isolated thermophilic fungi from self-heated wood chip piles stored at a paper factory. The dominant ones were *Chaetomium thermophile* var. *Coprophile*, *Chaetomium thermophile* var. *dissitum*, *Humicola grisea* var. *thermoidea*, *Humicola lanuginosa*, *Sporotrichum thermophile*, *Talaromyces emersonii*, *Talaromyces thermophilus*, and *Thermoascus aurantiacus*, besides the thermotolerant *Aspergillus fumigatus*. He also observed that fewer species were present in fresh chips and in chips from unheated piles. Ramnath et al. (2014) analyzed microflora from *Eucalyptus* sp. wood chips intended for pulping and isolated thermotolerant *Aspergillus fumigatus* and *Phanerochaete chrysosporium*.

4.3.3 NUCLEAR REACTOR EFFLUENTS

During the process of nuclear power generation, large quantities of water are used and subsequently discharged into river estuaries or oceans. Thermal effluent refers to discharges from nuclear reactors and provides habitats with temperatures from ambient to 70°C, pH values that are not extreme, and environmental parameters that may be conducive to the growth of thermophilic and thermotolerant fungi. Tansey and Fliermans (1978) described the diverse populations of thermophilic and thermotolerant fungi found in foam, water, microbial mats, plant debris, air, and soil associated with reactor effluent habitats at the Savannah River Plant. By direct microscopic examination and quantitative culturing, they established that except for the population of *Dactylaria gallopava* (syn. *Ochroconis gallopava*), populations of thermotolerant and thermophilic fungi were not significantly different in elevated and ambient temperature sites. *D. gallopava*, a thermotolerant pathogenic fungus causing encephalitis in young chickens and turkeys, was present in the mirobial mats, in foam, and in soils at the edges of the cooling water effluents that have temperatures above 43°C. Rippon et al. (1980) isolated 52 species of thermotolerant and thermophilic fungi from 17 samples of water, foam, microbial mat, soil, and air associated with the thermally enriched cooling canal of a nuclear power station. They grouped 11 species as opportunistic Mucorales or opportunistic *Aspergillus* sp. One veterinary pathogen (*D. gallopava*) was also isolated. The authors further reported that soil samples near the cooling canal reflected an enrichment of thermophilous organisms, the previously mentioned opportunistic Mucorales and *Aspergillus* spp. which were found in a greater number than that usually encountered in a mesophilic environment. Air and soil samples taken at various distances from the power station, however, produced no greater abundance of these thermophilous fungi, as would be expected from a thermally enriched environment. Their results indicated that there was no significant dissemination of thermophilous fungi from the thermally enriched effluents to the nearby environment. Ellis (1980) reported the isolation of thermophilic fungi from Hazelwood Power Station cooling pond (Morwell, Victoria, Australia) effluents. He isolated five species of thermophilic fungi: *Chaetomium thermophile, Aspergillus fumigatus, Humicola grisea* var. *thermoidea, Humicola insolens*, and *Thermomyces lanuginosus*.

4.3.4 MANURE

Manure is the organic material derived from animal, human, and plant residues containing plant nutrients in complex organic form. Agricultural operations generate large quantities of manure, which if applied directly to agricultural land may have adverse effects on soil, air, and water quality through contamination, odor and gas emission, and nutrient leaching (Larney and Hao, 2007). Manure is also a reservoir of antimicrobial-resistant bacteria. The composting of manure results in a product that, besides being nutrient rich, is comparatively free of microbial pathogens and phytotoxins. Recently, Holman et al. (2016) isolated *Remersonia thermophila, Talaromyces thermophilus*, and *Thermomyces lanuginosus* from a composting manure. The process of production of organic manure by composting is an unwitting exploitation of thermophilic microorganisms, including those from the fungal world.

4.3.5 STORED PEAT

Peat is an accumulation of partially decayed vegetation or organic matter that is unique to natural areas known as peatlands. One of the most common components of peat vegetation is *Sphagnum* moss (peat moss). In certain parts of the world, peat is used as an important source of fuel. Due to microbial metabolism in moist peat, conditions akin to the requirement of elevated temperatures for the growth of thermophilic fungi are frequent in this unique habitat. Investigation of the isolation of thermophilic fungi from peat and their taxonomy was carried out by Kuster and Locci (1964). In their study, *Humicola* was recorded for the first time from peat, and its various thermophilic strains were classified as *H. insolens*, *H. stellatus*, and *H. lanuginosus*. They also isolated thermophilic *Paecilomyces* (which is otherwise said to be thermotolerant) as commonly occurring in peat. However, confusion prevailed with regard to the identification of this genera, as after detailed examination, they put this strain in the related genera *Talaromyces* and *Thermoascus*. They further reported that *Mucor pusillus* and *Aspergillus fumigatus* were the most frequently isolated species from peat under thermophilic conditions.

4.3.6   RETTING GUAYULE

Guayule (*Parthenium argentatum*), native to the southwest United States and Mexico, is a flowering shrub belonging to the Asteraceae family. The name *guayule* is derived from the Nahuati word *ulli* or *olli*, meaning "rubber." The plant can be used as secondary source of latex for the production of rubber. During the Second World War, the demand for guayule rubber increased when Japan cut off America's Malaysian latex source. Thus, studies on the latex-bearing guayule plant showed that the extractability and quality of rubber improved if, before milling, this shrub was chopped and stored in a mass ("rets") that would self-heat to a high temperature, ranging between 65°C and 70°C. Allen and Emerson (1949) isolated several thermophilic fungi that would grow up to 60°C and were capable of decomposing the resin. They established that the observed improvement from retting resulted mainly from utilization and reduction in the amount of contaminating resin in crude rubber by the thermophilic fungi. They further demonstrated that for optimal development of these fungi in the retting guayule, its moisture content, porosity, and nutrient contents were crucial factors. Additionally, the size of the heap of chopped plant material was also an important factor in reducing the loss of heat produced by microbial metabolism for the successful colonization of thermophilic microflora.

4.3.7   STORED GRAINS

Plant products upon which man has depended for centuries and has stored in ever-increasing quantities are the various grains and the flour and other foodstuffs derived directly from them. Thermophilic fungi are regarded as storage fungi; they are present on grains while in the field but do not damage grains prior to storage (Tansey and Brock, 1978). During storage, heating of stored grains progresses, resulting in a distinct succession of microbial species that culminate in large populations of thermotolerant and thermophilic fungi, causing loss of grain quality in terms of germination, discoloration, and damage. Stored grains constitute an excellent substrate for the development and colonization of thermophilic fungi.

Thermophilic fungi have been isolated from standing parts of plants by Apinis (1963, 1972) and Apinis and Chesters (1964), whereas Clarke (1966) described them from harvested grains. A large population of thermotolerant fungi, especially *Absidia corymbifera*, *Aspergillus fumigatus*, and *Aspergillus candidus* and thermophilic *Humicola lanuginosa* and *Mucor pusillus*, has been reported from stored grains (Clark, 1967; Clarke, 1969; Lacey, 1971). A detailed investigation of thermophilous fungi from moist, stored barley grain in three concrete stave silos in Cumberland and

Lincolnshire, England, was carried out by Mulinge and Apinis (1969). They isolated *Absidia* spp., *A. candidus, Aspergillus flavus, Aspergillus fumigatus, Aspergillus terreus, Dactylomyces crustaceus, Eurotium amstelodami, Monascus* spp., and *Mucor pusillus* most frequently from both grain husks and dehusked surface sterilized grains. They further reported that the mycelium of thermophilous fungi was found to be mostly confined to husk tissue of the healthy grain at the outset of the storage. However, after prolonged storage a number of these species, such as *A. candidus, Aspergillus terreus,* and *Dactylomyces crustaceus,* invaded the grain tissue, especially when self-heating of grain took place.

Oso (1974) isolated seven strongly thermophilic fungi from stacks of oil palm kernels in parts of the western state of Nigeria. These were *Humicola lanuginose, Mucor pusillus, Talaromyces emersonii, Chaetomium thermophile* var. *coprophile, Chaetomium thermophile* var. *dissitum, Thermoascus aurantiacus,* and *Thermoascus crustaceus.* Of the 148 rice samples analyzed by Kalyanasundaram and Rao (1988) from south India, 84% harbored storage fungi internally. The fungi included several species of thermotolerant *Aspergillus* and *Penicillium.* Thermophilic fungi have also been isolated from cocoa beans (Broadbent and Oyeniran, 1968; Hansen and Welty, 1970). Wareing (1997) isolated a number of thermotolerant (*Aspergillus candidus, Aspergillus fumigatus, Aspergillus flavus,* and *Paecilomyces varioti*) and thermophilic (*Thermomyces lanuginosus, Rhizomucor pusillus, Thermoascus aurantiacus,* and *Thermoascus crustaceous*) fungi from shipments of food-aid grain, and from large bag stacks of maize stored in sub-Saharan Africa.

4.3.8 MUNICIPAL WASTE

The disposal of municipal wastes and refuse has presented a serious problem in big cities, and the odors that characterize the outskirts of such big cities confirm the severity of the problem and the health hazard. Municipal wastes, besides other things, contain lignocellulosic materials and large quantities of plastic and inorganic (glass and metal) materials. Various reports have suggested that municipal refuse contains more than 50% paper (Kaiser et al., 1968). Studies have indicated that a range of microorganisms are capable of degrading lignocellulosic materials present in the municipal waste (Han and Srinivasan, 1968; Stutzenberger et al., 1970). It is now well established that many thermophilic fungi can utilize plasticizers present in the plastics as a carbon source; however, they are unable to use the polyethylene itself. One of the fundamental properties of composting is a thermogenesis resulting from microbial activity. The highest temperature attained in these studies of municipal waste composting usually exceeded 60°C, which would allow the development of thermophilic microflora in the composts. During investigations on the process of composting municipal solid waste, Kane and Mullins (1973) isolated six thermophilic fungi: *Aspergillus fumigatus, Chaetomium thermophile, Humicola lanuginose, Mucor pusillus, Thermoascus aurantiacus,* and *Torula thermophila.* Municipal refuse decomposed by certain thermophilic fungi, such as *Humicola* sp., *Mucor* sp., and *Sporotrichum* sp., have been found to be richer in N, P, and K (Sen et al., 1979). While reviewing thermophilic fungi from different substrates, Subrahmanyam (1999) reported that cellulolytic thermophilic fungi like *Thermomucor* sp., *Thermoascus aurantiacus*, and *Myceliophthora thermophile*, have been isolated from municipal waste composts from different parts of India.

4.3.9 COMPOSTS

Lignocellulosic materials are the earth's most abundant renewable organic carbon source and are composed of 65%–72% utilizable polysaccharides (hemicellulose and cellulose) and 10%–30% lignin. Compost refers to the organic matter that has been decomposed in a solid-state fermentation, carried out through a succession of microbial population, and is often rich in organic nutrients. The compost is beneficial for the soil in several ways, including as a soil conditioner, a fertilizer, an addition to humic acid, and a natural pesticide for soil. During the process of composting, microorganisms break down organic matter and convert it into carbon dioxide, water, heat, and humus, the relatively stable organic end product. Three phases of composting are recognized under optimal conditions:

1. The mesophilic, or moderate-temperature phase, which lasts for a couple of days, breaking down the soluble and readily degradable compounds. The heat produced by mesophilic microorganisms causes the compost temperature to rapidly rise, paving the way for the development of thermophilic microorganisms.
2. The thermophilic, or high-temperature phase, which can last from a few days to several months. As the temperature rises above 40°C, the mesophilic microorganisms become less competitive and are replaced by thermophilic, or heat-loving, microorganisms. At temperatures above 55°C, several microorganisms that are human or plant pathogens are killed, because temperatures more than about 65°C kill many forms of microbes and limit the rate of decomposition. During this phase, high temperatures hasten the breakdown of major structural molecules in plant proteins, such as fats, and complex carbohydrates like cellulose and hemicellulose.
3. A several-month cooling and maturation phase, during which the compost temperature gradually decreases and mesophilic microorganisms once again take over for the final phase of "curing" or maturation of the remaining organic matter.

During the 1930s, Waksman and his coworkers investigated, for the first time, the role of thermophilic fungi, bacteria, and actinomycetes in the self-heating of mushroom compost (Waksman and McGrath, 1931; Waksman and Nissen, 1932; Waksman and Cordon, 1939). Since then, the majority of all known thermophilic fungi have been isolated from mushroom compost and play an important role in the preparation of a suitable compost for the cultivation of edible mushroom (Emerson, 1968; Fergus and Sinden, 1969; Hayes, 1969; Eicker, 1977; Fergus, 1978; Fermor et al., 1985). Furthermore, thermophilic fungi are believed to contribute significantly to the quality of compost (Seal and Eggins, 1976; Eicker, 1977; Ross and Harris, 1983; Gerrits, 1988). The effect of these fungi on the growth of mushroom mycelium and mushroom yield has been described at three distinct levels (Wiegant, 1992).

First, they decrease the concentration of ammonia in the compost, which otherwise would counteract the growth of the mushroom mycelium. Second, they immobilize nutrients in a form that apparently is available to the mushroom mycelium. And third, they may have a growth-promoting effect on the mushroom mycelium, as has been demonstrated for *Scytalidium thermophilum* and several other thermophilic fungi (Wiegant et al., 1992). The effectiveness of *Scytalidium thermophilum* in compost preparation for *Agaricus bisporus* has been shown by Straatsma et al. (1994), who obtained a twofold increase in the yield of mushrooms on inoculated compost compared with that on the pasteurized control. They also isolated 22 species of thermophilic fungi from mushroom compost (Table 4.3). Salar and Aneja (2006) isolated 18 species of thermophilic and thermotolerant fungi from the mushroom compost (Table 4.3), and these represent most of the known thermophilic taxa. Seven species were isolated during phase I of composting, which included *Absidia corymbifera*, *Rhizomucor pusillus*, *Rhizomucor miehei*, *Talaromyces emersonii*, *Talaromyces thermophiles*, and *Thermomyces lanuginosus*. Fungi isolated from phase II compost were *Chaetomium thermophile*, *Emericella nidulans*, *Thermoascus aurantiacus*, *Myriococcum albomyces*, *Humicola insolens*, *Malbranchea sulfurea*, *Torula thermophila*, *Stilbella thermophile*, and *T. lanuginosus*. We also visually observed two basidiomycetous species from phase II compost. Lee et al. (2014) isolated three species of thermophilic fungi from compost, which were identified as *Myriococcum thermophilum*, *Thermoascus aurantiacus*, and *T. lanuginosus*, all of which could grow above 50°C. As claimed by the authors, this was the first report on the isolation of thermophilic fungi from compost in Korea.

The above account of the occurrence of habitats of thermophilic fungi strongly argues in favor of the hypothesis that the first living organisms on the earth were thermophiles. Thermophilic fungi are ubiquitous in their occurrence in thermogenic as well as nonthermogenic environments. Natural geothermal areas are found in all parts of the globe associated with tectonic activity, but are usually concentrated in small places. The best-known and biologically most studied geothermal areas are in Iceland, North America (Yellowstone National Park), New Zealand, Japan, Italy, and the erstwhile Soviet Union. Other than geothermal, there are very few stable hot natural habitats; solar-heated ponds and soils are, of course, common. Similarly, transient ecosystems, such as biological self-heating in composts, hay, litter, or manure, may cause quite high temperatures and even spontaneous ignition. Several man-made constant hot environments, like hot water pipelines and heat exchangers in homes and factories, burning coal refuse piles and other types of mining heap, as well as thermophilic waste treatment plants, have been created. Several well-known thermophiles have been isolated from such man-made systems, and some have not even been found elsewhere. However, culture-independent approaches using molecular techniques may allow for the recovery of hitherto unknown thermophilic fungal species, which may have implications in several biotechnology-based industries.

TABLE 4.3
Thermophilous Fungi Reported from Mushroom Compost around the World[a]

Fungus	References	Detected in the Present Study	Log_{10} (CFU/g)
Zygomycetes			
Absidia corymbifera[b]	2, 5, 14	+	2.9
Rhizomucor miehei[b]	6, 14	+	3.1
Rhizomucor pusillus[b]	2–7, 10, 11, 13, 14	+	3.4
Ascomycetes			
Chaetomium thermophile	2–7, 10, 13	+	3.7
Corynascus thermophilus	9, 14	–	
Emericella nidulans	2, 4, 14	+	3.0
Talaromyces emersonii[b]	14	+	2.9
Talaromyces thermophilus[b]	5–7, 14	+	2.9
Thermoascus aurantiacus	3, 14	+	3.6
Thermoascus aurantiacus var. levispora	14	–	
Thermoascus crustaceus	14	–	
Thielavia terrestris	14	–	
Myriococcum albomyces	8, 13, 14	+	3.5
Myriococcum thermophilum	14	–	
Chaetomium sp.		–	
Basidiomycetes			
Coprinus cinereus	14	–	
Basidiomycetes 1		+	
Basidiomycetes 2		+	
Hyphomycetes			
Aspergillus fumigatus[b]	2–7, 10–14	+	3.9
Hormographiella aspergillata	14	–	
Humicola insolens		+	3.9
Malbranchea sulfurea	5, 6, 14	+	3.1
Paecilomyces variotii	14	–	
Torula thermophile (= *Scytalidium thermophilum*)	2–7, 10–14	+	4.2
Stilbella thermophila	4, 7, 11, 13, 14	+	2.8
Thermomyces lanuginosus[b]	2–7, 10, 11, 13, 14	+	3.7
Unidentified taxon		+	2.8

Note: 1 = Anonymous (1992); 2 = Basuki (1981); 3 = Bilai (1984); 4 = Cailleux (1973); 5 = Chang and Hudson (1967); 6 = Eicker (1977); 7 = Fergus (1964); 8 = Fergus (1971); 9 = Fergus and Sinden (1969); 10 = Fermor et al. (1979); 11 = Hayes (1969); 12 = Olivier and Guillaumes (1976); 13 = Seal and Eggins (1976); 14 = Straatsma et al. (1994); – = not detected; + = isolated from compost.

[a] The nomenclature of reference 1 was followed.

[b] Isolated from phase I compost only; the rest of the fungi were isolated from phase II compost.

REFERENCES

Abdel-Hafez, S.I.I. 1982. Thermophilic and thermotolerant fungi in the desert soils of Saudi Arabia. *Mycopathologia* 80:15–20.

Abdullah, S.K., and Al-Bader, S.M. 1990. On the thermophilic and thermotolerant mycoflora of Iraq. *Sydowia* 42:1–7.

Aguilar, A. 1996. Extremophile research in the European Union: From fundamental aspects to industrial expectations. *FEMS Microbiological Reviews* 18:89–92.

Allen, P.J., and Emerson, R. 1949. Guayule rubber, microbiological improvement by shrub retting. *Industrial & Engineering Chemistry* 41:346–365.

Aneja, K.R., and Kumar, R. 1994. *Chaetomium senegalense*—A new record from India. *Proceedings of the National Academy of Sciences* 64(B):229–230.

Anonymous. 1992. *Catalogue of the Culture Collections.* 10th ed. Egham, Surrey, UK: International Mycological Institute.

Apinis, A.E. 1963. Occurrence of thermophilic fungi in certain alluvial soils near Nottingham. *Nova Hedwigia* 5:57–78.

Apinis, A.E. 1972. Mycological aspects of stored grain. In *Biodeterioration of Materials*, ed. A.H. Walters and E.H. Hueck-Van der Plas, 493–498. New York: Halstead Press, John Wiley & Sons.

Apinis, A.E., and Chester, C.G.C. 1964. Ascomycetes of some salt marshes and sanddunes. *Transactions of the British Mycological Society* 47:419–435.

Apinis, A.E., and Pugh, G.J.F. 1967. Thermophilous fungi of birds' nests. *Mycopathologia et Mycologia Applicata* 33:1–9.

Barns, S.M., Delwiche, C.F., Palmer, J.D., and Pace, N.R. 1996. Perspectives on archaeal diversity, thermophily, and monophyly from environmental rRNA sequences. *Proceedings of the National Academy of Sciences of the United States of America* 93: 9188–9193.

Basuki, T. 1981. *Ecology and productivity of the paddy straw mushroom [Volvariella volvacea (Bull ex Fr.) Sing].* PhD thesis, University of Wales, Cardiff.

Bilai, V.T. 1984. Thermophilic micromycete species from mushroom composts. *Mikrobiology Zhurnal (Kiev)* 46:35–38.

Broadbent, J.A., and Oyeniran, J.O. 1968. A new look at mouldy cocoa. In Proceedings of 1st International Biodeterioration Symposium, Southampton, UK, September 9–14, 1968, 693–702.

Brown, R.A., Pilling, R.L., and Young, D.C. 1994. *Preservation of wood chips.* U.S. Patent 5342629.

Cailleux, R. 1973. Mycoflore du compost destine a la culture du champignon de couche. *Revue de Mycologie* 37:14–35.

Chang, Y., and Hudson, H.J. 1967. The fungi of wheat straw compost I. Ecological studies. *Transactions of the British Mycological Society* 50:649–666.

Chen, K.Y., Huang, D.J., and Liu. C.C. 2003. The mycoflora of hot spring soil in northern Taiwan. *Taiwania* 48:203–211.

Chen, M.Y., Chen, Z.C., Chen, K.Y., and Tsay, S.S. 2000. Fungal flora of hot springs of Taiwan (1): Wu Rai. *Taiwania* 45(2):207–216.

Clark, F.E. 1967. Bacteria in soil. In *Soil Biology*, ed. A. Burges and F. Raw, 15–49. New York: Academic Press.

Clarke, D.D. 1966. Production of pectic enzymes by *Phytophthora infestans. Nature* 211:649.

Clarke, J.H. 1969. Fungi in Stored Products. *PANS Pest Articles & News Summaries* 15: 473–481.

Cooney, D.G., and Emerson, R. 1964. *Thermophilic Fungi: An Account of Their Biology, Activities, and Classification.* San Francisco: W.H. Freeman and Co.

Dix, N.J., and J. Webster. 1995. *Fungal Ecology*. London: Chapman & Hall.

Eggins, H.O., and Malik, K.A. 1969. The occurrence of thermophilic cellulolytic fungi in a pasture land soil. *Antonie van Leeuwenhoek* 35:178–184.

Eicker, A. 1972. Occurrence and isolation of South African thermophilic fungi. *South African Journal of Science* 68:150–155.

Eicker, A. 1977. Thermophilic fungi associated with the cultivation of *Agaricus bisporus*. *Journal of South African Botany* 43:193–207.

Ellis, D.H. 1980. Thermophilous fungi isolated from some Antarctic and sub-Antarctic soils. *Mycologia* 72:1033–1036.

Ellis, D.H., and Keane, P.J. 1981. Thermophilic fungi isolated from some Australian soils. *Australian Journal of Botany* 29:689–704.

Emerson, R. 1968. Thermophiles. In *The Fungi*, ed. G.C. Ainsworth and A.S. Sussman, 105–128. New York: Academic Press.

Evans, H.C. 1971. Thermophilous fungi of coal spoil tips. II. Occurrence, distribution and temperature relationships. *Transactions of the British Mycological Society* 57:255–266.

Fatma, F.M. 2003. Distribution of fungi in the sandy soil in Egyptian beaches. *Pakistan Journal of Biological Sciences* 6:860–866.

Fergus, C.L. 1964. Thermophilic and thermotolerant molds and actinomycetes of mushroom compost during peak heating. *Mycologia* 56:267–283.

Fergus, C.L. 1971. The temperature relationships and thermal resistance of a new thermophile *Papulaspora* from mushroom compost. *Mycologia* 63:426–431.

Fergus, C.L. 1978. The fungal flora of compost during mycelium colonization by the cultivated mushroom *Agaricus brunnescens*. *Mycologia* 70:636–644.

Fergus, C.L., and Sinden, J.W. 1969. A new thermophilic fungus from mushroom compost: *Thielavia thermophila* spec. nov. *Canadian Journal of Botany* 47:1635–1637.

Fermor, T.R., Randle, P.E., and Smith J.F. 1985. Compost as a substrate and its preparation. In *The Biology and Technology of the Cultivated Mushroom*, ed. P.B. Flegg, D.M. Spencer, and D.A. Wood, 81–109. Chichester, UK: John Wiley & Sons.

Fermor, T.R., Smith J.F., and Spencer, D.M. 1979. The microflora of experimental mushroom composts. *Journal of Horticulture Science* 54:137–147.

Flannigan, B., and Sagoo, G.S. 1977. Degradation of wood by *Aspergillus fumigatus* isolated from self-heated wood chips. *Mycologia* 69:514–523.

Gerrits, J.P.G. 1988. Compost treatment in bulk for mushroom growing. *Mushroom Journal* 182:471–475.

Gregory, P.H., and Lacey, M.E. 1963. Mycological examination of dust from mouldy hay associated with farmer's lung disease. *Journal of General Microbiology* 30:75–88.

Guiraud, P., Steiman, R., Seigle-Murandi, F., and Sage, L. 1995. Mycoflora of soil around the Dead Sea II: Deuteromycetes (except *Aspergillus* and *Penicillium*). *Systematic and Applied Microbiology* 18:318–322.

Han, Y.W., and Srinivasan, V.R. 1968. Isolation and characterization of a cellulose-utilizing bacterium. *Applied Microbiology* 16:1140–1145.

Hansen, A., and Welty, R.E. 1970. Microflora of raw cacao beans. *Mycopathologia et Mycologia Applicata* 44:309–316.

Hayes, W.A. 1969. Microbiological changes in composting wheat straw/horse manure mixtures. *Mushroom Science* 7:173–186.

Hedger, J.N. 1974. The ecology of thermophilic fungi in Indonesia. In *Biodegradation and Humification*, ed. G. Kilbetus, O. Reisinger, A. Mourey, and J.A. Cancela Da Fonseca, 59–65. Sarreguemines, France: Rapport du Ie Colloque International-Nancy Université de Nancy, Pierron Editeur.

Hedger, J.N. 1975. *Ecology of Thermophilic Fungi in Indonesia*. Sarreguemines, France: Pierron, pp. 59–65.

Hemida, S.K. 1992. Thermophilic and thermotolerant fungi isolated from cultivated and desert soils, exposed continuously to cement dust particles in Egypt. *Zentralblatt für Mikrobiologie* 147:277–281.

Holman, D.B., Hao, X., Topp, E., Yang, H.E., and Alexander, T.W. 2016. Effect of co-composting cattle manure with construction and demolition waste on the archaeal, bacterial, and fungal microbiota, and on antimicrobial resistance determinants. *PloS One* 11(6):e0157539.

Hubalek, Z. 1967. Influence of pH on the occurrence of fungi in birds' nests. *Zeitschrift für Allgemeine Mikrobiologie* 16:65–72.

Isachenko, B.L., and Malchevskaya, N.N. 1936. Biogenic spontaneous heating of peat. *Doklady Akademii Nauk SSSR* 13:337–380.

Johri, B.N., Satyanarayana, T., and Olsen, J. 1999. *Thermophilic molds in biotechnology.* Netherlands: Springer.

Johri, B.N., and Thakre, R.P. 1975. Soil amendments and enrichment media in the ecology of thermophilic fungi. *Proceedings of the Indian National Science Academy* 41:564–570.

Kaiser, E.R., Zeit, C.D., and McCaffery, J.B. 1968. Municipal incinerator refuse and residue. In *Proceedings of the National Incinerator Conference.* New York: American Society of Mechanical Engineers.

Kalyanasundaram, I., and Rao, G.J. 1988. Storage fungi in rice in India. *Kavaka* 14:67–76.

Kambura, A.K., Mwirichia, R.K., Kasili, R.W., Karanja, E.N., Makonde, H.M., and Boga, H.I. 2016. Diversity of fungi in sediments and water sampled from the hot springs of Lake Magadi and Little Magadi in Kenya. *African Journal of Microbiology Research* 10 (10):330–338.

Kane, B.E., and Mullins, J.T. 1973. Thermophilic fungi in a municipal waste compost system. *Mycologia* 65:1087–1100.

Korniłowicz-Kowalska, T., and Kitowski, I. 2013. *Aspergillus fumigatus* and other thermophilic fungi in nests of wetland birds. *Mycopathologia* 175:43–56.

Kumar, R. 1996. *Taxophysiological studies on thermophilous fungi from northern Indian soils.* PhD thesis, Kurukshetra University, Kurukshetra, India.

Kuster, E., and Locci, R. 1964. Studies on peat and peat microorganisms II. Occurrence of thermophilic fungi in peat. *Archiv für Mikrobiologie* 48:319–324.

Lacey, J. 1971. The microbiology of moist barley storage in unsealed silos. *Annals of Applied Biology* 69:187–212.

Larney, F.J., and Hao, X. 2007. A review of composting as a management alternative for beef cattle feedlot manure in southern Alberta, Canada. *Bioresource Technology* 98:3221–3227.

Lee, H., Lee, Y.M., Jang, Y., Lee, S., Lee, H., Ahn, B.J., Kim, G.H., and Kim, J.J. 2014. Isolation and analysis of the enzymatic properties of thermophilic fungi from compost. *Mycobiology* 42:181–184.

Maheshwari, R., Kamalam, P.T., and Balasubramanyam, P.V. 1987. The biogeography of thermophilic fungi. *Current Science* 56:151–155.

Meadow, J.F. 2012. *Geothermal soil ecology in Yellowstone National Park.* PhD thesis, Montana State University, Bozeman.

Miehe, H. 1907. *Die Selbsterhitzung des Heus. Eine. Biologische studie.* Jena: Gustav Fischer.

Miller, J.D., Schneider, M.H., and Whitney, N.J. 1982. Fungi on fuel wood chips in a home. *Wood and Fiber* 14:54–59.

Mishustin, E.N. 1950. *Thermophilic Microorganisms in Nature and in Practice.* Moscow: Doklady Akademii Nauk SSSR.

Mouchacca, J. 1995. Thermophilic fungi in desert soils, a neglected extreme environment. In *Microbial Diversity and Ecosystem Function*, ed. D. Allsopp et al., 265–288. London: CAB International and Royal Holloway, University of London.

Mulinge, S.K., and Apinis, A.E. 1969. Occurrence of thermophilous fungi in stored moist barley grain. *Transactions of the British Mycological Society* 53:361–370.

Neill, W. 1971. *The Last of the Ruling Reptiles: Alligators, Crocodiles, and Their Kin*. New York: Columbia University Press.

Ogundero, V.W. 1979. Amylase and cellulase activities of thermophilic fungi causing deterioration of tobacco products in Nigeria. *Mycopathology* 69:131–135.

Olivier, J.M., and Guillaumes, J. 1976. Etude ecologique des composts de champignonnieres. I. Evolution de la microflore pendant l' incubation. *Annals of Phytopathology* 8:283–301.

Oso, B.A. 1974. Thermophilic fungi from stacks of oil palm kernels in Nigeria. *Zeitschrift für allgemeine Mikrobiologie* 14:593–601.

Pan, W.-Z., Huang, X.-W., Wei, K.-B., Zhang, C.-M., Yang, D.-M., Ding, J.-M., and Zhang, K.Q. 2010. Diversity of thermophilic fungi in Tengchong Rehai National Park revealed by ITS nucleotide sequence analyses. *Journal of Microbiology* 48:146–152.

Rajasekaran, A.K., and Maheshwari, R. 1993. Thermophilic fungi: An assessment of their potential for growth in soil. *Journal of Bioscience* 18:345–354.

Rajavaram, R.K., Bathini, S., Girisham, S., and Reddy, S.M. 2010. Incidence of thermophilic fungi from different substrates in Andhra Pradesh (India). *International Journal of Pharma and Bio Sciences* 1:0975–6299.

Ramnath, L., Bush, T., and Govinden, R. 2014. Method optimization for denaturing gradient gel electrophoresis (DGGE) analysis of microflora from *Eucalytpus* sp., wood chips intended for pulping. *African Journal of Biotechnology* 13:356–365.

Redman, R.S., Litvintseva, A., Sheehan, K.B., Henson, J.M., and Rodriguez, R.J. 1999. Fungi from geothermal soils in Yellowstone National Park. *Applied and Environmental Microbiology* 65:5193–5197.

Reysenbach, A.L., Longnecker, K., and Kirshtein, J. 2000. Novel bacterial and archaeal lineages from an in situ growth chamber deployed at a Mid-Atlantic Ridge hydrothermal vent. *Applied Environmental Microbiology* 66:3798–3806.

Rippon, J.W., Gerhold, R., and Heath, M. 1980. Thermophillic and thermotolerant fungi isolated from the thermal effluent of nuclear power generating reactors: Dispersal of human opportunistic and veterinary pathogenic fungi. *Mycopathologia* 70:169–179.

Ross, R.C., and Harris, P.J. 1983. The significance of thermophilic fungi in mushroom compost preparation. *Scientia Horticulturae* 20:61–70.

Salar, T.K., and Aneja, K.R. 2006. Thermophilous fungi from temperate soils of northern India. *Journal of Agricultural Technology* 2:49–58.

Sandhu, D.K., and Singh, S. 1981. Distribution of thermophilous microfungi in forest soils of Darjeeling (Eastern Himalayas). *Mycopathologia* 74:79–85.

Sarquis, M.I.D.M., and Oliveira, P.C.D. 1996. Diversity of microfungi in the sandy soil of Ipanema Beach, Rio de Janeiro, Brazil. *Journal of Basic Microbiology* 36:51–58.

Satyanarayana, T., Johri, B.N., and Saksena, S.B. 1977. Seasonal variation in mycoflora of nesting materials of birds with special reference to thermophilic fungi. *Transactions of the British Mycological Society* 62:307–309.

Seal, K.J., and Eggins, H.O.W. 1976. The upgrading of agricultural wastes by thermophilic fungi. In *Food from Wastes*, ed. G.G. Birch, K.J. Parker, and J.T. Wargan, 58–78. London: Applied Science Publishers.

Sen, S., Abraham, T.K., and Chakrabarty, S.L. 1979. Biodeterioration and utilization of city wastes by thermophilic microorganisms. *Indian Journal of Experimental Biology* 17:1284–1285.

Sharma, N., Vyas, G., and Pathania, S. 2013. Culturable diversity of thermophilic microorganisms found in hot springs of northern Himalayas and to explore their potential for production of industrially important enzymes. *Scholars Academic Journal of Biosciences* 1:165–178.

Sigler, L., Aneja, K.R., Kumar, R., Maheshwari, R., and Shukla, R.V. 1998. New records from India and redescription of *Corynascus thermophilus* and its anamorph *Myceliophthora fergusii*. *Mycotaxon* 68:185–192.

Singh, S., and Sandhu, D.K. 1986. Thermophilous fungi in Port Blair soils. *Canadian Journal of Botany* 64:1018–1026.

Skirnisdottir, S., Hreggvidsson, G.O., Hjörleifsdottir, S., Marteinsson, V.T., Petursdottir, S.K., Holst, O., and Kristjansson, J.K. 2000. Influence of sulfide and temperature on species composition and community structure of hot spring microbial mats. *Applied Environmental Microbiology* 66:2835–2841.

Steiman, R., Guiraud, P., Sage, L., Seigle-Murandi, F., and Lafond, J.L. 1995. Mycoflora of soil around the Dead Sea I: Ascomycetes (including *Aspergillus* and *Penicillium*), basidiomycetes, zygomycetes. *Systematic and Applied Microbiology* 18:310–317.

Stetter, K.O. 1999. Extremophiles and their adaptation to hot environments. *FEBS Letters* 452:22–25.

Straatsma, G., Samson, R.A., Olijnsma, T.W., Op Den Camp, H.J., Gerrits, J.P., and Van Griensven, L.J. 1994. Ecology of thermophilic fungi in mushroom compost, with emphasis on *Scytalidium thermophilum* and growth stimulation of *Agaricus bisporus* mycelium. *Applied and Environmental Microbiology* 60:454–458.

Stutzenberger, F.J., Kaufman, A.J., and Lossin, R.D. 1970. Cellulolytic activity in municipal solid waste composting. *Canadian Journal of Microbiology* 16:553–560.

Subrahmanyam, A. 1999. Ecology and distribution. In *Thermophilic Moulds in Biotechnology*, ed. B.N. Johri, T. Satyanarayana, and J. Olsen, 13–42. Dordrecht, Netherlands: Kluwer Academic.

Sundaram, B.M. 1977. Isolation of thermophilic fungi from soils. *Current Science* 46: 106–107.

Taber, R.A., and Pettit, R.E. 1975. Occurrence of thermophilic microorganisms in peanuts and peanut soil. *Mycologia* 67:157–161.

Tansey, M.R. 1971. Isolation of thermophilic fungi from self-heated industrial wood chip piles. *Mycologia* 63:537–547.

Tansey, M.R. 1973. Isolation of thermophilic fungi from alligator nesting material. *Mycologia* 65:594–601.

Tansey, M.R., and Brock, T.D. 1978. Microbial life at high temperature ecological aspects. In *Microbial Life in Extreme Environments*, ed. D. Kushner, 159–216. London: Academic Press.

Tansey, M.R., and Fliermans, C.B. 1978. Pathogenic species of thermophilic and thermotolerant fungi in reactor effluents of the Savannah River Plant. In *Energy and Environmental Stress in Aquatic Systems*, ed. J.H. Thorp and J.W. Gibbons, 663–690. U.S. Department of Energy Symposium Series CONF-771114. Springfield, IL: National Technical Information Service.

Tansey, M.R., and Jack, M.A. 1976. Thermophilic fungi in sun-heated soils. *Mycologia* 68:1061–1075.

Thakre, R.P., and Johri, B.N. 1976. Occurrence of thermophilic fungi in coal mine soils of Madhya Pardesh. *Current Science* 45:271–273.

Velonakis, E., Dimitriadi, D., Papadogiannakis, E., and Vatopoulos, A. 2014. Present status of effect of microorganisms from sand beach on public health. *Journal of Coastal Life Medicine* 2:746–756.

Waksman, S.A., and Cordon, T.C. 1939. Thermophilic decomposition of plant residues in composts by pure and mixed cultures of microorganisms. *Soil Science* 47:217–224.

Waksman, S.A., and McGrath, J.M. 1931. Preliminary study of the chemical processes involved in the decomposition of manure by *Agaricus campestris*. *American Journal of Botany* 18:573–581.

Waksman, S.A., and Nissen, W. 1932. On the nutrition of the cultivated mushroom and the chemical changes brought about by their organisms in manure compost. *American Journal of Botany* 19:514–537.

Waksman, S.A., Umbreit, W.W., and Cordon, T.C. 1939. Thermophilic actinomycetes and fungi in soils and in composts. *Soil Science* 47:37–61.

Ward, D.M., Ferris, M.J., Nold, S.C., and Bateson, M.M. 1998. A natural view of microbial biodiversity within hot spring cyanobacterial mat communities. *Microbiology and Molecular Biology Reviews* 62:1353–1370.

Ward, J.E., and Cowley, G.T. 1972. Thermophilic fungi of some central South Carolina forest soils. *Mycologia* 64:200–205.

Wareing, P.W. 1997. Incidence and detection of thermotolerant and thermophilic fungi from maize with particular reference to *Thermoascus* species. *International Journal of Food Microbiology* 35:137–145.

Whittaker, R.H. 1975. *Communities and Ecosystems*. New York: Macmillan.

Wiegant, W.M. 1992. Growth characteristics of the thermophilic fungus *Scytalidium thermophilum* in relation to production of mushroom compost. *Applied and Environmental Microbiology* 58:1301–1307.

Wiegant, W.M., Wery, J., Buitenhuis, E.T., and de Bont, J.A.M. 1992. Growth promoting effect of thermophilic fungi on the mycelium of the edible mushroom *Agaricus bisporus*. *Applied and Environmental Microbiology* 58:2654–2659.

Yee, T.L., Tajuddin, R., Nor, N.M.I.M., Mohd, M.H., and Zakaria, L. 2016. Filamentous ascomycete and basidiomycete fungi from beach sand. *Rendiconti Lincei* 27:603–607.

Zakaria, L., Yee, T.L., Zakaria, M., and Salleh, B. 2011. Diversity of microfungi in sandy beach soil of Teluk Aling, Pulau Pinang. *Tropical Life Sciences Research* 22:71–80.

Section II

Taxonomy, Biodiversity, and Classification

5 Bioprospecting of Thermophilic Fungi

5.1 INTRODUCTION

Bioprospecting refers to the exploration of the biodiversity of commercially valuable biochemical and genetic resources for achieving economic and conservation goals (Firn, 2003). This holds considerable promise for the development of novel compounds for food production and processes, consumer goods, public health, and environmental and energy uses. To serve these purposes, the existing diversity of microorganisms can act as a resource reservoir from which individual species with special traits can be exploited (Bull et al., 2000; Egorova and Antranikian, 2005). Most potential bioprospecting (searching for new biologically active chemicals in organisms) is currently focused on the study of extremophiles and their potential use in industrial processes.

Over the past few decades, the extremes at which life thrives have continued to challenge our understanding of biochemistry, biology, and evolution. As more new thermophiles are brought into the laboratory culture, they have provided a multitude of potential applications for biotechnology. More recently, innovative culturing approaches, environmental genome sequencing, and whole genome sequencing have provided new opportunities for the biotechnological exploration of thermophiles, particularly fungi. A great deal of research work on the occurrence of thermophilic fungi and their role as candidates for biotechnological applications has been carried out. Thermophilic fungi have been exploited for the bioconversion of organic matter, the production of enzymes, antibiotics, organic acids, biorefining, phenolic compounds, extracellular polysaccharides, and lipids (Kumar and Aneja, 1999). Thermostable enzymes and microorganisms have been topics of much research during the past two decades, but the interest in thermophiles and how their proteins are able to function at elevated temperatures actually started as early as the 1960s by the pioneering work of Brock and Freeze (1969). The use and development of molecular biology techniques, allowing genetic analysis and gene transfer for recombinant production, led to dramatically increased activities in the field of thermostable enzymes during the 1990s. This also stimulated the isolation of a number of microbes from thermal habitats in order to access enzymes that could significantly increase the window of opportunity for enzymatic bioprocesses. One of the early successful commercialized examples was the analytical use of a thermostable enzyme, Taq-polymerase, in polymerase chain reactions (PCRs) for the amplification of DNA, and a number of other DNA-modifying enzymes from thermophilic sources have since then been commercialized in this locale (Fujiwara, 2002; Satyanarayana et al., 2005; Podar and Reysenbach, 2006). Another area of interest has been the prospecting for industrial enzymes for use in technical products and processes, often on a large scale.

Enzymes can be advantageous as industrial biocatalysts, as they rarely require toxic metal ions for functionality, hence creating the possibility to use more environmentally friendly processing (Comfort et al., 2004). Thermostable enzymes offer robust catalyst alternatives, able to withstand the often relatively harsh conditions of industrial processing.

The sustainability of life at high temperatures has raised considerable interest in more systematic studies on fungal microorganisms. Man's quest for products from nature that will improve the quality of life, cure illness, and preserve food dates back thousands of years. This record of discovery on bioactive organisms and compounds, coupled with the knowledge that there is a largely undiscovered diversity of microorganisms, supports the argument for bioprospecting. Of special relevance to biodiversity-rich developing countries found in the tropical regions are those organisms that are able to survive under high temperature conditions, known as thermophiles. Tropical countries represent a greatly underexploited pool of potentially useful biological material. Thus, it is very much warranted to focus our research on the isolation and characterization of thermophilic fungi from specific physiological growth niches. The significance of such studies has further increased with the establishment of a biodiversity action plan and new international policy on patenting. Bioprospecting of these microorganisms will further add to the existing database of the culture collections of microorganisms and enhance their exploitation, besides yielding more information leading to unraveling the mechanism of their survival at high temperatures.

5.2   BIODIVERSITY PERSPECTIVE

Approximately half of the 17 ecosystem services within the 16 biomes distinguished as making up the earth's ecosystem are dependent, either directly or indirectly, on life and activities of microorganisms (Johri and Satyanarayana, 2005). India is an abode to biodiversity hot spots, including the well-known Western Ghats and northeastern hill region. India is bestowed with rich biodiversity locales for thermophilic microorganisms, including the thermal springs in the central Himalayan region, with Manikaran, Tatapani, and Vashisht being the most recognized. There is a tremendous scope of microbial diversity in human life and societal development. Although ecosystems with high temperatures are not as prevalent as temperate or low-temperature habitats, a variety of high-temperature, natural, and man-made habitats exist, including volcanic and geothermal areas, with temperatures often more than boiling, sun-heated litter and soil or sediments reaching 70°C, and biologically self-heated environments, such as compost, hay, sawdust, and coal refuse piles. In thermal springs, the temperature is above 60°C, and it is kept constant by continual volcanic activity. Besides temperature, other environmental parameters, such as pH, available energy sources, ionic strength, and nutrients, influence the diversity of thermophilic microbial populations (Satyanarayana et al., 2005).

Fungi constitute one of the largest and most diverse kingdoms of eukaryotes and are important biological components of terrestrial ecosystems. In general, fungi have been reported to represent prolific sources of various compounds, including several bioactive molecules. However, the potential of thermophilic fungi to produce bioactive compounds is poorly understood. Thus, prospecting has always tended to enrich the dreams and build up the hopes of some sectors of the community. It is further speculated that there are many exciting, very valuable chemicals awaiting discovery in organisms, but they lie hidden among a much larger number of chemicals that have little human value at present. The action plan for bioprospecting the microbial diversity as envisaged by Johri and Satyanarayana (2005) must focus on the following:

1. Isolation, characterization, and conservation of microorganisms from natural and man-made ecosystems
2. Bioprospecting of the gene pool for biologically active molecules with applications in the agriculture, health, industry, and environmental sectors
3. Establishment of a network of regional conservation centers
4. Creation of centers of excellence with state-of-the-art facilities at universities for teaching, research, and training
5. Creation of linkages between the national centers and international conservation centers through an appropriate networking system

5.3 CULTURABLE MICROBIAL DIVERSITY

The most fascinating aspect of the microbial world is its extraordinary diversity depicting almost every possible shape, size, physiology, and lifestyle. Fungi constitute an enormous group of microorganisms, with estimates of about 1.5 million species existing in various habitats (Hawksworth, 1991). However, not more than 90,000 fungal species have been described so far. It is thus not surprising that there exists a huge gap between the culturable and uncultivable or yet to be discovered fungal species. The culturable fungal diversity is thus a small proportion of the existing gene pool. Many mycologists have tried to answer the question "Where are the missing fungi?" by exploring habitats that are yet to be studied for the presence of such fungi (Suryanarayanan and Hawksworth, 2004). In this context, Darwin's prophecy has come true, when he had stated, "The time will come, … though I shall not live to see it, when we shall have very fairly true genealogical trees of each great kingdom of nature."

To date, approximately 60 thermophilic fungi have been documented from various habitats. Considering the diversity of thermogenic habitats, however, this number markedly underestimates the diversity of these fungi. Culture-dependent methods have helped to grow and develop these fungi as candidates for various biotechnological applications. Most research on these species has been conducted in the last 30 years and mainly focused on isolation and characterization. The use of several media (see Chapter 1) has further aided in the recovery of these fungi, as some species are recovered only on a particular medium (Langarica-Fuentes et al., 2014). Additionally, in view of the specific needs of microorganisms growing in diverse habitats, a variety of new culture techniques have been developed. Two of these techniques are discussed here, and both employ water collected from the natural environment, rather than culture media. The first is the *extinction culture technique*, in which samples are first microscopically examined and then diluted to a density of 1 to 10 cells, that is, diluted to extinction, similar to the most probable number assay. Multiple

1 mL cultures are grown in 48-well microtiter dishes under appropriate incubating conditions. The samples are then stained and examined microscopically to monitor for growth. The turbid cultures show signs of viable microbial growth and can be transferred to fresh medium for obtaining pure culture. In the second, the *microdroplet culture technique*, cells from mixed assemblages are encapsulated in a gel matrix. This matrix is then emulsified to generate "microdroplets" of gel, each containing a single cell. The gel matrix is porous enough to deliver nutrients to the gel-encapsulated microbes during incubation in continuous cultures. After incubating at appropriate conditions, the microdroplets are examined by flow cytometry, a technique in which fluorescently labeled cells are sorted and counted by laser beams. The microbeads containing colonies owing to the growth of the initial inoculum are separated from free-living cells and bare microdroplets. Microbeads with microbial growth can then be isolated for further culture and analysis.

As emphasized earlier, the culture-dependent methods do not display the whole spectrum of microbial diversity present in a particular ecosystem. Using the culture-dependent methods, only a miniscule fraction (<5%) of microbes are detected, and this has been referred to as the great plate count anomaly (GPCA). The inconsistency between the number of microbial cells observed microscopically and the number of colonies that can be cultivated from the same sample is due to two reasons. First, some of the cells observed under the microscope are actually nonviable, and second, the required conditions for their cultivation in the lab have not yet been developed. This has led to the depiction of such potentially viable microbes as being "nonculturable," as they show motility or the presence of dividing cells when observed directly or get stained with dyes that distinguish between the live and dead cells. In the following section, we consider the ways to culture the viable but nonculturable (VBNC) microbes.

5.4 BIOPROSPECTING THE UNCULTIVABLE

Traditionally, microbial populations have been studied by obtaining isolates in pure culture for the study of their morphology, physiology, and genetics. Unfortunately, it was not until it was realized that only a small fraction (about 1%) of microbes have been cultured in the laboratory that the need for developing alternative methods of assessing microbial diversity in different physiological niches was realized. It is now well known that a vast majority of microbes have not been grown under laboratory conditions, and hence remain obscure for their ecological functions and unexploited for their biotechnological applications (Kellenberger, 2001).

Although it is generally recognized that a pure culture is the "gold standard" for microbial analysis, most microbiologists agree that to understand natural microbial populations, one must also consider the diversity of life-forms found within each specific habitat. While the culture-based approaches have been useful in identifying several species of thermophilic and thermotolerant fungi from environmental samples, they are known to detect a small proportion of all fungi present in specific ecological niches. Most of the culture-independent studies are based on 16s rRNA gene sequence analysis. These studies have made it clear that a large proportion of these yet to be cultured microbes belong to new genotypes, classes, and divisions in all three domains of life. However, culture-independent techniques used for the analysis of thermophilic fungi are scanty. According to a recent estimate by Blackwell (2011) and Hawksworth (2012), only about 100,000 fungal species are know from approximately 3 million total fungal species on the earth. As has been stated earlier, thermophilic fungi constitute only the "tip of the iceberg" from such a large diversity of fungal species. Recently, several culture-independent approaches, such as PCR of the internal transcribed spacer (ITS) region, phospholipid fatty acid analysis (PLFA), denaturing gradient gel electrophoresis (DGGE) of PCR-amplified DNA fragments combined with the sequencing bands, terminal restriction fragment length polymorphism (T-RFLP) analysis, and more recently, next-generation sequencing, have given impetus to assessing the diversity of microbes in general (Anderson and Cairney, 2004; Bonito et al., 2010; Hultman et al., 2010; Lindahl et al., 2013) and thermophilic fungi in particular (Sharma et al., 2008; Pan et al., 2010; Morgenstern et al., 2012; Langarica-Fuentes et al., 2014). The prospect of accessing the huge and diverse gene pool of uncultivable microbes, including thermophilic fungi, led to unraveling a "golden treasure" in the past few years, with publications describing innovative techniques to exploit the metagenome for novel biocatalysts and bioactive compounds (Sharma et al., 2005). The metagenome from diverse habitats has been exploited for isolation of novel genes by PCR amplification using primers against the conserved domains of known genes or by preparation of the metagenomic libraries containing small inserts of 2–15 kb in plasmid vectors or large inserts of 40–130 kb in cosmid, fosmid, or bacterial artificial chromosome vectors. Conventional and 454 pyrosequencing methods for metagenome sequencing have recently been employed to replace biological cloning (Figure 5.1). The 454 pyrosequencing method is considered an efficient strategy for metagenomic studies.

Recently, the use of quantitative or real-time PCR of the ITS region between the 16S and 23S rRNA genes in prokaryotes has been used. Whereas in eukaryotes the rRNA cistron consists of the 18S, 5.8S, and 28S rRNA genes transcribed as a unit by RNA polymerase I, in thermophilic fungi, Langarica-Fuentes et al. (2014) amplified the ITS1-5.8S-ITS2 region of the fungal rRNA gene complex using the universal fungal primers ITS1 (5′- CCGTAGGTGAACCTGCGG-3′) and ITS4 (5′-TCCTCCGCTTATTGATATGC-3′). Sequences are used to interrogate the National Center for Biotechnology Information (NCBI) nucleotide database using the BLAST algorithm (http://www.ncbi.nlm.nih.gov/blast/). The ITS region is widely used to identify fungal species because of its variability in both sequence and length. Currently, ~172,000 full-length fungal ITS sequences are deposited in GenBank, and 56% are associated with a Latin binominal, representing ~15,500 species and 2,500 genera. An important fraction of the sequences lacking binominals is from environmental samples (O'Brien et al., 2005; Buée et al., 2009).

Another technique, which employs DGGE, relies on the fact that although the PCR fragments are about the same size, they differ in nucleotide sequence. In this technique, the identification of phylotypes begins with the extraction of DNA from a microbial community and the PCR amplification of the gene of choice, mainly that which encodes small-subunit rRNA. As the majority of amplified DNA fragments have about the same molecular weight, when visualized by agarose gel electrophoresis, they appear identical. In DGGE, a gradient of DNA denaturing agents to separate the fragments based on the condition under which they become single

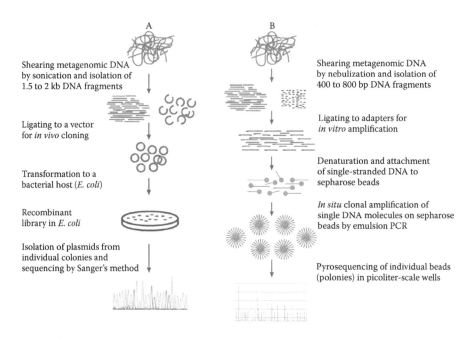

FIGURE 5.1 Conventional (A) and 454 pyrosequencing (B) methods for metagenome sequencing. In the conventional shotgun method, the metagenomic DNA is sheared by sonication and cloned in a plasmid vector. The cloned plasmid library is transformed into *Escherichia coli*. Recombinant plasmids from individual colonies are isolated and sequenced by Sanger's method. In 454 pyrosequencing technology, smaller fragments of metagenomic DNA are obtained by nebulization and ligated with adapters. One of the adapters carries a 5′ biotinylated strand. The ligation mixture is denatured and adhered to sepharose beads containing streptavidin. DNA and beads are taken in ratio, where each bead gets a single DNA molecule. The single DNA molecule is amplified *in situ* by emulsion PCR, and the amplified strands are tethered to the same bead within the emulsion. Each bead is considered a PCR colony (polony) containing about 1000 copies derived from the single DNA molecule. The polonies are then sequenced in picoliter-scale wells by pyrosequencing chemistry. (Reprinted from Rajendhran, J., and Gunasekaran, P., *Microbiol. Res.*, 166, 99–110, 2011. With permission from Elsevier.)

stranded is used. When a fragment is denatured, it stops mitigating through the gel matrix. Further, individual DNA fragments can then be cut out of the gel and cloned, and the nucleotide sequences determined.

5.5 BIOPROSPECTING AND CONSERVATION OF FUNGAL DIVERSITY

Biotechnology, as understood and practiced in the developing countries, ranges from traditional food fermentation to improved methods of industrial fermentation and development of novel products. Biotechnology development involves bioprospecting the increasing use of "newly discovered" local genetic resources and/or local knowledge that has not been properly documented or legally protected within the internationally accepted intellectual property framework.

While in the past culture collections were essentially seen and run as centers of conservation and distribution of microbiological material, now biological resource centers (BRCs) are conceived of as the source of all essentials, including legal advice and services for research and development in biological sciences. The concept of BRCs originated as early as 1946, at the United Nations Educational, Scientific, and Cultural Organization (UNESCO), to establish centers of microbial resources in developing countries and to conserve the vast microbial diversity. In 1999, the Organization for Economic Cooperation and Development (OECD) initiated the development of this concept in the twenty-first century.

Culture collections are important interfaces between providers and users of microbiological resources. Culture collections face primarily scientific and technical challenges. Several important culture collections of microbial resources exist all over the world (Table 5.1). As thermophilic fungi constitute a very small group of organisms growing at elevated temperatures, no exclusive culture collection exist anywhere in the world for their preservation. Curators of these culture collections receive strains from many sources and supply them to those who need these strains for various purposes. Because of the advances made in biotechnology, these collections increasingly offer other related services. Some of these services include

- Supply of cultures of bacteria, fungi, yeast, actinomycetes, algae, animal cell lines, plant cells, viruses, plasmids, and vectors
- Training related to the growth and culture of concerned organisms
- Identification of strains for the industry or academia
- Strain conservation
- Strain improvement
- Contract research
- Quality control

5.5.1   MICROBIAL STRAIN DATA NETWORK

This is an international information and communication network for microbiologists and biotechnologists that aims to provide support activities, such as electronic mail, bulletin boards, databases, and training. The secretariat of the Microbial Strain Data Network (MSDN) is located in Cambridge, United Kingdom. Several databases are linked through an electronic gateway to remote host computers in France, Germany, Canada, the United Kingdom, Japan, the Netherlands, and so forth. One of the distributed information centers, in Pune, India, provides regional support and virus expertise.

5.5.2   CLASSIFICATION OF MICROORGANISMS ON THE BASIS OF HAZARD

Various classification systems exist, such as the definitions for classification by the World Health Organization (WHO), U.S. Public Health Service (USPHS), Advisory Group on Dangerous Pathogens (ACDP), European Federation of Biotechnology (EFB), and European Community (EC). In Europe, the EC Directive on Biological Agents (93/88/EEC) sets a common baseline, which has been strengthened and

TABLE 5.1

Some of the World's Major Microbial Culture Collections

Country	Culture Collection
America	American Type Culture Collection (ATCC), 10801 University Boulevard, Manassas, VA 20110–2209
	Carolina Biological Supply Company, 2700 York Road, Burlington, NC 27215
	Agricultural Research Service Culture Collection (NRRL), National Center for Agricultural Utilization Research, 1815 North University Street, Peoria, IL 61604
United Kingdom	National Collection of Type Cultures (NCTC), PHLS Central Public Health Laboratory, 61 Colindale Avenue, London, NW9 5HT, UK
	CABI Bioscience, CABI Europe—UK Egham Bakeham Lane, Egham, Surrey TW20 9TY
	Culture Collection of Algae and Protozoa, 36 Story's Way, Cambridge University, Cambridge, CB3 ODT, UK
	European Culture Collections Organization (ECCO), ECACC/CAMR, Culture Collections, Public Health England Porton Down, Salisbury SP4 0JG UK
Germany	Deutsche Sammlung von Mikrooganismen und Zellkulturen (DSMZ) GmbH, Mascheroder Weg 1b, 38124, Braunschweig, Germany
Netherlands	Centraalbureau voor Schimmelculture (CBS), Oosterstaat 1, Baarn, The Netherlands
	NICMM Culture Collection, Post Box 8039 4330 EA, Middleburg, The Netherlands
Japan	Institute for Fermentation (IFO), Osaka, 17-85 Juso-Honmachi 2-chrome, Yodogawa-ku, Osaka, 532 Japan
	NITE Biological Resource Center (NBRC), Department of Biotechnology, National Institute of Technology and Evaluation, 2-5-8, Kazusakamatari, Kisarazu, Chiba 292-0818, Japan
France	Pasteur Culture Collection of Cyanobacterial Strains, Unite de physilogie Micro Bienne, Institut Pasteur, 28 Rue due Docteur Roun, 75724 Paris Cedex 15, France
India	Microbial Type Culture Collection and Gene Bank (MTCC), CSIR—Institute of Microbial Technology Sector 39-A, Chandigarh 160036, India
	Indian Type Culture Collection (ITCC), Division of Plant Pathology, Indian Agricultural Research Institute, New Delhi 110 012, India
Indonesia	Biotechnology Culture Collection (BTCC), Research Center for Biotechnology, Indonesian Institute of Science (LIPI), J1, Raya Bogor Km 46 Cibinong 16911, Indonesia
Thailand	BIOTEC Culture Collection, National Centre for Genetic Engineering and Biotechnology, 113 Phaholyothin Road, Klong 1, Klong Luang, Pathum Thani 12120, Thailand
Vietnam	Vietnam Type Culture Collection (VTCC), Center of Biotechnology, Vietnam National University (VNU), Hanoi, E2 Building, 144 Xuan Thuy Road, Hanoi, Vietnam

expanded in many of the individual member states. The definition and minimum handling procedures of pathogenic organisms are set by appropriate authorities in each country and are often the same or similar for all EC countries; in the United Kingdom, the ACDP lists four hazard groups with corresponding containment levels. On the basis of their potential to cause disease and their pathogenicity in humans, microorganisms are classified into four risk groups (Anon, 1995) as follows:

Risk Group 1: A biological agent that is most unlikely to cause human disease.

Risk Group 2: A biological agent that may cause human disease and which might be a hazard to laboratory workers but is unlikely to spread in the community. Laboratory exposure rarely produces infection, and effective prophylaxis or treatment is available.

Risk Group 3: A biological agent that may cause severe human disease and present a serious hazard to laboratory workers. It may present a risk of spreading in the community, but there is usually effective prophylaxis or treatment.

Risk Group 4: A biological agent that causes severe human disease and is a serious hazard to laboratory workers. It may present a high risk of spreading in the community, and there is typically no effective prophylaxis or treatment.

5.5.3 INTERNATIONAL DEPOSITORY AUTHORITIES

With the rapid advances in recombinant DNA techniques, the need for appropriate protection of inventions including or using microorganisms has been recognized. To give a particular status to microbiological resources and write a specific global treaty, an agreement on *Trade-Related Aspects of Property Intellectual Rights* (TRIPS) was signed among the countries party to the treaty. The TRIPS agreement establishes the patentability of microbiological resources. To comply with the demand of international recognition of patent protection in biotechnology, the founding members of the Budapest Union signed the *Budapest Treaty* on the International Recognition of the Deposit of Microorganisms for the Purposes of Patent Procedure (WIPO, 1977). One of the important features of this treaty is that a patent deposit that has been made with one International Depository Authority (IDA) is sufficient and valid for all other states that have signed the Budapest Treaty. Nonmember countries may also accept deposits according to the Budapest Treaty. To date, 59 states worldwide are party to the treaty.

5.5.3.1 Responsibilities of an IDA

As per the Budapest Treaty, an IDA has the following obligations:

- To be objective and impartial
- To be available on the same terms and conditions to any depositor
- To test the viability of the biological materials promptly after receipt
- To comply with the requirements of secrecy
- To accept and store the biological material for the full duration of time specified by the treaty (5 years after the most recent request for the furnishing of a sample of the material; at least for 30 years); conclusively, it must have a permanent existence
- To comply with the bureaucratic requirements specified in the treaty
- To furnish samples of deposited material for trials and examinations to persons entitled to receive them

5.5.3.2 Distribution of IDAs and the Biological Material Accepted

Until now, 35 institutions all over the world have acquired the status of an IDA under the Budapest Treaty. Out of these, 23 are located in the Europe, 3 in North America, 1 in Australia, and 8 in Asia. It is clear that there is a lack of IDAs in South America and Africa. The main reasons favoring a given country having its own IDA are the export and import restrictions that might exist for certain kinds of biological material and the possible communication problems that might arise due to the language predisposition.

The kind of biological material accepted by an IDA is at its own discretion, which mainly depends on the specific expertise of the collection and the equipment available. There are patent depositories that accept a broad range of different kinds of biological material, whereas others restrict the biological material to one or two kinds. Moreover, an IDA decides the acceptance of risk level of the material. Most IDAs accept organisms of Risk Groups 1 and 2 (Table 5.2).

5.5.3.3 Guide to the Deposit of Microorganisms under the Budapest Treaty

The World Intellectual Property Organization (WIPO) has issued a guide that offers assistance for the depositor or a requesting party. In this guide, the general requirements for a deposit and the furnishing of samples are provided. The specific requirements of individual IDAs and intellectual property offices are described in detail. The respective checklist and the model form the complete publication. The guide is regularly updated, and the changes are available online: http://www.wipo.int/portal/en/index.html.

5.5.3.4 Code of Practice for IDAs

For the discussion of problems arising out of the procedure for patent deposit, a code of practice for IDAs exists. Here a guideline for existing and future IDAs was established on the basis of an initiative of the European Culture Collections Organization (ECCO). The ECCO is supported by most of the existing IDAs, the European Patent Office, and the WIPO. Its aim is to harmonize the procedures involved in a patent deposit as per the Budapest Treaty. The main objective is to make a discussion forum for the existing and newly arising problems, and it is intended that it be updated, clarified, or extended if necessary. It deals with practical problems, such as the following:

- Will mixed cultures be accepted?
- Will unofficial statements be issued to the depositor?
- Can a deposit be withdrawn?
- Which test procedure or methods and criteria for viability testing are to be issued?
- When is the payment of a deposit effected?
- How is contamination of a deposited culture handled?
- Who is responsible for the purity and authenticity of the deposited culture?
- May requests for information about a deposited culture or the related deposit documents be answered by the IDA representative?

TABLE 5.2
Some Important IDAs and the Kind of Biological Material Accepted by Them

Sr. No.	IDA	Kinds of Biological Material Accepted	Highest Risk Level
1	ATCC	Bacteria; fungi; yeast; algae; protozoa; embryos; human, animal, and plant cell cultures; bacteriophages; animal and plant viruses; seeds; RNA and DNA	4 3 for GEMs
2	ABC	Animal and human cell lines	2
3	BCCM	Bacteria, fungi, yeast, plasmid DNA, RNA, and animal and human cell lines (1)	2 (1) 3
4	CCY	Yeast	1
5	CBS	Bacteria, fungi, yeast, bacteriophages, plasmid DNA	2
6	CCAP	Algae, free-living protozoa, microorganisms	3 2 for GEMs
7	IMI	Fungi, bacteria, nematodes	2
8	DSMZ	Bacteria; archaea; fungi; yeast; bacteriophages; plant viruses; plasmid DNA; animal, human, and cell cultures; murine embryos	2
9	IPOD	Bacteria, fungi, yeast, bacteriophages, plasmid DNA, animal and plant cell cultures, embryos, algae, seeds	2
10	NCAIM	Bacteria, fungi, yeasts	2
11	NCTC	Bacteria	3
12	NPMD	Bacteria, archaea, yeasts, fungi, bacteriophages	2
13	KCLRF	Animal, human, and plant cell cultures	1
14	MTCC	Bacteria, bacteria-containing plasmids, fungi, yeasts, bacteriophages, plasmids in a host and/or as isolated DNA preparations	1, 2
15	NRRL	Bacteria, yeast	1
16	VKM	Bacteria, fungi, yeast	1

Note: GEMs = Genetically Engineered Microorganisms.

5.5.3.5   Future Development of the IDA Network Worldwide

The number of IDAs is constantly increasing. There were only 25 IDAs in 1994; now this number has increased to 35 in 2004. Sixty-nine culture collections registered with the World Federation for Culture Collections (WFCC) offer a service for depositing a patent. India is the only developing country that has the honor of having an IDA at Chandigarh (MTCC). Most of these IDAs are located in Europe; these IDAs are not evenly distributed the world over. In the case of developing countries, the need for establishing the IDAs stems from the fact that biotechnology in these countries has started budding, and therefore establishment of IDAs in Africa and South America is fully justified and the WFCC incorporate the countries in these regions. Before an institution tries to obtain the status of an IDA, however, a well-functioning culture collection should exist first. The expertise to handle the patent strain should be there. It is mandatory that any country that wishes to establish an IDA must be a signing party to the Budapest Treaty.

5.5.4 CULTURE TRANSPORTATION

Microorganisms are isolated, cultured, characterized, preserved for the long term, stored, and transported between laboratories. They are shipped by various means: by mail, courer, or hand, from one laboratory to another within countries and often across borders and continents. They are sent for identification, reference, research, or production purposes from colleague to colleague, from and to culture collections. All these actions involving dispatch must be carried out safely and in compliance with various legislation and regulations that control these matters. The existing legislation is regularly updated or amended.

The importance of a laboratory's health and safety procedures stretches beyond its walls to all those who may come into contact with the substances and products dispatched from that laboratory. A microorganism in transit might put carriers, postal staff, freight operators, and recipients at risk; some organisms are relatively hazard-free, while others are quite dangerous. It is essential that safety and shipping regulations are followed to ensure safe transit. There are several other pieces of legislation that restrict the distribution of microorganisms of which a microbiologist must be aware.

Before a consignment containing microorganisms is offered for transport, the decision as to whether it is an infectious substance is crucial, as is the destination of the consignment. In order to choose the correct type of packaging and the correct mode of transport or carrier (postal mail or courier), shippers of biological material must have a sound knowledge of all relevant packaging and transport regulations. They must have recurrent training according to the latest International Air Transport Association (IATA) Dangerous Goods Regulations (DGR) if infectious substances are to be transported by air. Air transport plays the dominating role when living biological materials are transported over long distances. Furthermore, the DGR for air transport are most user-friendly making sure the responsible shipper is on the safe side and in conformity with international law. It is self-evident that the respective national or regional regulations for road transport have to be observed.

Infectious substances are by definition dangerous goods, and the DGR for transport fully apply so that they do not pose a risk for the people involved in the transportation chain, animals, or the environment. This usually does not apply to microorganisms classified in Risk Group 1. For the latter, consequently other regulations for packaging and transport are in place and have to be observed. They can usually be transported by postal mail services when packed in accordance with the respective packaging regulations laid down by the Universal Postal Union (UPU). Postal services usually differentiate between perishable (active) and nonperishable (dried, freeze-dried) biological substances. Shippers should be aware that any biological material is excluded from transport in postal parcels; the UPU permits letter mail only.

The term *freight* is used in connection with courier transport only, in contrast to postal parcels. Registered letter mail is generally recommended because of individual treatment and possible tracking. Also, note that, in general, postal mail systems exclude any dangerous goods; infectious substances classified in the new shipping Category B might be sent by national postal mail (on the road). The new deregulated transport requirements for Category B cultures imply that administrative expenditures and costs have become much less problematic. However, there are still strict requirements on the shipper's responsibility, training, and packaging quality, as well as on the correct labeling and marking. Although recent changes relevant for shippers of infectious substances resulted in the definition of a new classification system with two new shipping categories (A and B), instead of risk group definitions, the existing risk group allocation of an organism does help the sender to classify the material for transport purposes.

Additionally, the regulations for shipping genetically modified organisms (GMOs) have undergone a revision resulting in a clearer instruction for the transport of safety level 1 GMOs. Principally, the new deregulated requirements apply to the majority of Risk Group 2 microorganisms, as the definition of this risk group conforms to the definition of the new category: such cultures can be shipped under the same requirements as diagnostic specimens.

The transportation chain begins in the packaging department of a culture collection, ends in the recipient's laboratory, and may include transport by hand or postal or courier transport, within countries or across borders and continents. Only a correctly labeled and documented shipment reaches its destination quickly and safely; therefore, the courier services require their customers to fulfill the regulations. It is the responsibility of all laboratories supplying infectious substances to nominate a person who will receive recurrent training and take on the responsibility for signing the shipping documents. The latter can only be signed by a trained person who is thoroughly conversant with the regulations, including the applicability, limitations (state or operator variations), classification, identification, packing, marking, labeling, and documentation.

5.5.5 THE PREMISES BEFORE DISPATCH OF CULTURES

The sender of a microorganism must be sure that the receiver is authorized to work with it and has adequate facilities.

- Do *not* supply to private persons.
- Do *not* supply to new customers or unknown recipients who have not specified their institution.
- Supply infectious substances *only* to recipients who have the appropriate laboratory safety level that corresponds with the risk group of the organism.
- Supply animal or plant pathogens or GMOs *only* to recipients having an appropriate laboratory and the relevant permits for work.
- When shipping outside the country, be sure that the microorganism does not fall under export restrictions, like the Biological and Toxin Weapons Convention, dual-use restrictions, and other national legislation (relevant national authorities are the National Export Office, Department of Commerce, or Foreign Office), and that, if applicable, quarantine requirements

are fulfilled by the receiver and import permits (from health authorities) are ready to be shipped together with the organism.

Further, it is important for culture collections to

1. Establish a well-organized shipping department with trained staff
2. Nominate a trained person who replaces the legal trained shipper in cases of absence
3. Have access to the latest IATA DGR, the latest regulations for road transport of dangerous goods, and all further relevant information sources
4. Develop a step-by-step checklist
5. Establish a computerized system for filling in the shipping documents in order to have a fast and reliable system that avoids mistakes

5.5.6 Organizations Dealing with Microbial Cultures

The following international organizations deal with microbial cultures:

- *World Federation for Culture Collections*: http://www.wfcc.info/
- *World Data Centre for Microorganisms*: http://www.wdcm.org/
- *Microbial Strain Data Network*: http://sdb.im.ac.cn/msdn.shtml
- *The Microbial Underground*: http://blogs.warwick.ac.uk/microbialunder ground/
- *CBS-KNAW COLLECTIONS*: http://www.westerdijkinstitute.nl/collections/

5.6 FUTURE PERSPECTIVES

Obtaining new biotechnological products from uncultivable microorganisms is a topic of debate in recent times. Both basic research and biotechnological developments require routinely applicable tools for the functional analysis, expression, and manipulation of genes. These techniques include procedures for genetic transformation and selection of the transformants, well-characterized molecular markers, and expression signals. As thermophilic fungi colonize, multiply, and survive in habitats having elevated temperatures, they represent a formidable pool of bioactive compounds and are a strategic source for new and successful commercial products. Recent technological advances made in genomics, proteomics, and combinatorial chemistry show that nature continues to preserve compounds in its metagenome having the essence of bioactivity or function within the host and the environment. Bioprospecting of microbial sources such as thermophilic fungi offers several advantages for their biocatalysts, besides being thermostable. However, studies on fungal distribution and mapping are challenging due to the lack of sufficient knowledge about their taxonomy and the lack of expert mycologists around the world. Nevertheless, the fungal world provides a fascinating and almost continual source of biological diversity, which is a rich source to exploit for human welfare.

Considerable progress has been made in the search for thermophilic fungi, yet their true diversity has not been fully explored. The future challenges for tapping uncultured fungi using culture-independent molecular methods warrant that there are likely to be a high number of fungal species in various habitats, many of which might be thermophilic. Besides bioprospecting fungi in general and thermophilic fungi in particular, their conservation is of great concern, as they play a significant role in human welfare. Several steps for their conservation have been suggested by Moore et al. (2001), including (1) conservation of their habitats, (2) *in situ* conservation of nonmycological reserves and ecological niches, and (3) *ex situ* conservation, particularly for saprophytic species growing in culture. In addition, the Slovak Republic has passed legislation to protect 52 species of fungi, which enables managers to prevent damage to their natural habitats (Lizon, 1999). Thus, research on microorganisms requires sound techniques for their stable preservation for the confirmation of results and for future use.

REFERENCES

Anderson, I.C., and Cairney, J.W. 2004. Diversity and ecology of soil fungal communities: Increased understanding through the application of molecular techniques. *Environmental Microbiology* 6:769–779.

Anon. 1995. *Advisory Committee on the Microbiological Safety of Food. Report on Verocytotoxin-Producing Escherichia coli*. London: HMSO.

Blackwell, M. 2011. The fungi: 1, 2, 3 … 5.1 million species? *American Journal of Botany* 98:426–438.

Bonito, G., Isikhuemhen, S.O., and Vilgalys, R. 2010. Identification of fungi associated with municipal compost using DNA-based techniques. *Bioresource Technology* 101: 1021–1027.

Brock, T.D., and Freeze, H. 1969. *Thermus aquaticus*, a nonsporulating extreme thermophile. *Journal of Bacteriology* 98:289–297.

Buée, M., Reich, M., Murat, C., Morin, E., Nilsson, R.H., Uroz, S., and Martin, F. 2009. 454 pyrosequencing analyses of forest soils reveal an unexpectedly high fungal diversity. *New Phytologist* 184:449–456.

Bull, A.T., Ward, A.C., and Goodfellow, M. 2000. Search and discovery strategies for biotechnology: The paradigm shift. *Microbiology and Molecular Biology Reviews.* 64: 573–606.

Comfort, D.A., Chhabra, S.R., Conners, S.B., Chou, C.-J., Epting, K.L., Johnson, M.R., Jones, K.L., Sehgal, A.C., and Kelly, R.M. (2004). Strategic biocatalysis with hyperthermophilic enzymes. *Green Chemistry* 6:459–465.

Egorova, K., and Antranikian, G. 2005. Industrial relevance of thermophilic Archaea. *Current Opinion in Microbiology* 8:649–655.

Firn, R.D. 2003. Bioprospecting—Why is it so unrewarding? *Biodiversity and Conservation* 12:207–216.

Fujiwara, S. 2002. Extremophiles: Developments of their special functions and potential resources. *Journal of Bioscience and Bioengineering* 94:518–525.

Hawksworth, D.L. 1991. The fungal dimension of biodiversity: Magnitude, significance, and conservation. *Mycological Research* 95:641–655.

Hawksworth, D.L. 2012. Global species numbers of fungi: Are tropical studies and molecular approaches contributing to a more robust estimate? *Biodiversity and Conservation* 21:2425–2433.

Hultman, J., Vasara, T., Partanen, P., Kurola, J., Kontro, M.H., Paulin, L., Auvinen, P., and Romantschuk, M. 2010. Determination of fungal succession during municipal solid waste composting using a cloning-based analysis. *Journal of Applied Microbiology* 108:472–487.

Johri, B.N., and Satyanarayana, T. 2005. *Microbial Diversity: Current Perspectives and Potential Applications*. New Delhi: I.K. International.

Kellenberger, E. 2001. Exploring the unknown. *EMBO Reports* 2:5–7.

Kumar, R., and Aneja, K.R. 1999. Biotechnological applications of thermophilic fungi in mushroom compost preparation. In *From Ethnomycology to Fungal Biotechnology*, ed. J. Singh and K.R. Aneja, 115–126. New York: Plenum Publishers.

Langarica-Fuentes, A., Zafar, U., Heyworth, A., Brown, T., Fox, G., and Robson, G.D. 2014. Fungal succession in an in-vessel composting system characterized using 454 pyrosequencing. *FEMS Microbiology Ecology* 88:296–308.

Lindahl, B.D., Nilsson, R.H., Tedersoo, L., Abarenkov, K., Carlsen, T., Kjøller, R., Kõljalg, U., Pennanen, T., Rosendahl, S., Stenlid, J., and Kauserud, H. 2013. Fungal community analysis by high-throughput sequencing of amplified markers—A user's guide. *New Phytologist* 199:288–299.

Lizon, P. 1999. Current status and perspectives of conservation of fungi in Slovakia. In Abstracts XIII: Congress of European Mycologists, Madrid, Spain, September 21–25, 1999, 77.

Moore, D., Nauta, M.M., Evans, S.E., and Rotheroe, M. 2001. Fungal conservation issues: Recognising the problem, finding solutions. In *Fungal Conservation*, ed. D. Moore, M.M. Nauta, and S.E. Evans. Cambridge: Cambridge University Press 1–6.

Morgenstern, I., Powlowski, J., Ishmael, N., Darmond, C., Marqueteau, S., Moisan, M.C., Quenneville, G., and Tsang, A. 2012. A molecular phylogeny of thermophilic fungi. *Fungal Biology* 116:489–502.

O'Brien, H.E., Parrent, J.L., Jackson, J.A., Moncalvo, J.M., and Vilgalys, R. 2005. Fungal community analysis by large-scale sequencing of environmental samples. *Applied and Environmental Microbiology* 71:5544–5550.

Pan, W.Z., Huang, X.W., Wei, K.B., Zhang, C.M., Yang, D.M., Ding, J.M., and Zhang, K.Q. 2010. Diversity of thermophilic fungi in Tengchong Rehai National Park revealed by ITS nucleotide sequence analyses. *Journal of Microbiology* 48:146–152.

Podar, M., and Reysenbach, A.L. 2006. New opportunities revealed by biotechnological explorations of extremophiles. *Current Opinion in Microbiology* 17:250–255.

Rajendhran, J., and Gunasekaran, P. 2011. Microbial phylogeny and diversity: Small subunit ribosomal RNA sequence analysis and beyond. *Microbiological Research* 166:99–110.

Satyanarayana, T., Raghukumar, C., and Shivaji, S. 2005. Extremophilic microbes: Diversity and perspectives. *Current Science* 89:78–90.

Sharma, K.K., Kapoor, M., and Kuhad, R.C. 2005. *In vivo* enzymatic digestion, *in vitro* xylanase digestion, metabolic analogues, surfactants and polyethylene glycol ameliorate laccase production from Ganoderma sp. kk-02. *Letters in Applied Microbiology* 41: 24–31.

Sharma, M., Chadha, B.S., Kaur, M., Ghatora, S.K., and Saini, H.S. 2008. Molecular characterization of multiple xylanase producing thermophilic/thermotolerant fungi isolated from composting materials. *Letters in Applied Microbiology* 46:526–535.

Suryanarayanan, T.S., and Hawksworth, D.L. 2004. Fungi from little explored and extreme habitats. In *The Biodiversity of Fungi: Aspects of the Human Dimension*, ed. S.K. Deshmukh and M.K. Rai, 33–48. Enfield, NH: Science Publishers.

WIPO (World Intellectual Property Organization). 1977. *Licencing guide for developing countries*. WIPO Publication No. 620(E). Geneva: WIPO.

6 Taxonomy and Molecular Phylogeny of Thermophilic Fungi

6.1 INTRODUCTION

Perhaps the most intricate task that microbiologists face today is to see that which cannot be seen with the naked eye but can live apparently everywhere on earth. In this context, the extraordinary diversity of microbial form is of paramount importance, particularly the diversity of fungi having different shapes, sizes, physiologies, and lifestyles. To understand these attributes and survey the level of diversity, a reliable classification system is needed that can place such a diversity in different groups. *Taxonomy* (Greek *taxis*, meaning "arrangement" or "order," and *nomos*, "law" or "to govern") refers to the science of biological classification. It may be defined as the classification, nomenclature, and identification of organisms according to published rules. Although the term *systematics* is often used interchangeably with *taxonomy*, many taxonomists define systematics in more general terms as the scientific study of organisms with the ultimate goal of characterizing and arranging them in an orderly manner. Therefore, in systematics scientists make use of several disciplines, like morphology, ecology, physiology, biochemistry, genetics, epidemiology, and molecular biology, to achieve the coveted goal.

The earliest classification system, developed by Carolus Linnaeus, was the natural system of classification, which was based on some common characteristics shared by a group of organisms. Although this system of classification had several limitations, it was definitely superior to the artificial system of classification, as knowledge of an organism's position in the scheme provided information about many of its characteristics. For example, the classification of humans as mammals indicates that they have hair, a self-regulating body temperature, and milk-producing mammary glands in the female (Willey et al., 2011). However, in the context of microbes and fungi in particular, the natural system of classification was not necessarily rooted in the evolutionary relatedness.

Presently, the classification and nomenclature of microorganisms is based on *polyphasic taxonomy*, an approach that encompasses phenotypic, phylogenetic, and genotypic characteristics. The phenetic system of classification of microorganisms is anchored in the mutual similarity of their phenotypic characteristics. Organisms sharing many such characteristics form a single phenetic group or a taxon. This system of classification remained popular among microbial taxonomists for a long period of time. However, the major drawback of this system lay in its unrevealed evolutionary relationship. In contrast, the phylogenetic system of classification displays and compares organisms based on evolutionary relationships. This system of classification is more practical, as it offers insights into the history of life on earth. Nevertheless, this system of classification also remained ineffective for most of the twentieth century on account of the lack of information on the fossil record. It was only when small-subunit rRNA nucleotide sequences to assess evolutionary relationships among microorganisms were made available that the dawn of a new era revealing the origin and evolution of the majority of life-forms, including microorganisms, unfolded. Now, this approach is widely accepted for the classification of different varieties of organisms, and presently, more than 0.5 million different 16S and 18S rRNA sequences are available in international public databases, such as GenBank and MycoBank. 16S and 18S rRNA sequences are considered the most conserved sequences and are helpful in tracing the evolutionary history of an organism. Similarly, genotypic classification sought to compare organisms on the basis of their genetic similarity. In this approach, individual genes or whole genomes can be compared and individuals displaying 70% homology are considered to belong to the same species. Techniques involving the study of DNA, RNA, and protein estimation have advanced our understanding concerning microbial evolution and taxonomy.

6.2 CLASSIFICATION AND TAXONOMIC RANKS

For a long time, humans have desired to classify living and nonliving things. Things that are similar in some way can be grouped and given a particular name that allows them to be referred to by groups. There are several ways of defining similarity in fungi. This leads to many useful ways of grouping them, with the groups created by one definition of similarity different from those created by another. Nevertheless, when it comes to classifying living organisms, it is essential that the established rules of taxonomy are followed. While classifying microorganisms, it is necessary to place them in hierarchical taxonomic ranks, with each rank or level sharing a common set of explicit features. The ranks are arranged in such a manner that they do not overlap each other, and each rank includes not only the traits that define the rank above it but also a new set of more restrictive traits. A more general scheme of classification for living organisms uses concepts such as species and genus, and in this scheme, species is the basic unit of classification. Taxon (plural taxa) refers to a taxonomic group belonging to any rank. In fungi, the seven principal obligatory ranks of taxa in descending order are kingdom, phylum, class, order, family, genus, and species. Some microorganisms are additionally given a subspecies designation. Microbial groups at each rank have a specific suffix (ending) that is indicative of that rank or level. For instance, order has a suffix *ales*, family *aceae*, and so forth. Further, a genus typically contains several species, and species is said to be of lower taxonomic rank than genus. Consequently, there are fewer genera than there are species, fewer families than there are genera, and so on. Thus, it is clear that this is a hierarchical scheme, and any hierarchical system of classification is an example of a taxonomic classification. Fungal names are derived from various sources, such as a place, an important characteristic, the substrate on which the fungus grows, the symptoms it produces on the plant, or the commemoration of a person. The binomial system of classification is used by fungal taxonomists for the classification of fungi. The Latinized italicized name consists of two parts. The first part is the generic name (genus), and the second part is the species name. For example, in *Phytophthora infestans* (a causative agent of late blight of potato), the species epithet is stable and must always be used (principle of precedence) in any subsequent change of the name of the genus or when assigning the organism to another genus.

As of now, of the predicted 600,000 species of fungi, only 44,368 have been catalogued (Mora et al., 2011). Seven major groups (phyla), 10 subphyla, 35 classes, 12 subclasses, and 129 orders are recognized within the Fungi kingdom (Hibbett et al., 2007). Although a comprehensive classification of fungi is beyond the scope of the present text, an insight into the inclusion of fungi in different kingdoms post–Linnaean era is of fundamental importance. Until the latter half of the twentieth century, fungi were classified in the Plant kingdom (subkingdom Cryptogamia) and separated into four classes: Phycomycetes, Ascomycetes, Basidiomycetes, and Deuteromycetes (Fungi Imperfecti). In the middle of the twentieth century, the three major kingdoms of multicellular eukaryotes, Plantae, Animalia, and Fungi, were recognized as being absolutely distinct. The crucial difference between these three kingdoms was the mode of nutrition; animals (whether single-celled or multicellular) ingest food, plants photosynthesize, and fungi absorb externally digested nutrients. An appraisal of the genomic studies indicates that plant genomes lack gene sequences that are crucial in animal development, animal genomes lack gene sequences that are crucial in plant development, and fungal genomes have none of the sequences that are important in controlling multicellular development in either of the two groups. Such fundamental genetic differences imply that animals, plants, and fungi are very distinct cellular organisms. Molecular analyses have indicated that plants, animals, and fungi diverged from one another almost 1 billion years ago. Earlier, formal recognition of fungal nomenclature was governed by the International Code of Botanical Nomenclature (ICBN). However, to do justice to mycology, in 2011 the name was changed to the International Code of Nomenclature for algae, fungi, and plants (ICN) (McNeill et al. 2012), and all nomenclature pertaining to fungi is now governed by the ICN.

Thermophilic fungi are classified as those having a maximum temperature for growth at or above 40°C. This group of fungi is of wide interest due to their potential for the production of thermostable enzymes that can be used in industrial high-temperature bioprocesses. Only a small fraction of the estimated 600,000 fungi (Mora et al., 2011) are considered to be thermophilic. The taxonomic classification and nomenclature of thermophilic fungi is in a state of disarray, often leading to misidentification and confusion (Mouchacca, 2000). Currently, most of the known thermophilic fungi have been placed in the orders Sordariales, Eurotiales, Mucorales, and Onygenales (Berka et al., 2011). Earlier, Straatsma et al. (1994) described two isolates belonging to Basidiomycota. In addition, *Myriococcum thermophilum* is included as a mitosporic basidiomycete by the National Center for Biotechnology Information (NCBI), while the Index Fungorum lists it as an agaricomycete. Morgenstern et al. (2012) analyzed sequences from 86 fungal genomes to produce a robust molecular phylogeny of thermophilic fungi with an aim to identify the fungal orders that harbor thermophilic species and to resolve the evolutionary relationships among the thermophilic and nonthermophilic species in these orders. In order to understand the phylogenetic distribution of thermophilic fungi and evolution of thermophily in fungi, we must first have a detailed knowledge of phylogeny.

6.3   WHAT IS PHYLOGENY?

The term *phylogeny* is derived from the German *Phylogenie*, introduced by Haeckel in 1866, and it refers to a diagrammatic hypothesis about the evolutionary relationships of a group of organisms. An understanding of the diversity, systematics, and nomenclature of microbes is increasingly important in several branches of life science. The molecular approach to phylogenetic analysis, pioneered by Carl Woese in the 1970s and leading to the three-domain model (Archaea, Bacteria, and Eukarya), has revolutionized the philosophy of microbial evolution. The advancement of technological innovations in modern molecular biology and computational science has led to a spurt in nucleic acid sequence information, bioinformatic tools, and phylogenetic inference methods. For a long time, phylogenetic analysis has played a key role in microbiology and the emerging fields of comparative genomics and phylogenomics.

A quotation of Nieremberg (1635) is quite relevant in the present context of phylogeny:

> Nature rises up by connections, little by little and without leaps, as though it proceeds by an unbroken web, it proceeds in a leisurely and placid uninterrupted course. There is no gap, no break, no dispersion of forms: they have, in turn, been connected, ring within ring. That very golden chain is universal in its embrace.

The realization that all organisms on earth are related by common descent (Darwin, 1859) was one of the most profound insights in scientific history. To date, most fungal phylogenies have been derived from single-gene comparisons, or from concatenated alignments of a small number of genes. The increase in fungal genome sequencing presents an opportunity to reconstruct evolutionary events using entire genomes. The phylogenetic classification of fungi is intended to group fungi based on their ancestral relationships, also known as fungal phylogeny.

The genes possessed by organisms in the present day have descended through the lineage of their ancestors. Consequently, finding relationships between those lineages is the only way to establish the natural relationships between living organisms. Earlier, phylogenetic relationships could be inferred from a variety of data, traditionally including fossils, comparative morphology, and biochemistry. But modern phylogenetic analysis derives information from the molecular data of a large number of species. Traditional methods of systematics based on the morphology of vegetative cells, sexual states, physiological responses to fermentation, and growth tests can assign fungal species to particular genera and families. However, higher-level relationships among these groups are less certain and are best elucidated using molecular techniques.

The taxonomic rearrangement of the fungi started long before molecular techniques were at hand. However, within the last four decades, molecular techniques have altered the understanding of the evolution and phylogeny of life, and more than 1000 fungal research papers reporting phylogenies have been published. The majority of these studies used 16S and 18S rDNA sequencing. Although sequences based on protein-coding genes are rarely used in fungal phylogenetics, they can resolve deep-level phylogenetic relationships (Liu et al., 1999). In fungi and for most other organisms, phylogenetic reconstruction based on a single gene is blurred, as individual genes contain a limited number of nucleotide sites, and hence limited resolution. Thus, to have a robust phylogenetic analysis, multigene concatenation is performed using data from all available genes defining various characteristics of the organism. Until now, more than a hundred fungal genomes have been sequenced and are available in public databases. The genomes of some thermophilic fungi will be discussed in the later part of this chapter.

6.4　PHYLOGENETIC ANALYSIS

6.4.1　Molecular Phylogeny of Thermophilic Fungi

Thermophilic fungi are distributed widely throughout the fungal kingdom and are not consigned to any phylum or order. The definition of thermophilic fungi has been debated and defined variously by several workers (Cooney and Emerson, 1964; Crisan, 1964; Maheshwari et al., 2000), as discussed in the first chapter of this book. Thermophilic fungi are important from a biotechnological perspective, as they produce a number of thermostable enzymes that are useful in industrial processes, such as bleaching in the paper and pulp industry, the bioremediation of polluted sites and wastewater effluents, and the production of second- and third-generation biofuels (Wesenberg et al., 2003; Gianfreda and Rao, 2004; Sigoillot et al., 2005; Turner, 2007). Phylogenetic analysis of these fungi may resolve the pending issues of their misidentification and taxonomic classification (Mouchacca, 2000).

With the advent of polymerase chain reaction (PCR) technology, the molecular revolution of fungal taxonomy, which commenced in the early 1990s, has grown into a mature discipline where multilocus data sets, extensive taxon sampling, and rigrous analytical approaches are standard (Hibbett et al., 2007). While 16S rRNA gene–based phylogeny is the most commonly accepted taxonomic scheme in prokaryotes, its relevance to eukaryotic phylogeny is at a crossroads. Currently, there is a heated debate as to what is the best approach for reconstruction genome phylogenies, particularly in fungi. The 16S–23S internal transcribed spacer (ITS) regions of the rRNA operon might be under minimal selective pressure during evolution, and therefore have more variation in the sequences than the coding regions of 16S and 23S rRNAs. The size of the ITS varies considerably for different species, and even among different operons within a single cell having multiple rRNA operons (Rajendhran and Gunasekaran, 2011). To assess fungal phylogeny, single-gene phylogenies, especially those based on 18S rDNA, have established many of the accepted relationships between fungal organisms (Fitzpatrick et al., 2006). The use of the 18S rDNA approach is advantageous in view of the vertical transmission of this gene, its wide distribution, and the fact that it has slowly evolving sites. However, the phylogenetic relationships of a diverse range of fungi, including thermophilic species, must be performed using multigene concatenation, as this approach attempts to maximize the informativeness and explanatory power of the characteristic data used in the analysis (Kluge, 1989). Therefore, in recent studies (Pan et al., 2010; Berka et al., 2011; Morgenstern et al., 2012; Zhou et al., 2014) on the molecular phylogeny of thermophilic fungi, the multigene concatenation approach has largely been followed to elucidate the phylogenetic relationships among the fungal orders harboring thermophilic fungi.

Considering the diversity of thermophilic fungi in various habitats, relatively few studies focusing on phylogenetic analysis have been undertaken. Pan et al. (2010) attempted to identify thermophilic fungi from geothermal sites near neutral and alkalescent thermal springs in Tengchong Rehai National Park using ITS sequencing combined with morphological analysis for the identification of thermophilic fungi to the species level. The authors isolated 102 strains belonging to *Rhizomucor miehei*, *Chaetomium* sp., *Talaromyces thermophilus*, *Talaromyces byssochlamydoides*, *Thermoascus aurantiacus* Miehe var. *levisporus*, *Thermomyces lanuginosus*, *Scytalidium thermophilum*, *Malbranchea flava*, *Myceliophthora* sp. 1, *Myceliophthora* sp. 2, *Myceliophthora* sp. 3, and *Coprinopsis* sp. Similarly, Berka et al. (2011) analyzed the genomes of two thermophilic ascomycete species, *Myceliophthora thermophila* and *Thielavia terrestris*. They reported that the two genomes are similar in organization. The whole genome sequencing of these fungi offers new industrial applications of the enzymes from these organisms and their potential development as fungal production hosts. The authors further emphasized that with a finished genome as a scaffold and modern sequencing technologies, resequencing these strains and identifying mutations becomes relatively simple but very helpful for identifying beneficial and deleterious genetic changes.

A detailed study on the molecular phylogeny of thermophilic fungi by Morgenstern et al. (2012) investigated sequences from 86 fungal genomes to construct a robust molecular phylogeny of thermophilic fungi. Considering the criterion that a thermophilic fungus is one that optimally grows at 45°C, their phylogenetic study suggested that the known thermophilic fungi belong to the orders Sordariales, Eurotiales, Mucorales, and Onygenales. They examined the phylogenetic relationships of a diverse range of fungi, including thermophilic and thermotolerant species, using concatenated amino acid sequences of marker genes mcm7, rpb1, and rpb2 obtained from genome sequencing projects. These markers were used to reconstruct a phylogeny of the fungi (Figure 6.1).

The authors further observed that maximum likelihood analysis of the combined amino acid sequence alignment resulted in a tree with high bootstrap support (BSS) values even in deep internal nodes. Only two nodes received less than 50% BSS, and five nodes received support between 50% and 70%. Of the remaining 78 nodes receiving support above 70%, 61 nodes received 100% BSS (Figure 6.1).

According to Morgenstern et al. (2012), thermophilic species in the Sordariales are found only in Chaetomiaceae. To assess this connotation, they included sequences from species belonging to the classes Sordariaceae, Lasiosphaeriaceae, and Chaetomiaceae of Sordariales (Figure 6.2).

They reported that the concatenated data set for the Sordariales contains 60 operational taxonomic units (OTUs), of which 21 are represented by all three genes,

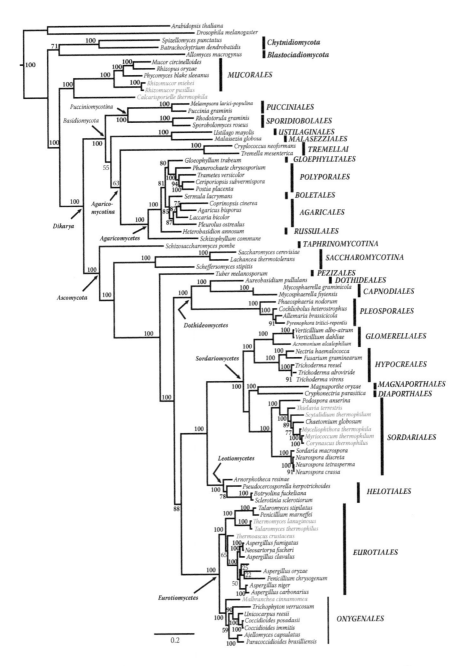

FIGURE 6.1 Maximum likelihood phylogeny inferred by amino acid sequence alignments. The amino acid sequences of Mcm7, Rpb1, and Rpb2 were used for concatenated analysis. Bootstrap values are indicated at each node and are given in bold font for values above 70. Thermophilic taxa are highlighted in red font. The tree is rooted with *Arabidopsis thaliana* and *Drosophila melanogaster*. (Reprinted from Morgenstern, I. et al., *Fungal Biol.*, 116, 489–502, 2012. With permission from Elsevier and British Mycological Society.)

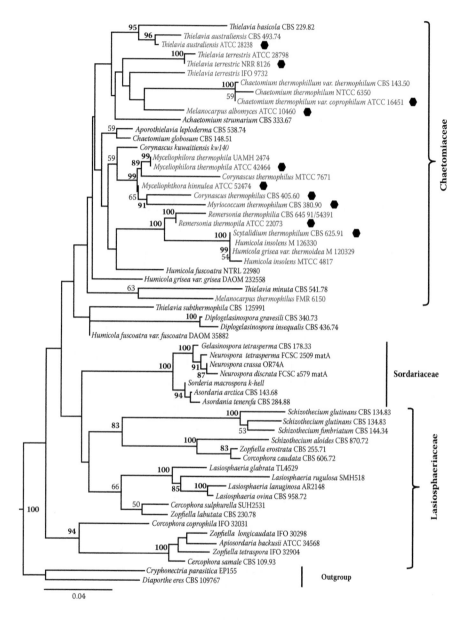

FIGURE 6.2 Maximum likelihood phylogeny for the Sordariales based on concatenated small-subunit, ITS/5.8S, and large-subunit nucleotide sequence alignments. Bootstrap values of 50 and higher are given next to each node; values above 70 are indicated in bold font. Thermophilic taxa are highlighted in red font. The black hexagons indicate thermophilic strains included in the temperature-dependent growth trial. Families are indicated to the right of vertical bars (monophyletic) or curly brackets (phyletic status unresolved). The tree is rooted with *Diaporthe eres* and *Cryphonectria parasitica* belonging to the Diaporthales. (Reprinted from Morgenstern, I. et al., *Fungal Biol.*, 116, 489–502, 2012. With permission from Elsevier and British Mycological Society.)

27 by two genes (mostly ITS and large subunit), and 12 by only one sequence. The tree reconstruction depicts the Chaetomiaceae, with the exclusion of *Thielavia subthermophila*, as monophyletic and the Lasiosphaeriaceae as paraphyletic (Figure 6.2). Furthermore, their study depicted that thermophilic fungi included in the order Eurotiales are not monophyletic; rather, they appear in three species' pairs and as single sequences throughout the tree (Figure 6.3).

In addition, the two *Thermoascus* species receive strong support as being monophyletic (100% BSS). Likewise, the species pairs of *Talaromyces byssochlamydoides* and *Talaromyces emersonii*, and *Talaromyces thermophilus* and *Thermomyces lanuginosus* received 100% BSS each. The two strains of *T. leycettanus* included in their analysis occupy different positions in the tree, indicating that they are not conspecific (Figure 6.3), which means they are not closely related to other *Talaromyces* isolates or any other thermophile. The results of their study confirmed that thermophily is not widespread in the fungal kingdom. Even the thermophilic taxa of uncertain taxonomic position did not fall outside one of the orders already known to harbor thermophiles, that is, the Sordariales, Eurotiales, Onygenales, and Mucorales.

Thermophilic fungi comprise less than 50 species, of which *Rhizomucor miehei*, *Rhizomucor nainitalensis*, *Rhizomucor pusillus*, *Rhizopus microsporus*, and *Rhizopus rhizopodiformis* belong to *Zygomycetes*. The genera Rhizomucor, Mucor, and Rhizopus are classified under the family Mucoraceae in the order Mucorales, which is a primitive and early divergent group of fungi (Zhou et al., 2014). Two species, Rhizomucor miehei and Rhizomucor pusillus, are true thermophilic fungi that are able to grow at 50°C or above. Zhou et al. (2014) sequenced the genome of R. miehei, which will facilitate future studies to better understand the mechanisms of fungal thermophilic adaptation and to explore the potential of R. miehei in the industrial-scale production of thermostable enzymes, such as proteases and lipases. The results of their study suggested that R. miehei is closer to Phycomyces blakesleeanus than to Mucor circinelloides and Rhizopus oryzae, although Rhizomucor, Mucor, and Rhizopus are classified under the same family, Mucoraceae.

Thermophilic fungi, as defined by their temperature preference to grow better at or above 45°C than at 25°C, have evolved independently in at least two lineages within the phylum Ascomycota, once each within the orders Sordariales and Eurotiales (Berka et al., 2011). Within the Sordariales, thermophily is limited to subgroups of the family Chaetomiaceae. Among fungi more broadly, thermophily also exists in the Zygomycota (Zhou et al., 2014); however, it appears to be rare or altogether absent in the phyla Basidiomycota and Chytridiomycota. The evolutionary trajectory of thermophily is obscured by chaotic taxonomy. Moreover, efforts on biosystematics have lagged behind research on thermophilic fungi useful in biotechnological industries, resulting in a body of literature for these organisms that lacks accurate taxonomic treatments (Mouchacca, 2000).

6.4.2 CONSTRUCTING PHYLOGENETIC TREES

A phylogenetic tree is a branching diagram or tree showing inferred evolutionary relationships among various biological species or lineages that are connected by nodes (Figure 6.4).

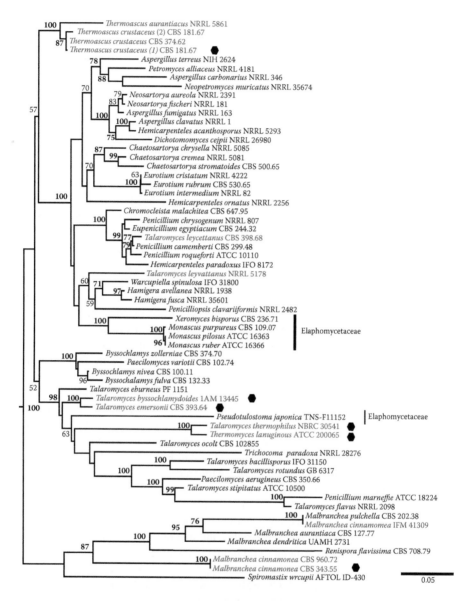

FIGURE 6.3 Maximum likelihood phylogeny for the Eurotiales based on concatenated small-subunit, ITS/5.8S, and large-subunit nucleotide sequence alignments. Bootstrap values of 50 and above are given next to each node; values above 70 are indicated in bold font. Thermophilic taxa are highlighted in red font. The black hexagons indicate thermophilic strains included in the temperature-dependent growth trial. Species belonging to the Elaphomycetaceae are indicated to the right of vertical bars. The tree is rooted with species belonging to the Onygenales. (Reprinted from Morgenstern, I. et al., *Fungal Biol.*, 116, 489–502, 2012. With permission from Elsevier and British Mycological Society.)

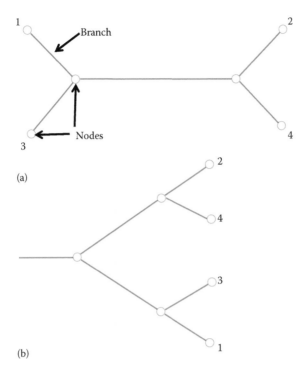

FIGURE 6.4 Phylogenetic tree depicting (a) unrooted tree and (b) rooted tree joining four taxonomic units. (Modified from Willey, J.M. et al., *Prescott's Microbiology*, 8th ed., McGraw-Hill, Singapore, 2011.)

Phylogenetic trees are central to the field of phylogenetics. In a phylogenetic tree, the tip of each branch represents the organism whose sequences have been analyzed. The tip of each branch is often called an OTU. The taxa joined together in the tree are implied to have descended from a common ancestor. The branching point (node) in a phylogenetic tree denotes a divergence event, while the length of the branches between the two nodes gives the number of molecular changes that have occurred during the evolutionary period.

The evolutionary relationships of organisms can be assessed by looking at their genome sequences. The fact that DNA is inherited from parents and is subject to change over time makes genomic analyses attractive as a means of determining genetic and hence evolutionary relatedness. In such analyses, homologous sequences from the genomes of representative specimens from different species and the results are displayed as sequences of the base letters A, C, G, and T, depicting how the four basic DNA components or bases (adenine, cytosine, guanine, and thymine) are arranged. For instance, if you have DNA sequences from representative specimens of different species, you can look for the differences between them. To illustrate this,

Microbe 1:   .....ACGTCATATTACCGAACTTA.....

Microbe 2:   .....ACATCACATTCCCGAACTTA.....

Microbe 3:   .....ACATTACATTCCCGAACTGA.....

FIGURE 6.5 Nucleotide sequences in three microbes and their pairwise comparison.

the sequences of three microbes are aligned and a pairwise comparison is made (Figure 6.5).

From Figure 6.5, it is clear that the DNA sequences of the three microbes are quite similar, with the three mismatches between the sequences of microbes 1 and 2 shown in red, and the two mismatches between the sequences of microbes 2 and 3 shown in blue. Thus, it can be inferred that with regard to this section of DNA, microbe 1 is closer to microbe 2 than it is to microbe 3, as it takes three changes to transform the sequence of microbe 1 into the sequence of microbe 2, but five changes to transform it into the sequence of microbe 3. This immediately suggests the possibility of defining a measure of distance between two sequences based on the number of differences in their bases. Therefore, as depicted, it would seem a natural extension to use the distances between sequences as proxies for measures of evolutionary relatedness between species. Therefore, in the above example, microbe 1 would be evolutionarily closer to microbe 2 than to microbe 3.

Phylogenetic trees consisting of a nontrivial number of input sequences are constructed using several computational phylogenetic methods. The two main categories are phenetic methods and cladistic methods. The phenetic methods, such as neighbor joining and the unweighted pair group method with arithmetic mean (UPGMA), are distance-based methods that measure the pairwise differences among sequences and build the tree totally from the resultant distance matrix. In contrast, the cladistic methods are character-based methods where all topologies are evaluated and the one that is chosen optimizes the evolution. The most commonly used cladistic methods include maximum parsimony and maximum likelihood. Phylogenetic tree construction methods can be assessed on the basis of several criteria, including efficiency, power, consistency, robustness, and falsifiability.

The prerequisite in constructing a phylogenetic tree is the choice of molecular markers. Either nucleotide or protein sequence data can be chosen, depending on the properties of the sequences and the purposes of the study. For evaluating very closely related organisms, a nucleotide sequence can be used, whereas for more widely divergent groups of organisms, a slowly evolving nucleotide sequence (rRNA) or protein sequence can be used. However, in most cases, protein sequences are preferable, as they are more conserved than codon-biased nucleotide sequences. There are five steps in constructing a phylogenetic tree. The first step is the alignment, which is also the most critical step in the procedure. The alignment can be done using online resources, such as ClustalW and T-Coffee. Only the correct alignment produces the phylogenetic tree. In this context, the program GBlocks can help to detect and eliminate the poorly aligned positions and divergent regions. The second step involves the examination of multiple alignment—whether it reflects the evolutionary process and is appropriate to continue with the tree building. Phylogenetic analysis can only be performed with those sequences that display a mixture of random and matched positions. The third step in the construction of a phylogenetic tree is to carefully choose the tree building method, as outlined above, that is, phenetic and cladistic methods. Calculate the probability of all possible topologies for each data partition. The fourth step involves combining data partitions and the application of the method performed by a computer. The last step is to manually examine the resulting tree with the highest overall probability at all partitions as the most likely phylogeny.

Furthermore, a phylogenetic tree may be unrooted or rooted. An unrooted tree (Figure 6.4a) shows the phylogenetic relatedness of nodes without depicting ancestry at all. It is clear from Figure 6.4a that organism 1 is more closely related to organism 3 than either organism 2 or 4, but it does not specify their common ancestor or the direction of change. The best way to root a tree is to use an outgroup (a species known to be very distantly related to all the species in the tree) genome and determine the point of the tree where the outgroup joins. On the contrary, a rooted tree (Figure 6.4b) always provides an evolutionary path and shows the development of the four species from this root. Each node with descendents represents the inferred most recent common ancestor of the descendents. Internal nodes are generally denoted as hypothetical taxonomic units, as they cannot be directly observed.

Although phylogenetic trees constructed on the basis of gene sequences or genomic data in various species offer evolutionary relationships, they have certain limitations:

1. First and foremost, they do not necessarily accurately represent the evolutionary history of the included taxa. In fact, they are literally scientific hypotheses that are subject to falsification by further studies.
2. The data on which they are based is noisy; that is, the analysis can be muddled by additional studies involving genetic recombination, horizontal gene transfer, and hybridization between species that were not nearest neighbors on the tree before hybridization took.
3. When using a single type of characteristic, for example, a single gene, a protein, or only morphological analysis while building a tree, great caution is required in inferring phylogenetic relationships among species because such trees constructed from another unrelated data source often differ from the first.
4. When extinct species are included in a tree, they represent the terminal nodes, as it is unlikely that they are direct ancestors of any extant species. Thus, skepticism might occur when extinct species are included in the trees that are wholly or partly based on DNA sequence data, as no DNA sequences long enough for use in phylogenetic analysis have yet been retrieved from material more than 1 million years old.

6.4.3 PHYLOGENY AND SYSTEMATICS

In the last two decades, molecular phylogenetics and phylogenomics have drastically revised and enriched fungal systematics by comparing single or multiple orthologous gene regions. The earlier traditional fungal phylogenetic systems based on morphology, physiology, and sexual states attracted several criticisms of their cladistic relationships. Presently, the most commonly accepted taxonomic scheme for fungi is based on 18s rDNA gene phylogeny, which arranges all fungi in a hierarchical taxonomic structure, from domain to species. However, studies based on multigene concatenation or whole genome analysis as revealed by several studies (Fitzpatrick et al., 2006; Hibbett et al., 2007; Wang et al., 2009) pave the way for constructing robust phylogenies for fungi. Gene concatenation attempts to maximize the informativeness and explanatory power of the characteristic data used in the analysis and has been justified on philosophical grounds. As more genes provide more phylogenetic information, a number of recent phylogenomic studies have attempted to infer the phylogenies of fungal organisms by combining large data sets of aligned genes. For example, Fitzpatrick et al. (2006), Hibbett et al. (2007), and Wang et al. (2009) built fungal phylogenies from large data sets of protein-coding genes of 42, 195, and 82 genomes, respectively. As a result of such studies, a huge database of their gene sequence data has been constructed and is available in the public domain. Further, *Ainsworth & Bisby's Dictionary of the Fungi* (Kirk et al., 2008) is one such taxonomic work that will influence fungal taxonomy in the twenty-first century; it contains a comprehensive kingdom-wide classification down to the level of genus. This dictionary contains more than 21,000 entries and provides the most complete listing available of genetic names of fungi, their families and orders, their attributes, and descriptive terms.

The recent trend in solving fungal systematic problems is to use the online fungal taxonomies that are proliferating the Internet. The following are some of the important online general classifications of fungi:

1. GenBank (www.ncbi.nlm.nih.gov/Taxonomy), a resource for a diverse community of researchers, including ecologists and molecular biologists
2. Tree of Life (www.tolweb.org/tree), which is widely used by teachers and students for basic studies and research
3. Myconet (www.fieldmuseum.org/myconet), a site on which the ascomycetous fungi are regularly updated
4. Index Fungorum (www.indexfungorum.org)
5. MycoBank (www.mycobank.org)

The integration of these and many more databases, particularly for taxonomic names, is likely to gain greater prominence for future fungal taxonomies. Further, if the classifications employed by these and other major taxonomic resources could be unified, this would promote communication and the awareness of fungal phylogeny, and also provide a framework for future revisions at all taxonomic levels. Therefore, classification of fungi based on monophyletic groups that can be recommended for general use is greatly warranted.

With the taxonomic advances in fungal nomenclature as supported by several phylogenetic and genome-wide studies, thermophilic fungi are described as being composed of 20 genera and 44 species (Oliveira et al., 2015). Using phylogenetic analysis, many taxonomic confusions regarding positions of taxa of a particular class or genus have recently been solved. For example, Morgenstern et al. (2012), while sequencing 86 fungal genomes, devised the phylogenetic classification of 22 thermophilic or thermotolerant fungi. In their classification scheme based on molecular phylogeny regarding uncertain taxonomic positions of some thermophilic fungi, they placed *Myriococcum thermophilum* and *Scytalidium thermophilum* into the Sordariales and *Thermomyces lanuginosus* in Eurotiales. Similarly, *Malbranchea cinnamomea* is assigned to Onygenales and *Calcarisporiella thermophila* belongs to Mucorales. Accordingly, the known thermophilic fungi appear to be limited to the orders Sordariales, Eurotiales, and Onygenales in Ascomycota and the Mucorales in the Mucoromycotina (basal fungi). Presently, no member of the thermophilic fungi appears to belong to the Basidiomycota.

The increased use of thermophilic fungi in biotechnological application requires further attention to their correct classification and nomenclature, a long pending problem that has been poorly addressed by fungal taxonomists. Further, as these fungi are an important source of thermostable enzymes, structural studies about their thermal stability and amenability to genetic manipulation will add to the existing knowledge of this important ecological group.

6.4.4 THERMOPHILIC FUNGAL GENOMES

Until now, a few hundred fungal genomes have been sequenced, including important human pathogens, plant pathogens, and model organisms (Zhou et al., 2014). The increase in fungal genome sequencing offers an opportunity to reconstruct evolutionary events using whole genomes. The genomes of several industrially useful fungi, such as *Aspergillus niger* (Pel et al., 2007) and *Trichoderma reesei* (Martinez et al., 2008,) have also been sequenced. As thermophilic fungi represent a potential reservoir of thermostable enzymes, sequencing of their genomes is advantageous from a biotechnological perspective; besides, their genomes are amenable to manipulation using classic and molecular genetics (Berka et al., 2011). Genomes of some thermophilic fungi have been sequenced in the recent past. Characteristic features of genomes of some thermophilic fungi are presented in Table 6.1. The following section highlights the genomes of some important thermophilic fungi.

Aspergillus fumigatus is a ubiquitous thermotolerant fungus commonly known as fungal weed in laboratories. It acts as both primary and opportunistic pathogens, causing aspergillosis in humans. The human respiratory tract is constantly exposed to its conidia because of the organism's prolific nature of conidial production. *A. fumigatus* is isolated from almost any substrate, including human habitats. Further, in immunocompromised individuals, the incidence of invasive infection may be as high as 50% and the mortality rate is often about 50%. Increasing incidences of asthma and sinusitis are often linked to the interaction of *A. fumigatus* and other airborne fungi with the immune system. Keeping in view the significant burden of invasive disease caused by *A. fumigatus* and its obscure basic biology, Nierman et al. (2005)

TABLE 6.1

General Features of Fungal Genomes of Some Thermophilic and Thermotolerant Fungi

Organism	Optimum Growth Temperature (°C)	Genome Size (Mb)	No. of Coding Genes	G + C Content (%)	No. of Chromosomes	Reference
Aspergillus fumigatus	40	29.4	9,926	49.9	8	Nierman et al. (2005)
Aspergillus niger	30	33.9	14,165	50.40	–	Pel et al. (2007)
Aspergillus nidulans	30	30.1	10,701	50.3	–	http://www.aspgd.org/
Myceliophthora thermophila	40–50	38.7	9,110	51.4	7	Berka et al. (2011)
Rhizomucor miehei	35–45	27.6	10,345	43.83	–	Zhou et al. (2014)
Thermomyces lanuginosus	45–55	23.3	5,105	52.14	–	McHunu et al. (2013)
Thielavia terrestris	45–50	36.9	9,813	54.7	6	Berka et al. (2011)
Chaetomium thermophilum	45–55	28.3	7,227	–	8	Amlacher et al. (2011)

sequenced the genome of its clinical isolate Af293. The complete genome of the organism comprises a 29.4 megabase (mb) genome sequence consisting of eight chromosomes containing 9926 predicted genes. The authors further reported the microarray analysis, which revealed a temperature-dependent expression of distinct sets of genes. With the availability of genome sequences, future understanding, including the management of allergic reactions caused by this fungus, will prove a boon to the clinician and public at large.

Genomes of two thermophilic fungi, *Myceliophthora thermophila* and *Thielavia terrestris*, were described by Berka et al. (2011). Both these fungi are important producers of thermostable hydrolyzing enzymes, such as cellulases, xylanases, and pectinases. Further, they characterize the biomass-hydrolyzing activity of recombinant enzymes, which suggests that these organisms are highly efficient in biomass decomposition at both moderate and elevated temperatures. The finished 38.7 mb genome of *M. thermophila* and the 36.9 mb genome of *T. terrestris* consist of seven and six complete telomere-to-telomere chromosomes, respectively. Berka et al. reported that the two genomes are fairly similar in organization, with the major difference occurring in chromosome (Ch) 1 of *T. terrestris*, which harbors most of the genes located on Ch2 and Ch4 of *M. thermophila*. Further, the protein-coding fractions of the genomes include 9,110 genes in *M. thermophila* and 9,813 genes in *T. terrestris* (Table 6.1), which are smaller than the average proteomes of other fungi in the class Sordariomycetes and their closely related mesophilic counterpart, *Chaetomium globosum*, which has 11,124 predicted genes in its 34.9 mb genome (Broad Institute, 2005). While comparing the GC content of *M. thermophila* and *T. terrestris* with *C. globusum* and other species within the class Sordariomycetes, the authors reported that although the genomes of the thermophilic species have a slightly lower genome-level GC content than *C. globosum*, they have a higher GC content in coding regions, which is reflected in the third position of codons (GC3). As G:C pairs are more thermally stable, this may suggest the potential mechanism of adaptability of protein-coding genes to elevated temperatures.

The genome sequence of the thermophilic zygomycete *Rhizomucor miehei* has recently been reported by Zhou et al. (2014). *R. miehei* has been extensively exploited for the production of various thermostable enzymes. Using a shotgun sequenced approach, Zhou et al. (2014) reported that the size of the *R. miehei* genome is 27.6 mb, with 10,345 predicted protein-coding genes contained in 10 chromosomes. The average G + C content of the genome is 43.8%, which is higher than that of zygomycete fungi (average 35.3%) but lower than that of most ascomycete fungi, including three thermophilic ascomycetes, *Thermomyces lanuginosus* (52.14%), *Thielavia terrestris* (54.7%), and *Myceliophthora thermophila* (51.4%) (Table 6.1). As G:C contents of the organism reflect more thermal stability, it is fairly surprising that the G + C content of the thermophilic *R. miehei* whole genome (43.8%) and its coding genes (47.4%) is close to that of the mesophilic ascomycetous fungi. However, the value is significantly higher than that of the mesophilic zygomycetes, such as *Rhizopus oryzae* (Zhou et al., 2014). Further, the genome information of *R. miehei* will facilitate future studies for us to better understand the mechanisms of fungal thermophilic adaptation and explore its potential in the industrial-scale production of thermostable enzymes.

The pioneering studies of Miehe on thermophilic fungi in 1907 led to the isolation of *Thermomyces lanuginosus*, which is explicitly associated with the composting of agricultural wastes, manure of mammals and bird droppings, and municipal refuge, where decomposition by mesophilic microorganisms paves the way for its colonization. Because of its exceptional capability to degrade plant biomass and produce high titers of thermostable hydrolyzing enzymes, the fungus finds application in various biotechnological processes. Considering the organism's industrial applications, its whole genome was sequenced by McHunu et al. (2013) using Roche 454 and Illumina paired-end sequencing strategies. The whole genome sequence of *T. lanuginosus* strain SSBP consists of 23.3 Mb. The proteome that was predicted from this assembly had a total of 5105 genes, which is comparatively less than that of other filamentous fungi, as well as thermophilic fungi (Table 6.1). The GC content of the whole genome is estimated to be 52.14%, whereas coding regions have a significantly higher GC content (55.6%). In spite of its low number of coding genes compared with that of other thermophilic fungi, *T. lanuginosus* is well adapted to high temperatures. The mechanism of thermal adaptation by this fungus stems from the fact that it has several DNA-related pathways. *T. lanuginosus* possesses a ubiquitin degradation pathway that plays a critical role in the responses to various stressors, such as nutrient limitation, heat shock, and heavy metal exposure (Staszczak, 2008). For adaptation to elevated temperature during composting, ubiquitin degradation is essential. Moreover, *T. lanuginosus* is also capable of histone acetylation and deacetylation and poly-ADP ribosylation, in addition to containing a high number of methylases, which play an important role in the condensation and packing of DNA (Nowak and Corces, 2004). All these events and repairing mechanisms impart thermoprotection and help the organism to thrive at elevated temperatures.

The genome of *Chaetomium thermophilum*, originally isolated by La Touche in 1948, has been sequenced by Amlacher et al. (2011). This fungus holds great promise for structural biology, has the exceptional ability to grow up to 60°C, and has been a subject of study for its nuclear pore complex (NPC). The fungus is known to produce a variety of thermostable enzymes, such as cellulose, amylase, xylanase, and lipase, and thus is a good candidate for biotechnological explorations. Using Roche 454 FLX and XLR platforms, Amlacher et al. (2011) sequenced a nuclear genome of 28.3 Mb contained in eight chromosomes. They identified a total of 7227 protein-coding genes. Later, Bock et al. (2014) further described the genome-wide proteome and transcriptome analysis of *Chaetomium thermophilum*. The refinement of the gene structure by their transcriptomics and proteomics approach in particular ensures individual gene expression studies and subsequent experimental characterizations. Their study implied that the individual proteins are the basis for the adaptation of a thermophilic lifestyle in thermophilic fungi.

Although genome sequencing of thermophiles has been initiated only recently, several of the thermophilic fungal genomes have not been sequenced to date. Genomic features of a few other thermotolerant and thermophilic fungi are presented in Table 6.1 with relevant references. Additionally, a variety of thermophiles may serve as model organisms that may help us to understand the possible mechanism of genomic diversification or environmental adaptation, which could be a driving force for the microbial evolution and associated thermophily.

6.5 FUTURE PROSPECTS

Currently, the 18s rDNA gene-based systematic scheme appears to be justified as a starting point for puckering the ideal classification based on natural and evolutionary relationships. As thermophiles are believed to share some characteristics with early life-forms, their further analysis would also play a critical role in the presumptions of these phylogenetic issues. Historically, the research and development efforts on thermophilic fungi have been focused on the identification and characterization of their enzymes from relatively few strains. The characterization of genes expressing these enzymes has largely been ignored for decades. This approach has produced advanced enzymes over time; however, recent studies based on genomic investigation have almost instantaneously yielded a diverse palette of novel, thermostable, high-efficiency gene products that can be combined and coordinated to improve existing enzyme concoctions or generate concoctions *de novo*.

Thermophilic fungi are also likely to have distinctive but as yet mostly uncharacterized roles in the global carbon cycle, as they possess optimal growth temperatures ($45°C–55°C$) well above those of the vast majority of microorganisms. Recently, researchers have been interested in developing a phylogenomic program to investigate the molecular basis of fungal thermophily (Bock et al., 2014) and to provide an empirical framework for linking genotype and phenotype to understand the molecular basis of fungal thermophily. The experimental findings of Bock et al. (2014) fit well with recent bioinformatics and deep sequencing studies, which also conclude that changes in protein primary structure lead to the thermostability of proteins and not the differential expression of thermoinducible genes. As genome annotations of several thermophilic fungi are available (Table 6.1), studies on the genomes of remaining thermophiles will further demonstrate their complex molecular structure. Further, various biotechnologically important hydrolytic enzymes from thermophilic fungi need to be cloned and expressed in homologous and heterologous systems for large-scale applications in various industries.

REFERENCES

Amlacher, S., Sarges, P., Flemming, D., van Noort, V., Kunze, R., Devos, D.P., Arumugam, M., Bork, P., and Hurt, E. 2011. Insight into structure and assembly of the nuclear pore complex by utilizing the genome of a eukaryotic thermophile. *Cell* 146:277–289.

Berka, R.M., Grigoriev, I.V., Otillar, R., Salamov, A., Grimwood, J., Reid, I., Ishmael, N. et al. 2011. Comparative genomic analysis of the thermophilic biomass-degrading fungi *Myceliophthora thermophila* and *Thielavia terrestris*. *Nature Biotechnology* 29: 922–927.

Bock, T., Chen, W.H., Ori, A., Malik, N., Silva-Martin, N., Huerta-Cepas, J., Powell, S.T. et al. 2014. An integrated approach for genome annotation of the eukaryotic thermophile *Chaetomium thermophilum*. *Nucleic Acids Research* 42(22):13525–13533.

Broad Institute. 2005. *Chaetomium globosum* Genome Database. http://www.broadinstitute .org/annotation/genome/chaetomium_globosum.

Cooney, D.G., and Emerson, R. 1964. *Thermophilic Fungi: An Account of Their Biology, Activities, and Classification*. San Francisco: W.H. Freeman and Company.

Crisan, E.V. 1964. Isolation and culture of thermophilic fungi. *Contributions from Boyce Thompson Institute* 22:291–301.

Darwin, C.R. 1859. *On the Origin of Species by Means of Natural Selection, or the Preservation of Favoured Races in the Struggle for Life*. 1st ed., issue 1. London: John Murray.

Fitzpatrick, D.A., Logue, M.E., Stajich, J.E., and Butler, G. 2006. A fungal phylogeny based on 42 complete genomes derived from supertree and combined gene analysis. *BMC Evolutionary Biology* 6:99; doi: 10.1186/1471-2148-6-99.

Gianfreda, L., and Rao, M.A. 2004. Potential of extra cellular enzymes in remediation of polluted soils: A review. *Enzyme and Microbial Technology* 35:339–354.

Hibbett, D.S., Binder, M., Bischoff, J.F., Blackwell, M., Cannon, P.F., Eriksson, O.E., Huhndorf, S., and James, T. 2007. A higher-level phylogenetic classification of the fungi. *Mycological Research* 111:509–547.

Kirk, P., Cannon, P.F., Minter, D.W., and Stalpers, J.A. 2008. *Ainsworth & Bisby's Dictionary of the Fungi*. 10th ed. Wallingford, UK: CAB International.

Kluge, A.G. 1989. A concern for evidence and a phylogenetic hypothesis of relationships among epicrates (Boidae, Serpentes). *Systematic Zoology*, 38 (1): 7–25.

Liu, Y.J., Whelen, S., and Hall, B.D. 1999. Phylogenetic relationships among ascomycetes: Evidence from an RNA polymerase II subunit. *Molecular Biology and Evolution* 16:1799–1808.

Maheshwari, R., Bharadwaj, G., and Bhat, M.K. 2000. Thermophilic fungi: Their physiology and enzymes. *Microbiology and Molecular Biology Reviews* 64:461–488.

Martinez, D., Berka, R.M., Henrissat, B., Saloheimo, M., Arvas, M., Baker, S.E., and Chapman, J. 2008. Genome sequencing and analysis of the biomass-degrading fungus *Trichoderma reesei* (syn. *Hypocrea jecorina*). *Nature Biotechnology* 26:553–560.

McHunu, N.P., Permaul, K., Rahman, A.Y.A., Saito, J.A., Singh, S., and Alam, M. 2013. Xylanase superproducer: Genome sequence of a compost-loving thermophilic fungus, *Thermomyces lanuginosus* strain SSBP. *Genome Announcements* 1:e00388-13.

McNeill, J., Barrie, F.R., Buck, W.R., Demoulin, V., Greuter, W., Hawksworth, D.L., Herendeen, P.S., Knapp, S., Marhold, K., Prado, J., Prud'homme van Reine, W.F., Smith, G.F., Wiersema, J.H., and Turland, N.J. 2012. International Code of Nomenclature for algae, fungi, and plants (Melbourne Code). Adopted by the Eighteenth International Botanical Congress Melbourne, Australia, July 2011. Koeltz, Konigstein.

Mora, C., Tittensor, D.P., Adl, S., Simpson, A.G.B., and Worm B. 2011. How many species are there on earth and in the ocean? *PloS Biology* 9:e1001127. doi: 10.1371/journal.pbio .1001127.

Morgenstern, I., Powlowski, J., Ishmael, N., Darmond, C., Marqueteau, S., Moisan, M.C., Quenneville, G., and Tsang, A. 2012. A molecular phylogeny of thermophilic fungi. *Fungal Biology* 116:489–502.

Mouchacca, J. 2000. Thermophilic fungi and applied research: A synopsis of name changes and synonymies. *World Journal of Microbiology and Biotechnology* 16:881.

Nieremberg I.E. 1635. *Historia naturae, maxime peregrinae, libris xvi*. Antwerp: Plantin Office.

Nierman, W.C., Pain, A., Anderson, M.J., Wortman, J.R., Kim, H.S., Arroyo, J., and Berriman, M. 2005. Genomic sequence of the pathogenic and allergenic filamentous fungus *Aspergillus fumigatus*. *Nature* 438:1151–1156.

Nowak, S.J., and Corces, V.G. 2004. Phosphorylation of histone H3: A balancing act between chromosome condensation and transcriptional activation. *Trends in Genetics* 20: 214–220.

Oliveira, T.B., Gomes, E., and Rodrigues, A. 2015. Thermophilic fungi in the new age of fungal taxonomy. *Extremophiles* 19:31–37.

Pan, W.Z., Huang, X.W., Wei, K.B., Zhang, C.M., Yang, D.M., Ding, J.M., and Zhang, KQ. 2010. Diversity of thermophilic fungi in Tengchong Rehai National Park revealed by ITS nucleotide sequence analyses. *Journal of Microbiology* 48:146–152.

Pel, H.J., de Winde, J.H., Archer, D.B., Dyer, P.S., Hofmann, G., Schaap, P.J., Turner, G. et al. 2007. Genome sequencing and analysis of the versatile cell factory *Aspergillus niger* CBS 513.88. *Nature Biotechnology* 25:221–231.

Rajendhran, J., and Gunasekaran, P. 2011. Microbial phylogeny and diversity: Small subunit ribosomal RNA sequence analysis and beyond. *Microbiological Research* 166:99–110.

Sigoillot, C., Camarero, S., Vidal, T., Record, E., Asther, M., Pérez-Boada, M., Martínez, M.J., Sigoillot, J.C., Asther, M., Colom, J.F., and Martínez, A.T. 2005. Comparison of different fungal enzymes for bleaching high-quality paper pulps. *Journal of Biotechnology* 115:333–343.

Staszczak, M. 2008. The role of the ubiquitin-proteasome system in the response of the lignolytic fungus *Trametes versicolor* to nitrogen deprivation. *Fungal Genetics and Biology* 45:328–337.

Straatsma, G., Samson, R.A., Olijnsma, T.W., Op Den Camp, H.J., Gerrits, J.P., and Van Griensven, L.J. 1994. Ecology of thermophilic fungi in mushroom compost, with emphasis on *Scytalidium thermophilum* and growth stimulation of *Agaricus bisporus* mycelium. *Applied and Environmental Microbiology* 60:454–458.

Turner, B.M. 2007. Defining an epigenetic code. *Nature Cell Biology* 9:2–6.

Wang, H.C., Moore, M.J., Soltis, P.S., Bell, C.D., Brockington, S.F., Alexandre, R., Davis, C.C., Latvis, M., Manchester, S.R., and Soltis, D.E. 2009. Rosid radiation and the rapid rise of angiosperm-dominated forests. *Proceedings of the National Academy of Sciences of the United States of America* 106:3853–3858.

Wesenberg, D., Kyriakides, I., and Agathos, S.N. 2003. White-rot fungi and their enzymes for the treatment of industrial dye effluents. *Biotechnology Advances* 22:161–187.

Willey, J.M., Sherwood, L.M., and Woolverton, C.J. 2011. *Prescott's Microbiology*. 8th ed. Singapore: McGraw-Hill.

Zhou, P., Zhang, G., Chen, S., Jiang, Z., Tang, Y., Henrissat, B., Yan, Q., Yang, S., Chen, C.-F., Zhang, B., and Du, Z. 2014. Genome sequence and transcriptome analyses of the thermophilic zygomycete fungus *Rhizomucor miehei*. *BMC Genomics* 15:294. doi: 10.1186/1471-2164-15-294.

7 Biodiversity and Taxonomic Descriptions

7.1 INTRODUCTION

Based on molecular taxonomic investigations, the major taxonomic divisions or phyla of fungi are continuously evolving. Thus, the most current classification of fungi utilizes recent molecular, multigene, phylogenetic studies and characteristics of their sexual and asexual reproductive structures, as well as spores to separate or delimit the groups. Environmental conditions on the earth with reference to air, land, or water vary considerably, affecting the growth and survival of living organisms. However, despite these variations, certain microorganisms, including fungi, cope with these adversities and can be found virtually inhabiting the most extreme environments. Fungal organisms have specific requirements for their survival and growth, and if any of these requirements are not met, or if they are significantly reduced or accelarated, their growth or survival is under threat. One such factor is the temperature, which is critical not only for fungi but also for all living organisms. At elevated temperature (55°C–60°C), only limited species of fungi are able to not only survive but also flourish at such an extreme condition. These fungi are referred to as thermophilic, as they are near the limits of cell functioning that limit enzyme activities or damage biomolecules (Rothschild and Mancinelli, 2001).

Due to climate change, global warming is becoming more eminent and accepted. It is further aggravated by human industrial activities, which at least offers the hope that it can be abridged by humans, in contrast with natural phenomena. The recent Paris Agreement on climate change restricts increases to a maximum of 2°C, and rapid, significant action is anticipated by many (Paterson and Lima, 2017). Nevertheless, climate change implies increases in temperature. Thus, more fungi that would tolerate or prefer high temperature can be expected to evolve. Occasionally, some organisms are able to not only withstand a single condition that would be harmful to most forms of life, but also tolerate multiple factors, which confer them the title of polyextremophiles. For example, a microorganism that lives in extremely hot and dry conditions would be referred to as a thermophilic xerophile. By researching extremophiles or, more specifically, the thermophile, as the subject of this text, we expand our understanding of the biodiversity and taxonomy of fungi that allow them to survive the hostile conditions in which they live.

Several thermophilic fungi have been isolated and identified during the last five decades, but their identification remains obscure or has not been carried out authentically. Compounding this unfortunate situation is that the ecologically important traits, such as cardinal temperature for growth and the ability to degrade particular substrates, differ significantly among species that are culturally and morphologically similar enough to be confused. The thermophilic fungi mostly belong to Zygomycetes, Ascomycetes, and Deuteromycetes (anamorphic fungi). No Myxomycetes or Basidiomycetes have been reported as a true thermophile. Of the currently less than 50 truly thermophilic fungi, more than 50% belong to Ascomycetes. While collecting information on the taxonomy of thermophilic fungi, we faced difficulties on account of the confusing nomenclature of these fungi. This confusion was compounded for several reasons: (1) the early taxonomic literature is scattered and often in languages other than English, (2) some species have been described repeatedly under different names, (3) the practice of interchangeably using the names of the sexual (teleomorph) and asexual (anamorph) stages of the same fungus, and (4) there are several instances in the literature of misidentifications of thermophilic fungi.

The terminology used for conidia and conidiophores of anamorphic fungi (Deuteromycetes) in this chapter is mainly that recommended by the proceedings of the Kananaskis conference (Kendrick, 1971). *Patterns of Development in Conidial Fungi*, by Cole and Samson (1979), was helpful in our studies of some of the hyphomycetous fungi. The latest classification of fungi, as outlined in *Ainsworth & Bisby's Dictionary of Fungi*, by Kirk et al. (2008), follows. The taxonomic descriptions of taxa included in this text are largely based on our previous publication (Salar and Aneja, 2007).

The reported thermophilic species in this text are grouped according to broad taxonomic categories. As observed by Mouchacca (2007), the nomenclature of zygomycetous taxa and anamorphic fungi is straightforward, as usually only one binomial is available or only one state is produced in culture, respectively. In contrast, for Ascomycetes regularly producing a conidial state in culture, the name of the sexual state (teleomorph) should be used to designate the organism even when a binomial is available for the anamorph; this prevents the practice of interchangeably using the name of either state of the same fungus (Mouchacca, 2007). When ascomycetous taxa produce the anamorph regularly and the teleomorph only under specific cultural conditions, the name of the anamorph could be preferentially selected. The goal is to introduce uniformity in name citations of fungi (one fungus, one name), particularly in the literature of applied research. In this text, maximum efforts were undertaken to trace updated information on the taxonomic position of these less than 50 strict thermophilic species. For each, information on the type of material, morphological features distinguishing it from related members of the genus, and salient ecological features, such as habitat and geographic distribution, is presented.

As noted, various aspects of the taxonomy of thermophilic fungi remain to be worked out. The difficulty encountered in properly classifying them, especially the deuteromycetous fungi, is that the genera involved are in need of revision. However, the number of known thermophiles is still so few that once their thermophilic character has been established, it will be a relatively easy matter, even for nonmycologists, to identify future isolates by means of the simple key provided here.

7.2 KEY TO THE IDENTIFICATION OF THERMOPHILIC FUNGI

7.2.1 ZYGOMYCOTA

1 Mitospores endogenous, formed in sporangia; zygospores formed by hyphal conjugation; saprobic or, if parasitic or predaceous, having mycelium immersed in the host tissue	**Zygomycetes**	**2**
2 Sporangiophore arising from trophocyst; sporangia multi-spored, thallus mycelial	**Mucorales**	**3**
3 Sporangia columellate, specialized sporangiola absent; zygospores smooth to warty, borne on opposed, tong-like or apposed, naked, or appandaged suspensors; polyphyletic	**Mucoraceae**	**4**
4a Sporangia (sub)globose; short apophysis may be present; sporangiophores unbranched, originating from distinct rhizoids; sporangiospores often striate or angular	*Rhizopus*	
I Sporangiospores distinctly striate	*Rhizopus microsporus*	
II Sporangiospores inconspicuously striate	*Rhizopus rhizopodiformis*	
b Sporangia (sub)globose, apophysis absent; sporangiophores branched, not originating from distinct rhizoids; sporangiospores not striate or angular	*Rhizomucor*	
Sporangiospores dumbbell-shaped	*Rhizomucor nainitalensis*	
I Zygospores always present at 37°C; homothallic	*Rhizomucor miehei*	
II Zygospores absent unless compatible matings are made; heterothallic	*Rhizomucor pusillus*	

7.2.2 ASCOMYCOTA

1 Meiospores endogenous, formed in asci by free cell formation	**Ascomycetes**	**2, 10**
2 Ascomata cleistothecial, composed of flattened, hyaline to pale brown hyphae; asci scattered, evanescent at a very early stage; ascospores ellipsoid, smooth, hyaline; anamorph composed of simple basipetal thallic or blastic chains of thick-walled conidia	**Eurotiales**	**3**
3 Ascomata brightly colored, composed of thick-walled pseudoparenchymatous tissue; asci often formed in chains; ascospores usually yellowish, usually bivalvate, sometimes ornamented	**Trichocomaceae**	**4, 7**
4 Ascomata mostly confluent, crust-like, reddish	*Thermoascus, Dactylomyces, Coonemeria*	**5**
5a Anamorph consisting of chlamydospores	*Thermoascus aurantiacus*	
b Anamorph *Paecilomyces* or *Polypaecilum*		**6**
6a **I** Anamorph *Paecilomyces crustaceus*	*Coonemeria crustacea*	
II Anamorph *Paecilomyces aegyptiaca*	*Coonemeria aegyptiaca*	

b Anamorph *Polypaecilum*　　　　　　　　　　　*Dactylomyces thermophiles*

7 Ascomata globose or subglobose, pale gray　　*Rasamsonia, Talaromyces*　　　**8**

8a Anamorph *Penicillium emersonii*; ascomata　　*Rasamsonia emersonii*
　　abundantly produced on agar media, ascospore
　　3.5–4 µm, subglobose to ovoid, yellow

b Ascospore size 2–3.8 µm, globose, hyaline to yellow　*Rasamsonia composticola*

c Anamorph *Paeciolomyces byssochlamydoides*;　　*Rasamsonia*
　　　　　　　　　　　　　　　　　　　　　　　　　byssochlamydoides

d Anamorph *Penicillium duponti*; ascomata rarely　　　　　　　　　　　　　　**9**
　　produced on agar media unless established
　　cultures are flushed with N$_2$

9a Asci subglobose, scattered　　　　　　　　　*Talaromyces duponti*

b Asci globose to subglobose, produced in short chains　*Talaromyces thermophilus*

10 Ascomata perithecial or cleistothecial, thick-　　**Sordariales**　　　**11, 20, 23, 24**
　　walled, fragmenting into well-defined plates;
　　asci persistent or evanescent; ascospores brown

11 Ascomata often hairy, ostiolate, or nonostiolate;　**Chaetomiaceae**　　　**12, 17**
　　ascospores usually small, with minute germ pores

12 Ascomata perithecial　　　　　　　　　　　*Chaetomium*　　　　**13, 15, 16**

13 Perithecia with terminal hairs dichotomously　　*Chaetomium thermophile*　　**14**
　　branched; ascospores subglobose, 7–8 µm long

14a Perithecia gregarious　　　　　　　　　　*Chaetomium thermophile*
　　　　　　　　　　　　　　　　　　　　　　　var. *coprophile*

b Perithecia scattered, only occasionally gregarious　*Chaetomium thermophile*
　　　　　　　　　　　　　　　　　　　　　　　var. *dissitum*

15 Perithecia with terminal hairs ribbon-like, some-　*Chaetomium senegalensis*
　　what twisted undulate; ascospores ovate or clavate,
　　bilaterally flattened, 9–11 × 7–8 × 6.7 µm

16a Perithecia with terminal hairs irregularly branched　*Chaetomium virginicum*
　　or suggestive of dichotomous branching; ascospores
　　almond-shaped, 9–11 µm long

b Asci clavate, ascospores 5.5–7.8 × 5.2–6.3 µm　*Chaetomium*
　　　　　　　　　　　　　　　　　　　　　　　mesopotamicum

c Asci club-shaped, ascospores 19–24 × 11–14 µm　*Chaetomium britanicum*

17 Ascomata cleistothecial; ascospores ellipsoidal or ovoid　*Thielavia, Chaetomidium*　　**18**

18a Peridium pseudoparenchymatous; ascoma wall brown　*Thielavia minor*

b Peridium composed of textura epidermoidea　　　　　　　　　　　　　　**19**

19a Ascoma wall light brown; asci globose, 10 µm　*Thielavia australiensis*

b scoma wall brownish to black; asci oval to　　*Thielavia terricola*
　　pyriform, 16–19 × 25–35 µm

c Ascomatal wall pseudiparenchymatous　　　*Chaetomidium pingtungium*

20 Ascomata not hairy, or with setae surrounding　**Ceratostomataceae**　　**21**
　　the ostiole; ostiolate or nonostiolate; ascospores
　　usually large, with conspicuous germ pores

21 Ascomatal wall composed of dark, flattened irregular, thick-walled cells; ascospores with two apicaf germ pores; anamorph *Myceliophthora* — *Corynascus* — 22

22a Heterothallic; ascospores 19–28 µm — *Corynascus thermophilus*

b Homothallic; ascospores 10.9–19 × 6–10 µm — *Corynascus sepedonium*

23 Ascomata nonstiolate, wall glabrous composed of polygonal cells, dark brown — **Familiae incertae sedis**

a Ascospores globose or elliptical, with a single apiculus, 10–15 µm — *Melanocarpus albomyces*

b Ascospores ovoid 7.5–9 × 6–7.5 µm — *Melanocarpus thermophilus*

24 Ascomata cleistothecial, ascoma wall made up of angular dark cells — **Microascaceae**

Ascospores greenish brown, with a subapical germ pore, 14.0–18.0 × 7.5–10.0 µm — *Canariomyces thermophila*

7.2.3 DEUTEROMYCETES (ANAMORPHIC FUNGI)

1 Asexual reproduction usually by conidia; teliomorph absent — **Deuteromycetes** — 2

2a Conidia absent; bulbils present — *Papulospora thermophila*

b Conidia present — 3

3a Conidiophores aggregated into synnematous conidiomata, with slimy conidial heads — *Remersonia thermophila*

b Conidiophores thick-walled — *Acremonium*

　I Conidia pyriform smooth 3–6 × 1–1.5 µm — *Acremonium albamensis*

　II Conidia ellipsoidal 3–4 × 1.3–1.7 µm — *Acremonium thermophilum*

c Conidiophores not aggregated into synnematous conidiomata, not with large slimy heads — 4

4a Conidia (endo)arthric — *Malbranchaea cinnamomea*

b Conidia not arthric — 5

5a Conidia hyaline, not darkly pigmented — *Chrysosporium*

　I Conidia aleurosporic, borne terminally or laterally on the conidiogenous cells, pyriform to clavate, rough — *Chrysosporium tropicum*

　II Conidia formed terminally or laterally from almost all parts of the aerial mycelium on conspicuous relatively narrow pegs, globose or elliptical, smooth — *Myceliophthora fergusii*

b Conidia hyaline to pale brown, produced singly or in short chains on ampulliform swellings, rough-walled in fresh isolates — *Myceliophthora thermophila*

c Conidia verrucose to spinulose — *Myceliophthora hinnulea*

d Conidia brown to black — 6

6a Conidia only in chains — *Scytalidium*

I Conidia smooth, globose or oval-shaped	*Scytalidium thermophilum*
II Conidia thick-walled, ellipsoid to barrel-shaped	*Scytalidium indonesicum*
b Conidia mostly solitary, rarely in short chains	7
7a Conidia stellate	*Arthrinium pterospermum*
b Conidia not stellate	8
8a Conidia verrucose	*Thermomyces lanuginosus*
b Conidia smooth	9
9 Intercalary chlamydospores absent	*Thermomyces ibadanensis*

7.3 TAXONOMIC DESCRIPTIONS OF THERMOPHILIC TAXA

7.3.1 ZYGOMYCETES

Rhizomucor miehei (Cooney and Emerson) Schipper (1978) (Figure 7.1a–c)
Syn. = *Mucor miehei* Cooney and Emerson (1964)
Description: Heavy colony-bearing numerous sporangia on YpSs agar at 35°C–45°C. Colonies compact, at first white, later gray brown to beige brown. *Sporangiphores* 8 µm in width, loosely sympodially branched. *Sporangia* spherical, 30–60 µm in diameter, walls echinulate. *Columellae* spherical to oval, 20–45 µm in diameter. *Sporangiophores* nearly spherical, colorless, 4–6 × 3–5 µm. *Zygospores* in abundance, subspherical, warty, reddish brown or yellow when young, blackish at maturity, 30–50 µm in diameter, produced on homothallic mycelium.
Habitat diversity: On retting guayule, soil and sand, coal mines, hay, stored barley, and compost.
Geographic distribution: United States, India, Ghana, United Kingdom, and Saudi Arabia.
Description: Based on Schipper (1978).

Rhizomucor nainitalensis Joshi (1982)
Description: Growth at 48°C is very rapid, filling half of a petri dish in 2 days. At 38°C, "the growth of the mycelium takes place after three days but about one week is required to colonize the culture medium in a petri dish at 25°C" (Joshi 1982). *R. nainitalensis* appears very close to *Rhizomucor miehei*. It differs from *R. miehei* and *Rhizomucor pusillus* mainly by sporangiospores of varying shapes and sizes. *Sporangiospores* subglobose, ellipsoidal, oblong, reniform, dumbbell-shaped, 3–6 µm or more wide.
Habitat diversity: Decomposed oak log.
Geographic distribution: India.
Description: Based on Joshi (1982).

Rhizomucor pusillus (Lindt) Schipper (1978) (Figures 7.1d–f and 7.2a and b)
Syn. = *Mucor pusillus* Lindt (1886)
= *Mucor septatus* Bezold in Siebenmann (1889)
= *Rhizomucor parasiticus* Lucet & Cost. (1899)
= *Rhizomucor septatus* (Bezold) Lucet & Cost. (1901)
= *Rhizopus parasiticus* (Lucet & Cost.) Lendner (1908)
= *Mucor buntingii* Lendner (1930)
= *Mucor hagemii* Naumov (1935)

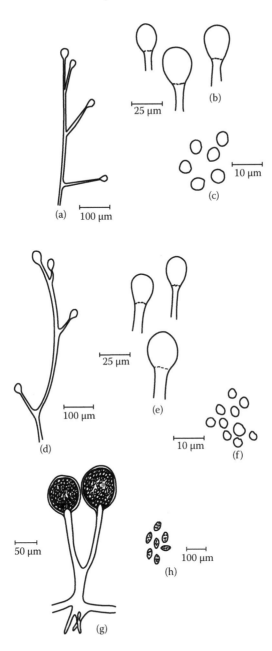

FIGURE 7.1 (a–c): *Rhizomucor mieheei*: (a) sporangiophores, (b) columellae, and (c) sporangiospores. (d–f): *Rhizomucor pusillus*: (d) sporangiophores, (e) columellae, and (f) sporangiospores. (g and h): *Rhizopus rhizopodiformis*: (g) sporangiophores with rhizoidal system and sporangia and (h) sporangiospores. ([a–f] redrawn from Schipper, M.A.A., *Stud. Mycol.*, 17, 53–71, 1978; [g and h] redrawn from Thakre, R.P., and Johri, B.N., *Curr. Sci.*, 45, 271–273, 1976.)

Description: Colony growth on YpSs at 45°C is very fast; initially white, later becoming grayish with whitish margin, color changes to grayish black at maturity; mycelial turf 2–3 mm high. *Sporangiophores* sympodially branched, 10–15 μm in diameter, colorless to yellow brown. *Sporangia* 50–80 μm in diameter, spinulose, rupturing at maturity, columellae subglobose to slightly elongate, 15–35 μm in diameter and up to 60 μm long. *Sporangiospores* colorless, globose to subglobose, 3–5 μm in diamete. Chlamydospores absent. *Zygospores* produced in compatible (heterothallic) isolates on YpSs and malt extract agar (MEA) at 30°C–40°C, reddish brown to black, warty, 45–70 μm in diameter.

Habitat diversity: Mainly on composting and fermenting substrates, like compost, municipal wastes, horse dung, composted wheat straw, guayule, hay, seeds of cacao, barley, oat, maize and wheat, sunflower seeds, groundnuts, pecans, sputum, bird nests, air, and soil.

Geographic distribution: United Kingdom, Chad, Czechoslovakia, South Africa, Indonesia, India, Japan, United States, Nigeria, and Australia.

Description: Based on our isolates IMI 339616, 318201, and 333312.

Rhizopus microsporus van Tieghem (Figure 7.2c and d)

Description: Growth at 45°C on YpSs is extremely rapid. Colonies cottony, turf high, at maturity grayish black, vegetative mycelium hyaline. This fungus is characterized by the production of stolons. Rhizoids pale brown, 60–170 × 4–7 μm. *Sporangiophores* unbranched, arise singly or in groups from reduced nodal regions, smooth, hyaline to pale brown, 400–1400 × 15–30 μm, up to 40 μm at the base. *Sporangia* black, usually moist, globose, smooth, 40–90 μm in diameter. Each sporangium possesses a broad oval columella. *Sporangiospores* numerous, pale brown, striate, oval, 5–8 μm in diameter. *Zygospores* rarely produced.

Habitat diversity: Soil, composting wheat straw, coal mine soils, fermenting plant materials, nesting materials, stored grains, air, and cow dung.

Geographic distribution: India, Sudan, Tanzania, Malaysia, and Australia.

Description: Based on our isolate IMI 339612.

Rhizopus rhizopodiformis (Cohn.) Zopf (Figure 7.1g and h)

Description: Growth very rapid at 45°C on Potato Dextrose Agar (PDA) filling a 90 mm petri dish in 24 h. Colony white at first, later grayish black. Vegetative hyphae hyaline, 3–7 μm in diameter, reduced stolons. *Rhizoids* pale brown, 120–150 × 12–14 μm. *Sporangiophores* arise singly, in twos or in groups from the reduced nodal regions. *Sporangia* black, spherical, smooth, 76–175 μm in diameter. *Sporangiospores* pale brown, spherical, smooth, 3–6 μm in diameter.

Habitat diversity: Coal mine soils, nesting material of birds, lung of pullet, stomach of pig, bread, wooden slats, soil, seeds of *Lycopersicon esculentum*, *Cucumis melo*, breeder cow, and oil palm effluents.

Geographic distribution: India, United Kingdom, South Africa, China, Ghana, Hong Kong, Indonesia, Malaysia, and Japan.

Description: Based on Thakre and Johri (1976).

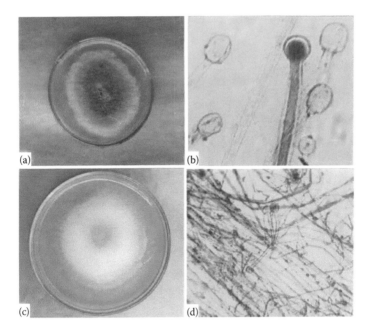

FIGURE 7.2   (a and b): *Rhizomucor pusillus*: (a) 3-day-old culture at 45°C on YpSs agar and (b) sporangiophores with young and dehisced sporangia showing columellae, 800×. (c and d): *Rhizopus microspores*: (c) 3-day-old culture at 45°C on YpSs agar and (d) sporangiophores with rhizoids and sporangia, 150×.

7.3.2 Ascomycetes

Thermoascus aurantiacus Miehe (1907) (Figures 7.3a–c and 7.4f–h)
Syn. = *Dactylomyces thermophilus* Sopp (1912) Cooney & Emerson (1964)
    = *Thermoascus isatschenkoi* Mal'chevskaya (1939) Cooney & Emerson (1964)
Description: Growth on YpSs at 45°C is very rapid and extends outward to fill a 90 mm petri dish in 2–3 days. At first, the hyphae are largely within the substratum or closely spread over the surface. Colony white in the early stages, later becoming very pale gray buff. *Hyphae* colorless, septate, 1.5–12 μm wide. *Conidiophores* erect, septate, irregularly branched, up to 1000 μm long, tapering from 10–12 μm in diameter at the base to 5 μm at the apices. *Phialides* flask-shaped, sometimes branched, irregularly arranged, 15–28 × 3–7 μm. *Conidia* colorless to pale brown, elliptical, produced in long chains, 3–8 × 2–6 μm in diameter. *Cleistothecia* reddish brown; irregular, globose, or somewhat angular; scattered or partly confluent; 0.6–1.5 mm in diameter; peridial wall dull; rough; composed of several layers of pseudoparenchymatous cells. *Asci* numerous, scattered, oval, eight-spored, evanescent, 10–15 × 8–12 μm. *Ascospores* oval or elliptical, hyaline, single-celled, smooth-walled, with a characteristic germ pore, 6–8 × 5–6 μm.

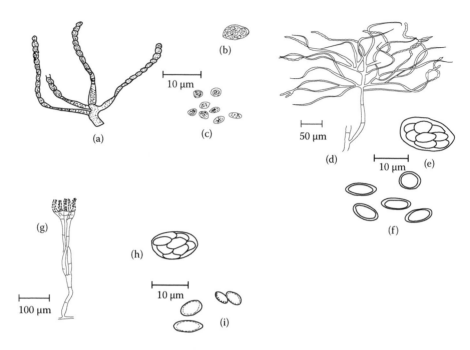

FIGURE 7.3 (a–c): *Thermoascus auranticus*: (a) conidiophores with chains of developing conidia, (b) ascus, and (c) ascospores. (d–f): *Coonemeria crustacea*: (d) conidiophores with long chains of conidia, (e) ascus, and (f) ascospores. (g–i): *Dactylomyces thermophilus*: (g) conidiophore tip with lobed or palamate annellophore, (h) ascus, and (i) ascospores. ([a–c] redrawn from Cooney, D.G., and Emerson, R., *Thermophilic Fungi: An Account of Their Biology, Activities, and Classification*, W.H. Freeman and Co., San Francisco, 1964; [d–f] redrawn from Stolk, A.C., *Antonie van Leeuwenhoek*, 31, 262–276, 1965; [g–i] redrawn from Apinis, A.E., *Trans. Br. Mycol. Soc.*, 50, 573–582, 1967.)

Habitat diversity: Heated hay, peat, cacao husks, mushroom compost, stored grains, coal mine soils, soil, air, self-heated wood chips, chaff, tobacco, and sawdust.

Geographic distribution: Germany, United States, India, Russia, Holland, Netherlands, South Africa, Italy, United Kingdom, Canada, Jordan, Australia, Indonesia, Egypt, and Japan.

Description: Based on our isolate maintained in our culture collection under accession no. T-167.

Coonemeria crustacea (Apinis & Chesters) Mouchacca (1997) (Figure 7.3d–f)
Anamorph = *Paecilomyces crustaceus*
Syn. = *Dactylomyces crustaceus* Apinis & Chesters (1964)
      = *Thermoascus crustaceus* (Apinis & Chesters) Stolk (1965)
Description: Colony growth on malt agar at 45°C is rapid. Hyphae hyaline, septate 2–12 μm in diameter. *Conidiophores* hyaline, septate, smooth, produced mainly from the submerged mycelium but also from trailing hyphae, conidiophores up to 1000 μm long, with a diameter of 7–12 μm at the base and tapering to 4–5 μm at the apex, the upper part bears irregularly arranged branches. *Phialides* either occur singly as side

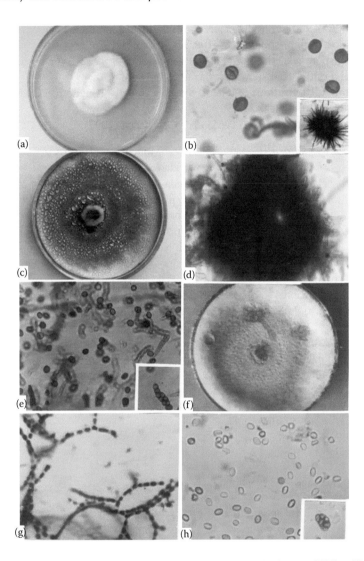

FIGURE 7.4   (a and b): *Chaetomium senegalensis*: (a) 5-day-old culture at 45°C on YpSs agar and (b) ascospores, 800×; inset showing perithecium, 250×. (c–e): *Melancarpus albomyces*: (c) 8-day-old culture at 45°C on YpSs agar showing exudates; (d) mature cleistothecium, 400×; and (e) ascospores; inset showing pyriforum ascus, 400×. (f–h): *Thermoascus aurantiacus*: (f) 5-day-old culture at 45°C on YpSs agar; (g) conidiophores with conidia in chains, 350×; and (h) ascospores, inset showing ascus, 800×.

branches or appear in irregular verticils of two or three at the end of the branches, 15–30 μm long, consisting of cylindrical basal part, 5–7 μm in diameter, tapering gradually to a long conidium-bearing tube, slightly bent away from the main axis, up to 12 μm long and about 3 μm in diameter. *Conidia* smooth, hyaline to pale brown, produced in conspicuous, long, diverging chains, cylindrical when young, ellipsoidal when mature, 6–10 × 3–6 μm. *Ascocarps* usually confluent or globose when borne separately, 300–900 μm in diameter. *Asci* subglobose, produced singly by means of croziers, mostly eight-spored, 16–20 × 13–15 μm. *Ascospores* oval, yellow to pale brown, ascospore wall shows fine echinulations, 6.5–8 × 5–6.5 μm.

Habitat diversity: Coal spoil tips, bagasse, and soil.

Geographic distribution: United States, United Kingdom, Ghana, Japan, Netherlands, and Indonesia.

Description: Based on Stolk (1965) and Mouchacca (1997).

Coonemeria aegyptiaca (Ueda & Udagawa) Mouchacca (1997)

Basionym = *Thermoascus aegyptiacus* Ueda & Udagawa (1983)

Anamorph = *Paecilomyces aegyptiaca* Ueda & Udagawa (1983)

Description: Growth of the fungus takes place between 25°C and 55°C. At 40°C, colonies fill the petri plate within 4 days, producing numerous superficial ascocarps, often forming a crusty mass, vinaceous to reddish brown, initially a simple coiled hyphae. *Conidia* fairly abundant, grayish yellow and not affecting colony color. *Cleistothecia* superficial, subglobose, orange brown, 250–550 μm wide. *Asci* borne singly on croziers, scattered in the ascomatal cavity, eight-spored, ovate, 14–18 × 11–15 μm, evanescent. *Ascospores* single-celled, ellipsoid to ovoid, yellowish to pale reddish orange, 6.0–8.5 × 4.0–5.5 μm, thick-walled and nearly smooth. *Conidiophores* erect, arising more commonly from aerial hypae, hyaline, smooth-walled, 50–300 × 5–7 μm; apical parts irregularly branched and bearing terminal verticils of two to four phialides, usually without any metulae. *Phialides* solitary or irregularly verticillate, cylindric, 12–30 × 3–6 μm. *Conidia* sometimes ovoid to subglobose, formed in long divergent or tangled chains, continuous, hyaline but fulvous in mass, cylindrical to elliptical, 4.5–11 × 3–4 μm.

Habitat diversity: Sludge and soil.

Geographic distribution: Egypt and Iraq.

Description: Mouchacca (1997).

Dactylomyces thermophilus Sopp (1912) (Figure 7.3g–i)

Anamorph = *Polypaecilum* sp. (Apinis) 1967

Syn. = *Thermoascus thermophilus* (Sopp) von Arx (1975)

    = *Penicillium thermophilus* (Sopp) Biourge

    = *Penicillium thermophilum* (Sopp) Saccardo

Description: *Hyphae* hyaline, branched, septate, 2–15 μm thick. *Conidiophores* smooth, dichotomously branched, robust and hyaline; conidiophores are usually twisted and arise from either creeping surface hyphae or trailing aerial hyphae, up to 25 μm in diameter; conidiophore tips bear a lobed or palmate anellophore. *Conidia* produced in chains, nonseptate, smooth, cylindrical to oval, green or yellow green in mass, hyaline or subhyaline, 3–11 × 2.5–5.5 μm. *Ascocarps* superficial, firm, irregularly globose, red brown, more or less solitary, 150–600 μm in diameter. *Asci* subglobose to oval, evanescent, eight-spored, 12–15 μm in diameter. *Ascospores* hyaline, nonseptate, smooth, more or less oval, 5.5–8 × 3.5–6 μm.

Habitat diversity: Wood and bark of *Pinus*, and plant debris.

Geographic distribution: Sweden, Norway, and United Kingdom.

Description: Based on Apinis (1967).

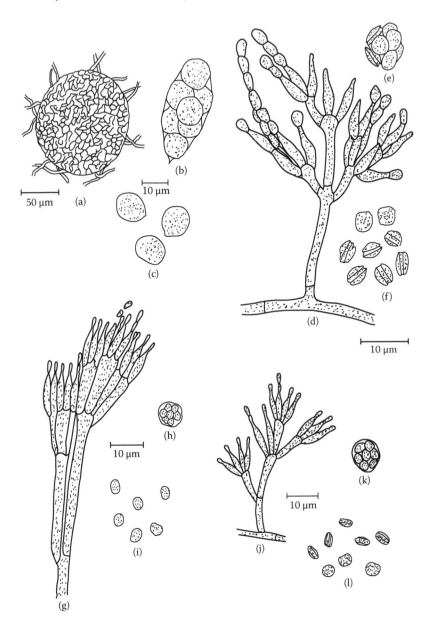

FIGURE 7.5   (a–c): *Melanocarpus albomyces*: (a) mature aspocarp, (b) ascus, and (c) ascospores. (d–f): *Talaromyces dupontii*: (d) penicillus, (e) a cluster of eight ascospores, and
(f) ascospores showing equatorial furrows and ridges. (g–i): *Rasamsonia emrsonii*:
(g) penicillus, (h) ascus, and (i) ascospores. (j–i): *Talaromyces thermphilus*: (j) penicillus,
(k) ascus, and (l) ascospores. ([a–f] redrawn from Cooney, D.G., and Emerson, R., *Thermophilic
Fungi: An Account of Their Biology, Activities, and Classification*, W.H. Freeman and Co., San
Francisco, 1964; [g–l] redrawn from Stolk, A.C., *Antonie van Leeuwenhoek*, 31, 262–276,
1965.)

Rasamsonia emersonii Stolk (1965), Houbraken & Frisvad comb. nov. (Figure 7.5g–i)

Basionym = *Talaromyces emersonii* Stolk (1965)

    = *Penecillium emersonii* Stolk (1965)

Syn. = *Geosmithia emersonii* (Stolk) Pitt (1979)

    = *Byssochlamys* sp. *fide* Cooney & Emerson (1964)

Description: Colonies on malt agar growing rapidly, attaining a diameter of 5–7 cm in 7 days at 45°C. Mycelium hyaline, partly submerged, partly aerial, appearing definitely funiculose, producing colorless to cream exudate in droplets. *Ascocarps* produced in dense yellow to reddish brown layer. *Hyphae* hyaline, 1–3 µm in diameter. *Ascocarps* often confluent, but when borne separately, they are globose to ellipsoid, 50–300 µm in diameter. *Asci* evanescent, subglobose to ellipsoid, eight-spored, 8–9 × 6–8 µm. *Ascospores* yellow, thick-walled, wall smooth, subglobose to ovoid, 3.5–4.0 × 3–3.5 µm. *Conidiophores* arising from the funiculose aerial mycelium, hyaline to pale brown, erect, septate, 35–150 × 3–4.5 µm; conidiophores usually with one or two branches, densely compacted. *Phialides* almost cylindrical with abruptly tapering tip, up to 2 µm in length; all elements of the conidiophores appear roughened. *Conidia* mostly cylindrical, occasionally ellipsoidal, smooth, hyaline to pale brown, 3.5–5.0 × 1.5–2.5 µm; produced in long, loosely adherent parallel chains, often forming loose columns.

Habitat diversity: Compost, soil, piles of wood chips, riverbanks, grassland, municipal waste, peat, coal spoil tips, sugarcane bagasse, palm oil kernels, Blesbok dung, outdoor air, rhizosphere of *Cassia tora*, and *Cassia occidentalis*.

Geographic distribution: Italy, Netherlands, United Kingdom, United States, Sweden, Canada, Japan, India, Nigeria, South Africa, and Indonesia.

Description: Based on Stolk (1965), Houbraken et al. (2012).

Rasamsonia composticola Y.Y. Su & L. Cai, sp. nov. (2013) (Figure 7.6a–h)

Description: Truly thermophilic, optimal growth temperature 45°C–50°C with minimum growth temperature of 30°C on MEA and PDA. Colonies on MEA 70–80 mm diameter after 5 days at 45°C, surface texture velutinous, white, sparse, with floccose buff aerial mycelia in center. *Asci* often confluent, yellow brown to orange brown, globose to ellipsoid, 100–250 µm in diameter. Coverings scanty, consisting of inconspicuous networks of hyphae, surrounded by twisted and yellowish hyphae, about 1 µm in diameter. Asci evanescent, globose to subglobose, eight-spored, 6.7–11 × 6–8 µm, formed in short chains. *Ascospores* hyaline, single-celled, smooth, globose, 2–3.8 µm diameter. *Conidiophores* arising from the subsurface, surface, or aerial mycelium on MEA; 30–200 × 2–5 µm; verrucose; hyaline; straight. Penicilli biverticillate, terverticillate, often asymmetric; metulae 6– 15 × 1.5–4 µm, often with enlarged apices, two to four per branch, usually with verrucose walls. Phialides three to six per metulae, 5.5–9 × 1.5–3 µm, cylindrical with long collula, 1–3 µm, verruculose. *Conidia* smooth, hyaline, cylindrical, slightly larger at one end, one-celled, 3–8.75 × 1.5–3.75 µm, formed in chains.

Habitat diversity: Compost from rice straw.

Geographic distribution: China.

Description: Based on Su and Cai (2013).

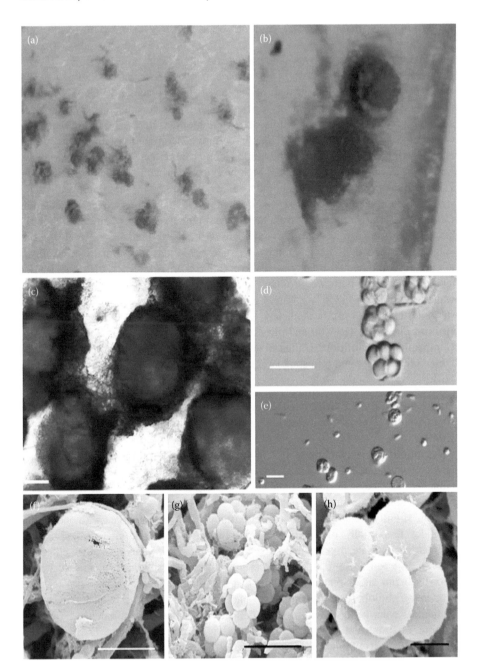

FIGURE 7.6 (a and b): *Rasamsonia composticola* ascocarps forming on MEA. (c): Ascocarps. (d and e): Asci and globose ascospores. (f): Scanning electron micrographs of ascocarp. (g and h): Scanning electron micrographs of asci and ascospores. Scale bars: 50 μm (c), 10 μm (d–g), 2 μm (h). (Reprinted from Su, Y.Y., and Cai, L., *Mycol. Prog.*, 12, 213–221, 2013. With permission from Springer.)

Rasamsonia byssochlamydoides Stolk & Samson (1972) Houbraken & Frisvad comb. nov.

Basionym = *Talaromyces byssochlamydoides* Stolk & Samson (1972)

Anamorph = *Paecilomyces byssochlamydoides* Stolk & Samson (1972)

Description: The optimum growth of the fungus takes place at 40°C–45°C, and it poorly grows at 37°C. It is mainly distinguished by its conspicuous *Paecilomyces* anamorph having cylindrical conidia. *Ascomata* globose, up to 200 μm in diameter, covering very scanty, always develops in culture concomitantly with the anamorph and as such prevents its confusion with the similar imperfect taxon of *Paecilomyces variotii* Bainier. *Asci* eight-spored, subglobose to broadly ellipsoidal, 9–12.5 × 6.5–7.5 μm. *Ascospores* globose to subglobose, 3.7–4.5 × 3.5–4.0 μm, thick-walled and smooth, often partially covered by remnants of a gelatinous covering material. *Conidiophores* arising directly from the agar or in the floccose mycelium, irregularly branched (*Paecilomyces* type), terverticillate or quarterverticillate, up to 300 μm in length; stipes broad, 3–6 μm, smooth to verrucose. Metulae in terminal whorls of two to eight, broad, unequal in length, 10–15 × 2.0–5.0 μm. Phialides, cylindrical base-tapering in a long narrowed collula, 10–15 (–20) × 2.5–3.5 μm. Conidia smooth walled, cylindrical, 4–8 × 1–2.5 μm. Chlamydospores are rarely produced, globose or subglobose, about 4 μm in diameter.

Habitat diversity: Dry soil and piles of peat.

Geographic distribution: United States, Germany, Japan, and Egypt.

Description: Stolk and Samson (1972), Awao and Otsuka (1974), Houbraken et al. (2012).

Talaromyces duponti Griffon and Maublanc (1911) (Figure 7.5d–f)

Anamorph = *Penicillium dupontii* Griffon & Maublanc(1911) Raper & Thom (1949)

Description: On yeast glucose agar at 45°C, growth is fairly rapid. Colony initially white, changing to dull grayish green, lavender, or pinkish brown. Hyphae delicately branched, septate, 2–3 μm in diameter. *Conidiophores* short, lateral, more or less perpendicular to the main hyphae, simple or branched, septate, 5–30 × 2–3 μm, tapering slightly toward the base. *Penicilli* irregular, monoverticillate with one to four phialides to partially or nearly regularly biverticillate. *Metulae* 5–7 × 2–3 μm. *Phialides* acuminate, divergent, 8–10 × 2 μm. *Conidia* pale yellow, ovoid to elliptical, smooth, 2–5 × 1.5–3 μm. *Cleistothecia* grayish tan, subspherical, mostly scattered, 400–1300 μm in diameter, peridium distinct, smooth papery, nonostiolate. *Asci* scattered, subglobose, eight-spored, evanescent, 9–10 μm in diameter. *Ascospores* lenticular, pale yellow, with a well-defined equatorial furrow, flanked by low, smooth, or somewhat jagged ridges; 3–5 × 2.5–3.5 μm.

Habitat diversity: Manure and damp hay, self-heated guayule shrub, leaf litter, soil, cigarette, and *Hordeum vulgare*.

Geographic distribution: United States, India, United Kingdom, France, South Africa, Netherlands, Jordan, and Nigeria.

Description: Based on Cooney and Emerson (1964).

Talaromyces thermophilus Stolk (1965) (Figure 7.5j–l)

Anamorph = *Penecillium duponti* Griffon et Maublanc, emend. Emerson apud Raper and Thom (1949)

Description: Colonies on malt agar grow rapidly at 45°C. Mycelium floccose to slightly funiculose, producing abundant conidial structures in gray-brown colors. *Hyphae* hyaline, septate, 2–3 μm in diameter. *Conidiophores* borne as perpendicular branches from aerial hyphae, smooth-walled, short, 2–30 × 2–3 μm, slightly inflated at apex, 4 μm in diameter. *Phialides* in small clusters of four in verticil, divergent, swollen at base, tapering abruptly to a long conidium-bearing tip, 7–10 × 2–2.5 μm. *Conidia smooth*, ovoid to ellipsoidal, 2.5–5 × 1.5–3 μm. *Ascocarps* pale gray, globose to subglobose, nonconfluent, 400–1300 μm in diameter. *Asci* produced in short chains, globose to subglobose, 9–10 μm in diameter. *Acospores* lenticular, ornated by two to three or more, somewhat jagged, irregular ridges, yellow, 3.5–5 × 2.5–3.5 μm.

Habitat diversity: Guayule shrub, fermented straw, dung, compost, and soils.

Geographic distribution: United States, Netherlands, India, United Kingdom, Japan, Australia, and Indonesia.

Description: Based on Stolk (1965).

Chaetomium thermophile **var.** *coprophile* Cooney and Emerson (1964) (Figures 7.7d–f and 7.8a–c)

Description: On YpSs agar at 45°C, the colonies appear white at first, producing typical concentric rings of growth; the rings of dark brown perithecia are separated by narrow zones of whitish hyphae. *Perithecia* superficial, more or less gregarious, globose or subglobose, 75–150 μm in diameter, ostiolate, but the ostiole is not easily observed, as the perithecia are densely clothed with dark, dichotomously branched hairs. Hairs smooth, 40–140 × 4.5–7 μm. *Asci* short stalked, cylindrical, produced in a basal tuft, bearing eight spores in a linear row, 45–55 × 6–8 μm. *Ascospores* dark or olive brown, globose or subglobose, 7–9 μm in diameter.

Habitat diversity: Decomposing wheat straw, horse dung, mushroom compost, vegetable detritus, and soil.

Geographic distribution: United States, United Kingdom, India, Netherlands, and Ghana.

Description: Based on our isolate IMI 348710.

Chaetomium thermophile **var.** *dissitum* Cooney and Emerson (1964) (Figures 7.7g–i and 7.8d–f)

Description: On YpSs agar at 45°C, the colonies appear colorless at first and then become dark brown. Perithecia initially appear in about 2 days as small, colorless knots on the hyphae. *Perithecia* superficial; gray, gray green, to dark brown; produced in diffuse manner; scattered; ostiolate but the opening is not easily observed as the perithecia are densely clothed with dense terminal hairs; mature perithecia 60–175 μm in diameter, usually globose, although subglobose or oval forms are also quit common. *Asci* short stalked, cylindrical, eight-spored, arranged in a single row, 50–60 × 7–8 μm. *Ascospores* olive brown in color, globose or subglobose, smooth-walled, 6.5–9.5 μm in diameter, with a single apiculus.

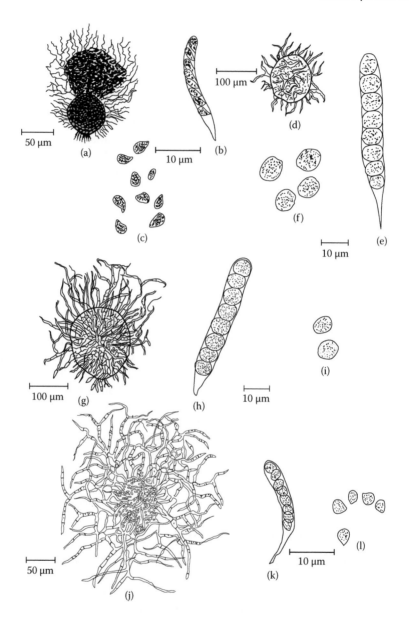

FIGURE 7.7 (a–c): *Chaetomium senegalensis*: (a) mature ascocarp, (b) ascus, and (c) asco-spores. (d–f): *Chaetomium thermophile* var. *coprophile*: (d) mature perithecium, (e) ascus, and (f) ascospores. (g–i) *Chaetomium thermophile* var. *dissitum*: (g) mature perithecium, (h) ascus, and (i) ascospores. (j–l): *Chaetomium verginicum*: (j) mature perithecium, (k) ascus, and (l) ascospores. ([a–c] redrawn from Ames, L.M., *A Monograph of the Chaetomiaceae*, U.S. Army Research and Development Series 2, Bibliotheca Mycologica Cramer, Lehre, Germany, 1963; [d–i] redrawn from Cooney, D.G., and Emerson, R., *Thermophilic Fungi: An Account of Their Biology, Activities, and Classification*, W.H. Freeman and Co., San Francisco, 1964.)

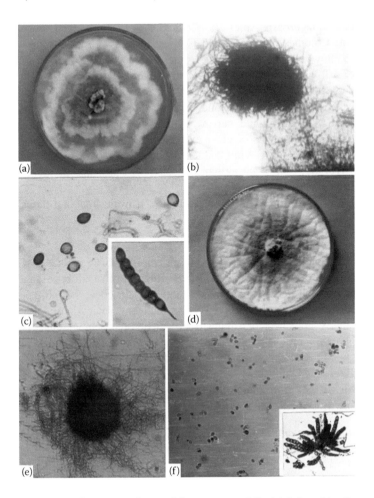

FIGURE 7.8 (a–c): *Chaetomium thermophile* var. *coprophile*: (a) 3-day-old culture at 45°C on YpSs agar with typical concentric rings of growth; (b) a mature perithecium, 350×; and (c) ascospore; inset showing young ascus, 1000×. (d–f): *Chaetomium thermophile* var. *dissitum*: (d) 5-day-old culture at 45°C on YpSs agar; (e) mature perithecium, 350×; and (f) ascospores; inset showing young asci, 400×.

Habitat diversity: Nesting materials, decomposing wheat straw, mushroom compost, and soil.

Geographic distribution: United States, United Kingdom, India, Netherlands, and Ghana.

Description: Based on our isolate maintained in our culture collection under accession number T-132.

Chaetomium senegalensis Ames (1963) (Figures 7.4a and b and 7.7a–c)
Description: Colonies on YpSs at 45°C appear white with a daily growth rate of 6–7 mm. Yellow exudate in mature colonies appears after 18–20 days. *Ascomata* dark, spherical or ovate, ostiolate, 110–230 μm in diameter, with a dark brown or black wall of angular, flattened, 6–10 μm cells. *Ascomatal hairs* narrow, delicate, often branched with bulbous base and tapering ends, punctulate or verrucose, 1.5–2 μm in width. *Asci* fasciculate, narrow, cylindrical, eight-spored, 50–70 × 7–9 μm. *Ascospores* uniseriate, ovate or clavate, bilaterally flattened, dark greenish, brown when mature, 6.5–12.5 × 4–7 μm, teardrop-shaped with a subapical germ pore.
Habitat diversity: On plant remains, seeds of *Capsicum annuum*, soil, and decomposing wheat straw.
Geographic distribution: Senegal, Netherlands, Kuwait, Iran, and India.
Description: Based on Aneja and Kumar (1994).

Chaetomium virginicum (1963) (Figure 7.7j–l)
Description: Colonies appear rich brown in color at 45°C, hyphae 2–4 μm wide, often anastomose to form a loose network. *Perithecia* brown, globose, attached to the substratum with undifferentiated rhizoids, terminal hairs cover the entire perithecium, giving it the appearance of a tumbleweed in miniature. Hairs dense, granular, intricate, irregularly branched, or suggestive of dichotomous branching. *Asci* stalked, long, cylindrical, eight-spored, 70 × 10 μm. *Ascospores* unistichous, light yellow brown to pale brown, almond-shaped, 8–11 μm.
Habitat diversity: On decomposing leaves.
Geographic distribution: United States.
Description: Based on Ames (1963).

Chaetomium mesopotamicum Abdullah & Zora (1993)
Description: This is a newly described species and has a growth temperature range from 30°C–52°C. It differs from *Chaetomium thermophilum* LaTouche and *C. virginicum* Ames by its *asci*, which are clavate and possess long highly branched terminal hairs. *Ascospores* globose to ovoid, olive to brown, 5.5–7.8 × 5.2–6.3 μm, provided with one apical germ pore.
Habitat diversity: Date palm plantation.
Geographic distribution: Iraq.
Description: Based on Abdullah and Zora (1993).

Chaetomium britannicum Ames (1963)
Description: Growth of the fungus and production of perithecia take place at 47°C. *Ascomata* ovoid to vase-shaped. Terminal and lateral hairs are very slender, grayish, straight to undulate. *Asci* club-shaped, eight-spored. *Ascospores* brown, large, 19–24 × 11–14 μm, irregularly oval, rounded on the ends, and have a single apical germ pore.
Habitat diversity: Mushroom compost and soil.
Geographic distribution: United Kingdom.
Description: Based on Ames (1963).

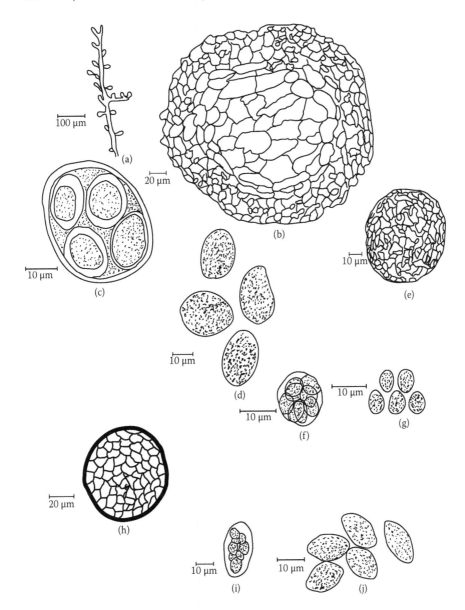

FIGURE 7.9 (a–d): *Corynascus thermophilus*: (a) young conidiophores with conidia, (b) median section through a young cleistothecium before ascus formation, (c) young ascus with wall intact containing ascospores, and (d) mature ascospores. (e–g): *Thielavia australiensis*: (e) mature ascocarp, (f) ascus containing eight ascospores, and (g) ascospores. (h–j): *Thielavia terricola*: (h) ascocarp, (i) ascus, and (j) ascospores. ([a–d] redrawn from von Arx, J.A., *On Thielavia and Some Similar Genera of Ascomycetes*, Netherlands, 1975; [e–g] redrawn from Tansey, M.R., and Jack, M.A., *Can. J. Bot.*, 53, 81–83, 1975; [h–j] redrawn from Emmons, C.W., *Bull. Torrey Bot. Club*, 57, 123–126, 1930.)

Thielavia minor (Rayss and Borut) Malloch and Cain (1973)

Description: Colony on YpSs agar broadly spreading, composed of white, cottony, aerial and submerged hyphae, 2–5 µm in diameter. *Ascocarps* spherical to globose, walls brown, nonostiolate, nonappendaged, brownish to almost black at maturity, 80–160 µm in diameter. *Asci* oval to globose, eight-spored, 15–18 × 10–14 µm. *Ascospores* broadly fusiform to elliptical, apiculate at both ends, olive brown, irregularly arranged, 10–12 × 13–15 µm.

Habitat diversity: Coal mine soils, *Elaeis guineensis* leaf, and groundnut kernels.

Geographic distribution: India, Zaire, and Zambia.

Description: Based on Malloch and Cain (1973).

Thielavia australiensis Tansey and Jack (1975) (Figure 7.9e–g)

Description: Colonies on YpSs agar at 40°C are sparse, colorless; mycelium of septate, branched hyphae; the agar is stained fulvous by a diffusible pigment. *Ascocarps* produced in abundance, scattered and immersed within the agar medium, globose, nonostiolate, glabrous, yellow brown (appearing black when containing mature ascospores), 20–200 µm in diameter, with a peridium of textura epidermoidea and produced from mature asci. *Asci* globose, irregularly distributed, eight-spored, evanescent, 10 µm in diameter. *Ascospores* brownish black, black in mass, ovoid, occasionally irregular in shape, with a single apical germ pore, 4–5 × 5–8 µm. *Conidia* are aleurioconidia of the form genus *Chrysosporium*, borne singly along the length of hyphae, single-celled, sessile, ovoid, smooth, colorless, 3–5 × 5–8 µm.

Habitat diversity: Nesting material of mallee fowl.

Geographic distribution: Australia.

Description: Based on Tansey and Jack (1975).

Thielavia terricola (Gilman and Abbott) Emmons (1930) (Figure 7.9h–j)

Description: Colonies on YpSs agar at 45°C broadly spreading; composed of white, cottony aerial and submerged hyphae; 1–6 µm in diameter; branches constricted at base; homothallic. *Ascocarps* spherical, arise from an ascogonial coil, nonostiolate, brownish to black at maturity, color largely due to masses of dark spores inside, 100–250 µm in diameter, outer wall of cleistothecium composed of uninucleate, rather thick-walled cells, somewhat carbonized, inner wall of cleistothecium composed of thin-walled, flattened cells. *Asci* oval to pyriform, 16–19 × 25–35 µm, deliquescing within the cleistothecium. *Ascospores* fusiform or elliptical, slightly apiculate at both ends, dark olivaceous to brown, 7–9 × 10–16 µm, with a wall much thickened at the end opposite the germ pore.

Habitat diversity: Soil, cow dung, compost, and *Ficus* sp.

Geographic distribution: United States, China, India, Canada, Kenya, Australia, United Kingdom, and Indonesia.

Description: Based on our isolate IMI 310851.

Chaetomidium pingtungium (Chen & Chen) Mouch., in Johri, Satyanarayana, & Olsen (eds.) 1999

Syn. = *Thielavia pingtungia* Chen & Chen (1996)

Description: Growth at 48°C fairly good, but no growth between 25°C and 30°C. The species is characterized by dark globose cleistothecia covered with brown thick-walled hairy appendages, *ascomatal hairs* echinulate, 2.5–4.0 µm wide and up to

350 μm long, ascomatal wall pseudoparenchymatous. *Asci* cylindrical, 40–52 × 7–9 μm, stipitate, fasciculate, eight-spored. *Ascospores* usually limoniform with pointed ends and elliptical in side view, uniseriate, dark brown, smooth, thick-walled, 8.5–10.0 × 6.5–8.5 μm. No anamorph stage developed in cultures.
Habitat diversity: Sugarcane field.
Geographic distribution: Taiwan.
Description: Based on Chen and Chen (1996).

Corynascus thermophilus (Fergus and Sinden) von Arx (1975) (Figure 7.9a–d)
Anamorph = *Myceliophthora fergusii* (van Klopotek) van Oorschot (1977)
Syn. = *Thielavia thermophila* Fergus and Sinden (1969)
　　 = *Chrysosporium fergusii* van Klopotek (1974a)
　　 = *Chaetomidium thermophilum* (Fergus & Sinden) Lodha (1978)
Description: Colonies at 45°C on starch yeast extract, cornmeal, and Czapek yeast extract agars white and floccose. Mycelium immersed and superficial, aerial hyphae become powdery with masses of conidia. *Hyphae* hyaline, irregularly three septate, 2–10 μm wide. *Cleistothecia* globose, glabrous, black at maturity, nonostiolate, 190–260 μm in diameter. *Asci* hyaline, globose to oval, irregularly arranged. The ascus wall disappears while the ascospores are hyaline, and they subsequently become pink light brown, and finally dark brown, four-spored, 37 × 40 μm. *Ascospores* brown, ellipsoidal, unicellular with a germ pore at each end, 19 × 28 μm in diameter.
Habitat diversity: Mushroom compost.
Geographic distribution: United States.
Description: Based on Fergus and Sinden (1969).

Corynascus sepedonium (Emmons) von Arx (1975)
Syn. = *Thielavia sepedonium* Emmons (1932)
　　 = *Thielavia lutescens* Kamyschko
Description: *Conidia* produced as solitary or catenulate aleurioconidia, globose, tuberculate, echinulate, bright yellow in en masse, 4–12 μm in diameter. *Ascocarps* globose, olivaceous to black, glabrous, 20–150 μm in diameter. *Asci* ovoid, irregularly disposed, not borne on croziers, stipitate at first but often not so as the ascocarps enlarge, eight-spored, 25–35 × 17–25 μm. *Ascospores* ellipsoidal, slightly flattened on one side, with a germ pore at each end, dark brown, 10.9–19 × 6–10 μm.
Habitat diversity: Dung, soil, human skin, pasture soil, hay, coal spoil tips, compost, cellulose material, *Litchi sinensis* leaf, seeds of *Triticum*, *Foeniculum vulgare*, and *Carpentaria acuminata*.
Geographic distribution: India, Kenya, United States, Uzbekistan, United Kingdom, Ghana, Egypt, Hungary, Australia, China, and Senegal.
Description: Based on von Arx (1975).

Melanocarpus albomyces (Cooney and Emerson) von Arx (1975) (Figures 7.4c–e and 7.5a–c)
Syn. = *Myriococcum albomyces* Cooney and Emerson (1964)
　　 = *Thielavia albomyces* (Cooney & Emerson) Malloch & Cain (1972)
Description: On YpSs agar, colonies appear white and cottony in the early stages, becoming grayish black by the fourth day and grayish-black pigment is excreted in the

agar medium. Two distinct types of hyphae: *aerial hyphae* septate and variable in thickness, 2–10 μm wide; *prostrate hyphae* constricted at the septa, forming branched, chain-like series of cylindrical or oval thick-walled cells that break apart easily. *Ascocarps* superficial, dark brown, scattered or gregarious, globose, glabrous wall, nonostiolate, 150–250 μm in diameter. *Asci* pyriform when young, at maturity irregularly oblong, eight-spored, ascus membrane simple, very thin, evanescent, 35–40 × 15–20 μm. *Ascospores* single-celled, smooth, dark brown, globose or elliptical, with a single apiculus, irregularly distributed in the ascus, 10–15 μm in diameter. **Habitat diversity:** Nesting material of chickens, decomposing wheat straw, soil, and grass compost.

Geographic distribution: United States, India, United Kingdom, and Saudi Arabia.

Description: Based on our isolate maintained in our laboratory in our culture collection under accession number T-178.

Melanocarpus thermophilus (Abdullah & Al-Bader) Guarro, Abdullah, & Al-Bader (1996)

Basionym = *Thielavia minuta* (Cain) Malloch & Cain var. *thermophila* Abdullah & Al-Bader

Description: Not much data on the growth profile of this fungus exist in the literature. This ascomycete does not produce the arthroconidial anamorph characteristic of the type species. *Asci* are eight-spored, with *ascospores* being ovoid, dark brown, 7.5–9.0 × 6.0–7.5 μm, each provided with a single germ pore.

Habitat diversity: Forest soil.

Geographic distribution: Iraq.

Description: Guarro et al. (1996).

Canariomyces thermophila Guarro & Samson in von Arx, Figueras, & Guarro (1988)

Description: Growth of the fungus takes place at 45°C, but no data on minimum and maximum growth temperature are available. *Ascomata* cleistothecial, ascoma wall made up of angular dark cells. *Asci* irregularly disposed. *Ascospores* greenish brown when mature with a subapical germ pore, 14.0–18.0 × 7.5–10.0 μm. No anamorph is reported.

Habitat diversity: Soil.

Geographic distribution: Africa.

Description: von Arx et al. (1988).

7.3.3 DEUTEROMYCETES (ANAMORPHIC FUNGI)

Papulospora thermophila Fergus (1971) (Figure 7.10e)

Description: Colonies on YpSs at 45°C white and downy, margins possess closely spaced indentations. *Mycelium* of two types, one very narrow, 1.5 μm in diameter, sparingly septate and moderately branched; the other wide, 5–7.5 μm in diameter, regularly septate and very branched. *Bulbils* produced on aerial hyphae and on submerged hyphae; those on the aerial hyphae are globose to subglobose and average 105 × 90 μm in diameter; those formed on submerged hyphae are irregular in shape

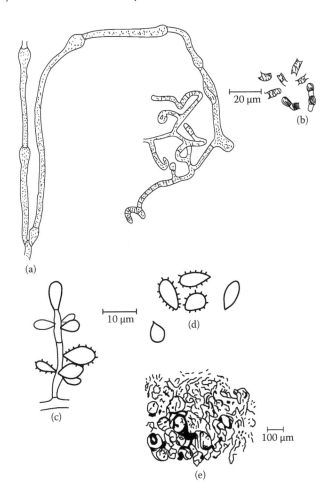

FIGURE 7.10 (a and b): *Malbranchea cinnamomea*: (a) Fertile hyphae with conidiophores showing swellings near the septum and terminal septate portions and (b) mature conidia showing thin-walled isthmuses. (c and d): *Myceliophthora thermophila*: (c) hyphae with mature aleuriophore and (d) aleuiospores. (e): *Papulospora thermophila*, bulbils on aerial mycelium. ([a and b] redrawn from Cooney, D.G., and Emerson, R., *Thermophilic Fungi: An Account of Their Biology, Activities, and Classification*, W.H. Freeman and Co., San Francisco, 1964; [c and d] redrawn from van Oorschot, C.A.N., *Persoonia*, 9, 404–408, 1977; [e] redrawn from Fergus, C.L., *Mycologia*, 63, 426–431, 1971.)

and average 125 × 113 μm in diameter. They are white at first, then become yellow, and finally orange at maturity. The outer three layers of the mature bulbils are formed of narrow and elongate cells, while the interior of globose cells is lighter in color. The bulbils begin to form after 24 h incubation at 45°C.

Habitat diversity: Mushroom compost and soil.

Geographic distribution: Switzerland, India, and Japan.

Description: Based on Fergus (1971).

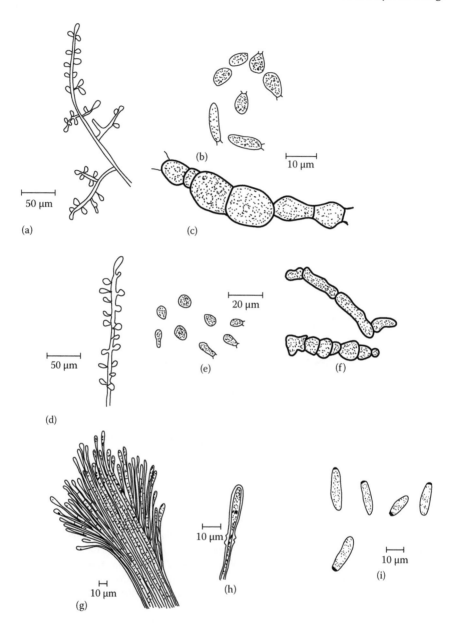

FIGURE 7.11 (a–c): *Chrysosporium tropicum*: (a) fertile hyphae with conidia, (b) conidia showing variation in shape and size, and (c) chain of globose chlamydospores. (d–f): *Myceliophthora fergusii*: (d) fertile hyphae bearing conidia, (e) conidia showing variation in shape and size, and (f) chlamydospores in chain of globose swelling. (g–i): *Remersonia thermophila*: (g) synnema with conidial head, (h) conidium attached to the annellate coniogenous cell, and (i) conidia.

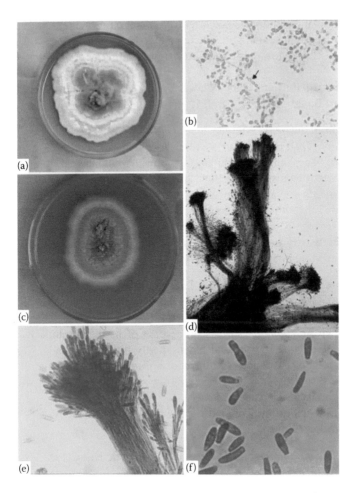

FIGURE 7.12 (a and b): *Malbranchea cinnamomea*: (a) 5-day-old culture at 45°C on YpSs agar and (b) arthroconidia with thin-walled isthmuses (arrow), 400×. (c–f): *Remersonia thermophila*: (c) 5-day-old culture at 45°C on YpSs agar; (d) erect and determinate synnemata, 150×; (e) synnmeta showing conidial head, 500×; and (f) hyaline, guttulate conidia, 1000×.

Remersonia thermophila (Fergus) Seifert, Samson, Bockhout, & Louis-Seize (1997) (Figures 7.11g–i and 7.12c–f)
Syn. = *Stilbella thermophila* Fergus (1964)
Description: Colonies on YpSs at 45°C are discrete. *Mycelium* mostly submerged, composed of brownish-gray, aggregated hyphae, superficial mycelium septate, smooth, 2–4 μm in diameter. *Synnemata* parallel, determinate, conspicuous, distinct, superficial, erect, grayish brown to pale brown; up to 450–600 μm long, 7–60 μm wide at the base, 18–120 μm wide at the convex apex. Constituent hyphae usually unbranched, smooth, septate, pale brown, 2–3 μm wide, compacted together at the base, then splaying out to form a wide convex conidial head. *Conidiogenous cells*

hyaline, terminal, cylindrical, percurrent proliferation, 15–37 × 5–7 µm. *Conidia* hyaline, solitary, ellipsoidal, guttulate, aseptate, thick-walled, 7–25 × 4–6 µm.

Habitat diversity: Mushroom compost, straw bedding used for pigs, horse dung, and soil.

Geographic distribution: United States, United Kingdom, and India.

Description: Based on our isolate IMI 364224.

Acremonium alabamense Morgan-Jones (1974)

Teleomorph: = *Thielavia terrestris* (Apinis) Malloch & Cain (1973)

Description: The fungus produces fast-growing colonies at high temperature. *Colonies* velvety, whitish, with yellowish to brownish runner hyphae that are 3–4.5 µm wide. *Conidiophores* are simple, short, 8–25 × 1–1.5 µm. *Conidia* are obovoid to pyriform, smooth, with a truncated base, 3–6 × 2–3 µm.

Habitat diversity: Alluvial soil and needles of *Pinus taeda.*

Geographic distribution: United States and United Kingdom.

Description: Morgan-Jones (1974).

Acremonium thermophilum Gams & Lacey (1972)

Description: Growth at 20°C is strong but slow; very good growth takes place between 25°C and 40°C and very weak at 47°C. The fungus is regarded as unique among known *Acremonium* species on account of its thermophilic habit and production of submerged hyphae and having partly pigmented walls. *Conidiophores* thick-walled with basitonous ramification. *Conidia* are ellipsoidal, 3.0–4.0 × 1.3–1.7 µm.

Habitat diversity: Sugarcane bagasse.

Geographic distribution: Trinidad.

Description: Gams and Lacey (1972).

Malbranchea cinnamomea (Libert) van Oorschot & de Hoog (1984) (Figures 7.10a and b and 7.12a and b)

Syn. = *Malbranchea sulfurea* (Miehe) Sigler and Carmichael (1976)

    = *Malbranchea pulchella* var. *sulfurea* (Miehe) Cooney and Emerson (1964)

    = *Thermoideum sulfureum* Miehe (1907)

    = *Geotrichum cinnamomeum* (Libert) Sacc. (1881)

Description: At 45°C, colonies on YpSs are robust; dense; thick; smooth or with a few outward radiating folds; velvetty with coarse, creamy yellow tufts of hyphae; sulfur yellow. The medium turns dark brown or black from diffused pigment. *Vegetative hyphae* hyaline, later becoming yellowish brown with prominent racket hyphae, swelling near the septum to a diameter of 9 µm or more. *Arthroconidia* borne on curved or loosely coiled lateral branches arising from broader vegetative hyphae, 3–6 µm in diameter. *Conidia* cylindrical, often curved, thick-walled, often with attached hyaline frill from the outer hyphal wall of the separating empty cell; hyaline, later yellow, tan, or yellowish green; 3–4.5 × 4–7 µm.

Habitat diversity: Guayule rets, composting heaps, wheat straw compost, stacked tobacco leaves, soil, peanut kernels, coal spoil tips, feces of Cape sparrow, deer dung, cattle, henhouse litter, snuff, air, and silage.

Geographic distribution: United States, Germany, United Kingdom, South Africa, Japan, Canada, Netherlands, Australia, Ghana, Egypt, India, and Indonesia.
Description: Based on our isolate IMI 361367.

Chrysosporium tropicum Carmichael (1962) (Figures 7.11a–c and 7.13a and b)
Description: Colony growth on YpSs at 45°C flat, granular in texture, dry, white to cream in color. *Mycelium* hyaline septate, smooth-walled, 2–4 μm in diameter. *Conidiophores* poorly differentiated. *Conidia* borne at the tips of the hyphae, directly along the sides or on short or long lateral branches, aleurosporic. Intercalary conidia also occur. Conidia hyaline, pyriform to clavate with a broadly truncate base and smooth or inconspicuous roughened walls, 3–5 × 4–9 μm.

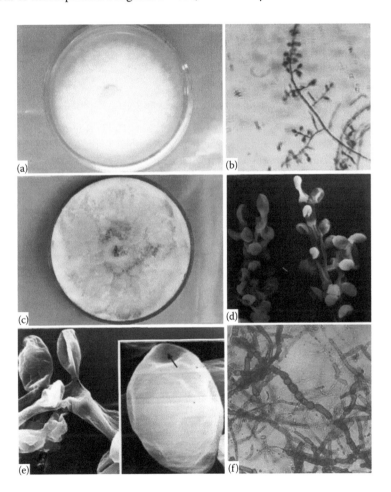

FIGURE 7.13   (a and b): *Chrysosporium tropicum*: (a) 3-day-old culture at 45°C on YpSs agar and (b) conidiophores with aleurosporic conidia, 400×. (c–f): *Myceliophthora fergusii*: (c) 5-day-old culture at 45°C on YpSs agar; (d) Scanning electron micrograph of conidiophores, 1800×; (e) Scanning electron micrograph of conidium borne on narrow peg (arrow), 4500×; inset showing a conidium with germ pore (arrow), 15,000×; and (f) chlamydospores in chain, 500×.

Habitat diversity: Deteriorating woolen fabric, dung, soil, and air.
Geographic distribution: New Guinea, India, United States, and Canada.
Description: Based on our isolates IMI 318198 and 339617.

Myceliophthora fergusi (van Klopotek) van Oorschot (1977) (Figures 7.11d–f and 7.13c–f)
Description: Growth on YpSs at 45°C very rapid, densely floccose, granular, creamy, colony margins irregular, feather-like in appearance and with age becomes mealy or dirty yellow as the aerial mycelium dies down and very old colonies have a dark brown reverse. Aerial mycelium hyaline, septate, smooth-walled, 2–4.5 μm in diameter. *Conidiophores* poorly differentiated from the mycelium. *Conidia* formed terminally and laterally from almost all parts of the aerial mycelium on conspicuous, relatively narrow pegs. Conidia hyaline, smooth, truncate, round (3.7–7.5 μm) or elliptical (5.5–11 × 4.5 μm). *Chlamydospores* in chains of globose swellings, dark brown, abundant in older cultures.
Habitat diversity: Soil.
Geographic distribution: India.
Description: Based on our isolates IMI 364225 and UAMH 8054.

Myceliophthora thermophila (Apinis) van Oorschot (1977) (Figure 7.10c and d)
Teleomorph: *Corynascus heterothallicus* (van Klopotek) von Arx (1981b)
Syn. = *Sporotrichum thermophile* Apinis (1963)
    = *Chrysosporium thermophilum* (Apinis) van Klopotek (1974b)
    = *Myceliophthora indica* Basu (1984)
Description: Growth at 45°C on Czapek Dox Agar (CDA) is very rapid. Colonies dry, thin, broadly spreading with a surface texture that varies from floccose or cottony to granular or powdery. The color is at first white, then pink, buff, and finally fulvous or cinnamon brown. *Hyphae* colorless, about 2 μm broad. *Aleuriospores* terminal or lateral on the hyphae or on short stalks, which may be ampulliform, occasionally catenate, mostly ovate or pyriform or strongly clavate with a broad or narrow scar, thick-walled, smooth or variously encrusted, orange brown, 4–8 × 2–4 μm.
Habitat diversity: Soil.
Geographic distribution: United States, Canada, India, United Kingdom, Japan, and Australia.
Description: Based on van Oorschot (1977).

Myceliophthora hinnulea Awao & Udagawa (1983)
Description: Growth of the fungus is extremely reduced at 20°C, optimal growth takes place at 40–45°C, and maximum slightly above 50°C. *Colonies* mainly dull to grayish brown. *Conidia* brownish, conspicuously verrucose to spinulose, 8.0–10.0 × 6.0–7.5 μm.
Habitat diversity: Cultivated soil.
Geographic distribution: Japan.
Description: Awao and Udagawa (1983).

Scytalidium thermophilum (Cooney and Emerson) Austwick (1976) (Figures 7.14c and d and 7.15k)
Syn. = *Torula thermophila* Cooney and Emerson (1964)

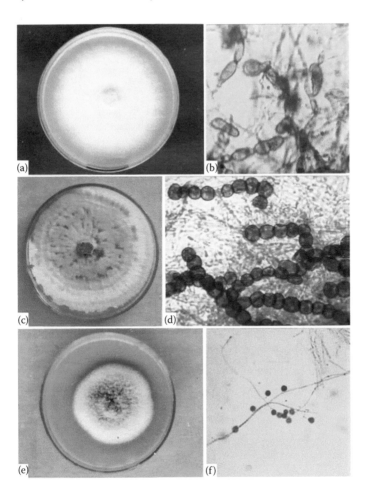

FIGURE 7.14   (a and b): *Humicola insolens*: (a) 5-day-old culture at 45°C on YpSs agar and (b) mature intercalary conidia in chains, 500×. (c and d): *Scytalidium thermophilum*: (c) 5-day-old culture at 45°C on YpSs agar and (d) chains of globose intercalary conidia, 800×. (e and f): *Thermomyces lanuginosus*: (e) 3-day-old culture at 45°C on YpSs agar and (f) conidia arising at right angles to the hyphae on conidiogenous cells, 500×.

= *Humicola insolens* Cooney & Emerson (1964) (Figures 7.14a and b and 7.15c and d)

= *Humicola fuscoatra* var. *longispora forma insolens* Cooney & Emerson (1967)

= *Humicola grisea* Traaen var. *thermoidea* Cooney & Emerson (1964) (Figure 7.15a and b)

= *Humicola insolens* Cooney & Emerson var. *thermoidea* Ellis (1982)

= *Scytalidium allahabadum* Narain, Srivastava, & Mehrotra (1983)

Description: Colonies on YpSs agar at 45°C are white at first but soon turn grayish to jet black as spore maturation proceeds. Hyphae colorless, prostrate, branched, septate, 2–5 μm wide. *Conidiogenous cells* small, 8.7 × 3.7 μm. *Conidia* dark brown,

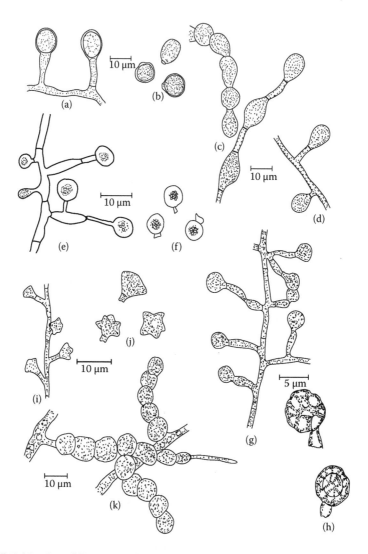

FIGURE 7.15 (a and b): *Humicola grisea* var. *thermoidea*: (a) Portion of fertile hyphae showing arrangement of conidia at right angle to the hyphae and (b) mature conidia. (c and d): *Humicola insolens*: (c) mature intercalary spores and chains of spores borne on lateral branches and (d) mature hyphae showing the development of spores on short lateral branches. (e and f): *Thermomyces ibadanensis*: (e) conidiophores showing conidium formation and (f) mature conidia showing appendages. (g and h): *Thermomyces lanuginosus*: (g) mature hyphae bearing conidia on short lateral branches and (h) mature conidia showing attachment piece and reticulate sculpturing. (i and j): *Thermomyces stellatus*: (i) fertile hyphae showing arrangement of conidia at right angle to the filament and (j) mature conidia showing characteristic stellate shape and the attachment piece. (k): *Scytalidium thermophilum*: Chains of mature spores showing the methods of development. ([a–d] and [g–k] redrawn from Cooney, D.G., and Emerson, R., *Thermophilic Fungi: An Account of Their Biology, Activities, and Classification*, W.H. Freeman and Co., San Francisco, 1964; [e and f] redrawn from Apinis, A.E., and Eggins, H.O.W., *Trans. Br. Mycol. Soc.*, 49, 629–632, 1966.)

smooth-walled, translucent, generally globose, 7–12.5 μm in diameter, or oval 11.2–14.6 × 7.5–10 μm, produced basipetally in chains on hyphal branches or developed intercalarily.
Habitat diversity: Nesting litter of chickens, mushroom compost, soil, horse dung, and wood chips.
Geographic distribution: United States, Japan, Indonesia, India, United Kingdom, and Netherlands.
Description: Based on our isolate IMI 361370.

Scytalidium indonesicum Hedger, Samson, & Basuki (1982)
Description: The growth of the fungus is very rapid at 45°C, filling the 90 mm petri dish in about 36 h. The fungus is distinguished by the production of intercalary conidia (chlamydospores). *Conidia* thick-walled, brown, ellipsoid to barrel-shaped, often with irregular outgrowths and also often constricted at the middle of the cell, 15–25 × 7–12 μm. On maturity, these conidia secede rather easily and appear irregular in shape. Dark brown and thick-walled similar but less wide conidia (arthroconidia) also develop in chains, 13–32 × 5–8 μm, and these do not secede easily.
Habitat diversity: Soil and *Dipterocarp* forest soil.
Geographic distribution: Indonesia, Java, and Sumatra.
Description: Hedger et al. (1982).

Arthrinium pterospermum (Cooke & Massee) von Arx (1981a)
Syn. = *Humicola stellata* Bunce (1961)
         = *Thermomyces stellatus* (Bunce) Apinis (1963) (Figure 7.15i and j)
Description: Colonies on YpSs agar at 37°C, dull black. Hyphae colorless, septate, about 2 μm in diameter. *Conidiogenous cells* short, up to 3 μm long, arising at right angle to the filaments. *Conidia* single on each conidiogenous cell, subglobose and colorless when young, becoming dark brown and stellate at maturity, 7.6 × 5.3 μm, separating from the conidiogenous cell and commonly retaining a short attachment piece.
Habitat diversity: Moldy hay and soil.
Geographic distribution: United Kingdom and United States.
Description: Based on Apinis (1963).

Thermomyces lanuginosus Tsiklinskya (1899) (Figures 7.14e and f and 7.15g and h)
Syn. = *Humicola lanuginosa* (Griffon and Maublanc) Bunce (1961)
         = *Sepedonium lanuginosum* (Miehe) Griffon and Maublanc (1911)
         = *Monotospora lanuginosa* (Griffon and Maublanc) Mason (1933)
         = *Acremoniella* sp. Rege (1927) Mason (1933)
         = *Acremoniella thermophila* Curzi (1930) Mason (1933)
Description: Colonies on YpSs at 45°C appear white at first, but soon turn gray, beginning at the center of the colony. Gradually, the colony turns purple brown, and the agar stains deep pink or wine color due to the secretion of diffusible substances. Mature colonies appear dull dark brown to black. *Hyphae* colorless, septate, 1.5–4 μm in diameter. *Conidiogenous cells* arise at right angle to the hyphae, 10–15 μm long, generally unbranched or rarely branched once or twice near the base, forming clusters, often septate. *Conidia* single on each conidiogenous cell, colorless, spherical, smooth-walled when young, at maturity turn dark brown and sculptured, 6–10 μm in

diameter, separating easily from the conidiogenous cell and commonly retain a short attachment piece.

Habitat diversity: Soil, moist oats, cereal grains, coal mine soils, coal spoil tips, mushroom compost, guayule rets, hay, manure, leaf mold peat, garden compost, horse, sheep and pig dung, cow, air, and various plant substances.

Geographic distribution: United States, United Kingdom, Nigeria, Ghana, India, Japan, Australia, and Indonesia.

Description: Based on our isolates IMI 361366 and 361369.

Thermomyces ibadanensis Apinis and Eggins (1966) (Figure 7.15e and f)
Description: Colonies on MEA velvety gray green, spreading with a narrow, colorless to green marginal zone. Hyphae hyaline, septate, branched, 2–3 μm in diameter. *Conidiophores* short, formed as simple or branched, clustered, more or less cylindrical or spindle-shaped side branches of the main hyphae, 6–14 μm long and 1.5–3 μm wide, cut off from the main hyphae by septa at their base, tip narrow, bearing a single conidium. *Conidia* hyaline at first, later turning brown, single-celled, smooth, thick-walled, about 4–8 μm in diameter with a single cylindrical, stalk-like appendage 1.5–8 × 1.5–2.5 μm.

Habitat diversity: Oil palm kernel stacks and soil.

Geographic distribution: Nigeria and India.

Description: Based on Apinis and Eggins (1966).

7.4   NOMENCLATURAL DISAGREEMENT AND SYNONYMIES

As discussed in the preceding section, thermophilic fungi form a small, physiologically distinct assemblage of little more than 40 species. The growth of these fungi at high temperature is a unique feature. The first thermophilic fungus, *Mucor pusillus*, was isolated from bread and described by Lindt (1886) more than a century ago. Tsiklinskaya discovered another thermophilic fungus, *Thermomyces lanuginosus*, in 1899 from potato inoculated with garden soil. Later, Hugo Miehe (1907) investigated the cause of thermogenesis of stored agricultural produce. He reported four thermophilic fungi, *Mucor pusillus*, *Thermoascus aurantiacus*, *Thermoidium sulfureum*, and *Thermomyces lanuginosus*, from self-heating hay. Miehe, who explained the self-heating of hay and other plant materials, was the first scientist to work extensively on thermophilic fungi. Following the work of Miehe, thermophilic fungi were isolated from various natural substrates (Noack, 1912; Allen, 1950). A detailed discussion on their habitat diversity is presented in Chapter 4. Cooney and Emerson (1964), for the first time, presented a comprehensive account of thermophilc fungi in which they provided taxonomic descriptions of the 13 species of thermophilc fungi known at that time. This monograph, in English, for the first time proved to be a gateway for mycologists and fungal taxonomists to explore this uncommon and fascinating group of fungi. Since then, a number of thermophilic fungi have been isolated and documented in the literature. A Russian compilation of descriptions and published illustrations of thermophilic fungi was prepared by Bilai and Zakharchenko (1987); 38 species were considered, but a few are not strict thermophiles. Finally, according to Abdullah and Al-Bader (1990), around 70 species detected in various substrates are now reported to be thermophilic or thermotolerant. Mouchacca (1997) reviewed the taxonomic status of thermophilic fungi. In this chapter, an attempt was made to provide detailed taxonomic descriptions, a simplified key to their identification, and the natural habitats of known thermophilic fungi. The geographic distribution of these fungi as available from literature has been included in the text.

While reviewing the literature, we faced difficulties on account of the confusion created by several investigators regarding thermophily. One such example is that of Crisan (1973), who presented a list of 55 thermophilous fungi; however, only half of these were thermophiles in the sense of Cooney and Emerson (1964). Further, difficulties in defining true thermophiles emanate from the absence of reliable growth data covering a wide range of temperatures for most taxa proposed to be thermophiles. Thus, such data are needed to ascertain the true nature of a few members of this physiologically distinct group of fungi.

The ability of thermophilic fungi to develop at high temperature is depicted by a few Zygomycetes, and several Ascomycetes and Deuteromycetes (Hyphomycetes). No Basidiomycetes were found to be thermophilic. Five thermophilic zygomycetous fungi discussed in this chapter comprise three *Rhizomucor* and two *Rhizopus* species. The genera *Rhizomucor*, *Mucor*, and *Rhizopus* are classified under the family Mucoraceae in the order Mucorales, which is a primitive and early divergent group of fungi. The genus *Rhizomucor* consists of *Mucor*-like fungi that produce nonapophysate sporangia and branched sporangiophores, but unlike *Mucor*, they form rhizoids. *Rhizomucor* species are evidently distinct from *Mucor* by virtue of their thermophilic nature and some morphological features (Zhou et al., 2014). Further, as reported by Mouchacca (1997), the validity of *Rhizomucor tauricus* and *Rhizomucor nainitalensis* is questioned. From an ecological perspective, the equivocal application of the now widely accepted (in spite of its limitations) definition as proposed by Cooney and Emerson (1964) led to the consideration of well-established thermotolerants as thermophiles. Ellis (1981) considered all *Rhizopus* able to grow at 45°C to be thermophilic, although they exhibit growth well below 20°C. These zygomycetes and some other true mesophilic fungi are also currently considered to be thermophiles in several publications focusing on biotechnological problems (Singh et al., 2016). Several authors also classify all fungi that develop in isolation plates incubated at 45°C as thermophiles (Mouchacca, 2007).

The thermophilic ascomycetous fungi comprise a group of 24 species belonging to 10 genera. Following Kirk et al. (2008), *Thermoascus*, *Rasamsonia*, *Dactylomyces*, *Coonemeria*, and *Talaromyces* belong to the family Trichocomaceae in Eurotiales. The remaining genera in Ascomycetes are representatives of the families Chaetomiaceae (*Chaetomium* and *Thielavia*), Ceratostomatatceae (*Corynascus*), Familiae incertae sedis (*Melanocarpus*), and Microascaceae (*Canariomyces*) of the order Sordariales. *Dactylomyces*, *Thermoascus*, and *Canariomyces* appear monospecific as far as thermophily is concerned. *Talaromyces* comprises two species, whereas six species and two varieties in this text represent *Chaetomium*. Most, but not all, thermophilic ascomycetes have a conidial anamorphic state. Species belonging to *Chaetomium* do not develop conidia of any kind, and thus do not possess an anamorphic state. Species belonging to *Thermoascus* regularly produce conidia, and our own culture of *Thermoascus* produced catenate condia and can be regarded as anamorphic, having a *Polypaecilum* stage (Salar and Aneja, 2007). For the time being, one representative species of *Thermoascus* has been retained. The genus *Coonemeria* is still a matter of debate, as stressed by Mouchacca (2000). The genus *Talaromyces* has the *Penicillium* anamorphic stage. Regarding the genus *Thielavia*, *T. pingtungia* has no conidial state, a feature characteristic of all known chaetomia. *T. australiensis* was reported to have an anamorphic state of the *Trichosporiella*. *T. terricola* is associated with *Acremonium alabamensis*, a distinctive anamorphic state of the teleomorph. The anamorphs of several species of *Corynascus* (von Arx, 1975) are placed in *Myceliophthora* (van Oorschot, 1977, 1980). This genus is appropriate for disposition of cellulosic sordariaceous anamorphs in which solitary conidia (aleurioconidia) are borne on swollen cells or are sessile.

Based on phenotypic and physiological characters, Houbraken et al. (2012) created a new genus, *Rasamsonia* gen. nov., to accommodate some species of *Talaromyces*. They opined that the majority of *Talaromyces* species are mesophiles; however, exceptions are species within the sections *Emersonii* and *Thermophila*. They further observed that section *Emersonii* includes *Talaromyces emersonii*, *Talaromyces byssochlamydoides*, *Talaromyces bacillisporus*, and *Talaromyces leycettanus*, all of which grow well at 40°C. *Talaromyces bacillisporus* is thermotolerant, *Talaromyces leycettanus* is thermotolerant to thermophilic, and *T. emersonii* and *T. byssochlamydoides* are truly thermophilic (Stolk and Samson, 1972). The sole member of the *Talaromyces* section *Thermophila*, *T. thermophilus*, grows rapidly at 50°C (Evans, 1971; Evans and Stolk, 1971; Stolk and Samson, 1972). Further, macro- and microscopical analysis of *T. emersonii*, *T. byssochlamydoides*, *Talaromyces eburneus*, *Geosmithia argillacea*, and *Geosmithia cylindrospora* showed that these species share various phenotypic and physiological characters. Based on phenotypic, physiological, and molecular data, Houbraken et al. (2012) proposed to transfer the species *T. emersonii*, *T. byssochlamydoides*, *Talaromyces eburneus*, *Geosmithia argillacea*, and *Geosmithia cylindrospora* to *Rasamsonia* gen. nov. in honor of Robert A. Samson, for his excellent work in fungal taxonomy. Recently, one more species of truly thermophilic fungi belonging to the genus *Rasamsonia*, *R. composticola*, has been described by Su and Cai (2013) from paddy straw compost in China. It is further clarified that *Rasamsonia* phenotypically resembles *Paecilomyces* and its teleomorph *Byssochlamys* and is placed within the order Eurotiales, family Trichocomaceae.

In Eurotiales, thermophilic fungi are contained in the genera *Talaromyces*, *Thermoascus*, *Thermomyces*, *Coonemeria*, and *Dactylomyces* (Stolk, 1965; Apinis, 1967; Stolk and Samson, 1972; Mouchacca, 1997). The study of Houbraken et al. (2012) shows that no thermophiles are accommodated in *Talaromyces*. *T. emersonii* and *T. byssochlamydoides* are pooled in *Rasamsonia*. Further, *T. thermophilus* is phylogenetically closely related to the type species of *Thermomyces*, *T. lanuginosus*, and not to the type of *Talaromyces*, *T. flavus*. It has been observed that *T. thermophilus* and *T. lanuginosus* share similar features, including thermophilicity, thick-walled chlamydospores, or chlamydospore-like conidia. However, further research is required to clarify the taxonomy of *Thermomyces* and the relationship between *T. thermophilus* and the other species in *Thermomyces* (Houbraken et al., 2012). According to Morgenstern et al. (2012), the species pairs *T. thermophilus* and *T. lanuginosus*, *Talaromyces byssochlamydoides*, and *T. emersonii* receive strong support as being monophyletic. However, for the time being *Thermomyces* is included in the Deuteromycetes in this text to avoid confusion, as its placement with regard to taxonomic classification is obscure (Mouchacca, 1997; Prakash and Johri, 2016).

Deuteromycetes (anamorphic fungi) comprises 14 species belonging to nine genera: Acremonium, Arthrinium, Papulospora, Remersonia, Malbranchea, Chrysosporium, Myceliophthora, Scytalidium, and Thermomyces. The imperfect state of Corynascus thermophilus is Myceliophthora fergusii, and it should not be confused with the teleomorph of Myceliophthora thermophila. Malbranchea cinnamomea is the only species of Malbranchea with thermophilic nature; other species of this genus are mesophiles.

Thermophilic fungi have received substantial attention in industry for their potential to produce thermostable enzymes and as production platforms tolerant of high temperature. The species commonly known as *Scytalidium thermophilum* is one of the most frequently encountered organisms in surveys of thermophilic fungi, and for the sake of convenience is included in Deuteromycetes, as reported in earlier literature. The current text also followed suit. There is evidence that it is ecologically and economically important, for example, in the context of commercial mushroom growing. *Scytalidium thermophilum* is a synonym of *Torula thermophila*, which is frequently reported from composts. Recently, Natwig et al. (2015) vehemently objected to the placement of this species and argued against placing this species in the genus *Scytalidium* or any other existing genus. They therefore proposed a new genus and combination, *Mycothermus thermophiles*, for this species.

Acremonium albamensis is reported to be the imperfect state of *Thielavia terricola*. The taxonomic status of *Thermomyces lanuginosus* as the first assessed thermophilic fungus is supported by the majority of the investigators. Its nomenclatural history involves genera such as *Acremoniella*, *Humicola*, *Monotospora*, and *Sepedonium*. The name *Remersonia thermophila* has been proposed for *Stilbella*, as the latter had an uncertain position and required a more appropriate genus (Seifert et al., 1997).

As suggested by Mouchacca (1997), strict restriction to nomenclatural rules governing citations of fungal binomials is fundamental. Many researchers use their own strains for the production of metabolites; this is, of course, welcome, as some strains of related isolates are highly productive. Therefore, the authors of applied research dealing with thermophiles should necessarily follow such regulations in order to stabilize names of strains used for applied research or in produced goods. This will definitely bring an end to the chaotic state prevailing, especially in publications relating to fungal taxonomy and biotechnology. Further, genome-wide studies of a number of thermophilic fungi will provide strong support in clarifying the taxonomic position and genes for degrading their natural substrate, for coping with different ecological conditions, and for mediating various stresses during thermophilic growth.

REFERENCES

Abdullah, S.K., and Al-Bader, S.M. 1990. On the thermophilic and thermotolerant mycoflora of Iraq. *Sydowia* 42:1–7.

Abdullah, S.K., and Zora, S.E. 1993. *Chaetomium mesopotamicum*, a new thermophilic species from Iraqi soil. *Cryptogamic Botany* 3:387–389.

Allen, M.B. 1950. The dynamic nature of thermophily. *Journal of General Physiology* 33:205–214.

Ames, L.M. 1963. *A Monograph of the Chaetomiaceae*. U.S. Army Research and Development Series 2. Lehre, Germany: Bibliotheca Mycologica Cramer.

Aneja, K.R., and Kumar, R. 1994. *Chaetomium senegalense*—A new record from India. *Proceedings of the National Academy of Sciences India, Section B* 64: 229–230.

Apinis, A.E. 1963. Occurrence of thermophilic fungi in certain alluvial soils near Nottingham. *Nova Hedwigia* 5:57–78.

Apinis, A.E. 1967. *Dactylomyces* and *Thermoascus*. *Transactions of the British Mycological Society* 50:573–582.

Apinis, A.E., and Chesters, C.G.C. 1964. Ascomycetes of some salt marshes and sand dunes. *Transactions of the British Mycological Society* 47:419–435.

Apinis, A.E., and Eggins, H.O.W. 1966. *Thermomyces ibadanensis* sp. nov. from oil palm kernel stacks in Nigeria. *Transactions of the British Mycological Society* 49:629–632.

Austwick, P.K.C. 1976. Environmental aspects of *Mortierella wolfii* infection in cattle. *New Zealand Journal of Agricultural Research* 19:25–33.

Awao, T., and Otsuka, S. 1974. Notes on thermophilic fungi in Japan. *Transactions of the Mycological Society of Japan* 15:7–22.

Awao, T., and Udagawa, S.I. 1983. A new thermophilic species of *Myceliophthora*. *Mycotaxon* 16:436–440.

Basu, M. 1984. *Myceliophthora indica* is a new thermophilic species from India. *Nova Hedwigia* 40:85–90.

Bilay, V.T., and Zakharchenko, B.A. 1987. *Opredelitel' Termofil'nykh Gribov*. Kiev: Kholodny Institute of Botany, Ukrainian Academy of Sciences.

Bunce, M.E. 1961. *Humicola stellatus* sp. nov., a thermophilic mould from hay. *Transactions of the British Mycological Society* 44:372–376.

Carmichael, J.W. 1962. *Chrysosporium* and some other aleurosporic hyphomycetes. *Canadian Journal of Botany* 40:1137–1174.

Chen, K.Y., and Chen, Z.C. 1996. A new species of *Thermoascus* with a *Paecilomyces* anamorph and other thermophilic species from Taiwan. *Mycotaxon* 50:225–240.

Cole, G.T., and Samson, R.A. 1979. *Patterns of Development in Conidial Fungi*. London: Pitman.

Cooney, D.G., and Emerson, R. 1964. *Thermophilic Fungi: An Account of Their Biology, Activities, and Classification*. San Francisco: W.H. Freeman and Co.

Crisan, E.V. 1973. Current concept of thermophilism and thermophilic fungi. *Mycologia* 65:1171–1198.

Curzi, M. 1930. Ricerche morphologiche e sperimentali su micromicete termofilo (*Acremoniella thermophila* Curzi). *Bolletino Stazione di Patologia Vegetale, Roma, N.S.* 10:222–280.

Ellis, D.H. 1981. Ultrastructure of thermophilic fungi IV. Conidial ontogeny in *Thermomyces*. *Transactions of the British Mycological Society* 77:229–241.

Ellis, D.H. 1982. Ultrastructure of thermophilic fungi V. Conidial ontogeny in *Humicola grisea* var. *thermoidea* and *H. insolens*. *Transactions of the British Mycological Society* 78:129–139.

Emmons, C.W. 1930. *Coniothyrium terricola* proves to be a species of *Thielavia*. *Bulletin of Torrey Botanical Club* 57:123–126.

Emmons, C.W. 1932. The development of the ascocarp in two species of *Thielavia*. *Bulletin of Torrey Botanical Club* 59:415–420.

Evans, H.C. 1971. Thermophilous fungi of coal spoil tips. I. Taxonomy. *Transactions of the British Mycological Society* 57:241–254.

Evans, H.C., and Stolk, A.C. 1971. *Talaromyces leycettanus* sp. nov. *Transactions of the British Mycological Society* 56:45–49.

Fergus, C.L. 1964. Thermophilic and therotolerant molds and actinomycetes of mushroom compost during peak heating. *Mycologia* 56:267–283.

Fergus, C.L. 1971. The temperature relationships and thermal resistance of a new thermophile *Papulaspora* from mushroom compost. *Mycologia* 63:426–431.

Fergus, C.L., and Sinden, J.W. 1969. A new thermophilic fungus from mushroom compost: *Thielavia thermophila* spec. nov. *Canadian Journal of Botany* 47:1635–1637.

Gams, W., and Lacey. 1972. *Cephalosporium*-like hyphomycetes. Two species of *Acremonium* from heated substrates. *Transactions of the British Mycological Society* 59:519–522.

Griffon, E., and Maublanc, A. 1911. Deux moisissures thermophiles. *Bulletin of the Society of Mycology, France* 27:68–74.

Guarro, J., Abdullah, S.K., Al-Bader, S.M., Figueras, M.J., and Gene, J. 1996. The genus *Melanocarpus*. *Mycological Research* 100:75–78.

Hedger, J.N., Samson, R.A., and Basuki, T. 1982. *Scytalidium indonesicum*. *Transactions of the British Mycological Society* 78:365.

Houbraken, J., Spierenburg, H., and Frisvad, J.C. 2012. *Rasamsonia*, a new genus comprising thermotolerant and thermophilic *Talaromyces* and *Geosmithia* species. *Antonie van Leeuwenhoek* 101:403–421.

Johri, B.N., Satyanarayana, T., and Olsen, J. 1999. *Thermophilic Moulds in Biotechnology*. Dordrecht, the Netherlands: Kluwer Academic Publishers.

Joshi, M.C. 1982. A new species of *Rhizomucor*. *Sydowia* 35:100–103.

Kendrick, W.B. 1971. *Taxonomy of Fungi Imperfecti*. Toronto: University of Toronto Press.

Kirk, P.M., Cannon, P.F., Minter, D.W., and Stalpers, J.A. 2008. *Ainsworth & Bisby's Dictionary of Fungi*. Wallingford, UK: CAB International.

Lendner, A. 1908. Les Mucorinees de la Suisse. *Materiaux Pour la Flore Cryptogamique Suisse* 3:1–177.

Lendner, A. 1930. *Bulletin de la Société Botanique de Genève* 21:260.

Lindt, W. 1886. Mitteilungen über einige neue pathogene Shimmelpilze. *Archiv für Experimentelle Pathologie und Pharmakologie* 21:269–298.

Lodha, B.C. 1978. Generic concepts in some Ascomycetes occurring on dung. In Taxonomy of Fungi: Proceedings of the International Symposium on Taxonomy of Fungi, Madras, India, 1973, 241–257.

Lucet, A., and Costantin, J. 1899. Sur une nouvelle Mucorinee pathogene. *Comptes Rendus Hebdomadaires des Séances de l'Académie des Sciences, Paris* 129:1033.

Lucet, A., and Costantin, J. 1901. Contribution a l'etude des Mucorinees pathogene. *Archives de Parasitologie* 4:362–408.

Mal'chevskaya, T.P. 1939. Nanchno-issled. *Lab. Razv. Sel'khoz Zhivot.* 13:26.

Malloch, D., and Cain, R.F. 1972. The Trichocomataceae: Ascomycetes with *Aspergillus*, *Paecilomyces*, and *Penicillium* imperfect states. *Canadian Journal of Botany* 50:2613–2628.

Malloch, D., and Cain, R.F. 1973. The genus *Thielavia*. *Mycologia* 65:1055–1077.

Mason, E.W. 1933. *Annotated account of fungi received at the Imperial Mycological Institute. List II (Fascicle 2)*. Kew: Imperial Mycological Institute.

Miehe, H. 1907. *Die Selbsterhitzung des Heus. Eine Biologische Studie*. Jena, Germany: Gustav Fischer.

Morgan-Jones, G. 1974. Notes on Hyphomycetes-V. A new thermophilic species of *Acremonium*. *Canadian Journal of Botany* 52:429–431.

Morgenstern, I., Powlowski, J., Ishmael, N., Darmond, C., Marqueteau, S., Moisan, M., Quenneville, G., and Tsang, A. 2012. A molecular phylogeny of thermophilic fungi. *Fungal Biology* 116:489–502.

Mouchacca, J. 1997. Thermophilic fungi: Biodiversity and taxonomic status. *Cryptogamie Mycologie* 18:19–69.

Mouchacca, J. 2000. Thermophilic fungi and applied research: A synopsis of name changes and synonymies. *World Journal of Microbiology & Biotechnology* 16:881–888.

Mouchacca, J. 2007. Heat tolerant fungi and applied research: Addition to the previously treated group of strictly thermotolerant species. *World Journal of Microbiology & Biotechnology* 23:1755–1770.

Narain, R., Srivastava, R.B., and Mehrotra, B.S. 1983. A new thermophilic species of *Scytalidium* from India. *Zentralblatt für Mikrobiologie* 138:569–572.

Natwig, D.O., Taylor, J.W., Tsang, A., Hutchinson, M.I., and Powell, A.J. 2015. *Mycothermus thermophilus* gen. et comb. nov., a new home for the itinerant thermophile *Scytalidium thermophilum* (*Torula thermophila*). *Mycologia* 107:319–327.

Naumov, N.A. 1935. *Opredelitel' Mukorovykh (Mucorales)*. Moscow.

Noack, K. 1912. Beiträge zur Biologie der thermophilen Organismen. *Jahrbücher für Wissenschaftliche Botanik* 51:593–648.

Paterson, R.R.M., and Lima, N. 2017. Thermophilic fungi to dominate aflatoxigenic/mycotoxigenic fungi on food under global warming. *International Journal of Environmental Research and Public Health* 14:199.

Pitt, J.I. 1979. *The Genus Penicillium and Its Teleomorphic States Eupenicillium and Talaromyces*. London: Academic Press.

Prakash, A., and Johri, B.N. 2016. *Thermomyces lanuginosus*: A true representative of thermophilic, fungal world. *Kavaka* 47:46–53.

Raper, K.B., and Thom, C. 1949. *A Manual of Penicillia*. Baltimore: Williams and Wilkins Co.

Rege, R.D. 1927. Biochemical decomposition of cellulosic materials with special reference to the action of fungi. *Annals of Applied Biology* 14:1–44.

Rothschild, L.J., and Mancinelli, R.L. 2001. Life in extreme environments. *Nature* 409: 1092–1101.

Salar, R.K., and Aneja, K.R. 2007. Thermophilic fungi: Taxonomy and biogeography. *Journal of Agricultural Technology, Bangkok* 3:77–107.

Schipper, M.A.A. 1978. On the genera *Rhizomucor* and *Parasitella*. *Studies in Mycology* 17:53–71.

Seifert, K.A., Samson, R.A., Boekhout, T., and Louis-Seize, G. 1997. *Remersonia*, a new genus for *Stilbella thermophila*, a thermophilic mould from compost. *Canadian Journal of Botany* 75:1158–1165.

Siebenmann, F. 1889. *Die Schimmelmykosen des menschlichen Ohres*. Wiesbaden, Germany: Bergmann.

Sigler, L., and Carmichael, J.W. 1976. Taxonomy of *Malbranchea* and some other hyphomycetes with arthroconidia. *Mycotaxon* 4:349–488.

Singh, B., Poças-Fonseca, M.J., Johri, B.N., and Satyanarayana, T. 2016. Thermophilic molds: Biology and applications. *Critical Reviews in Microbiology* 42:985–1006.

Sopp, J.O. 1912. *Monographie der Pilzgruppe Penicillium mit besonderer Berücksichtigung der in Norwegen gefundenen Arten*. Videnskabs-Selskabets i Christiania, Skrifter, 1 Math.-Naturv. Kl., N° 11, i–vi.

Stolk, A.C. 1965. Thermophilic species of *Talaromyces* Benjamin and *Thermoascus* Miehe. *Antonie van Leeuwenhoek* 31:262–276.

Stolk, A.C., and Samson, R.A. 1972. The genus *Talaromyces*—Studies on *Talaromyces* and related genera. II. *Studies in Mycology* 2:1–65.

Su, Y.Y., and Cai, L. 2013. *Rasamsonia composticola*, a new thermophilic species isolated from compost in Yunnan, China. *Mycological Progress* 12:213–221.

Tansey, M.R., and Jack, M.A. 1975. *Thielavia australiensis* sp. nov., a new thermophilic fungus from incubator bird (mallee fowl) nesting material. *Canadian Journal of Botany* 53:81–83.

Thakre, R.P., and Johri, B.N. 1976. Occurrence of thermophilic fungi in coal-mine soils of Madhya Pradesh. *Current Science* 45:271–273.

Tsiklinskaya, P. 1899. Sur les mucedinees thermophiles. *Annales de l'Institut Pasteur, Paris* 13:500–505.

Ueda, S., and Udagawa, S.I. 1983. *Thermoascus aegyptiacus*, a new thermophilic ascomycete. *Transactions of the Mycological Society of Japan* 24:135–142.

van Klopotek, A. 1974a. Revision der thermophilen *Sporotrichum* arten: *Chrysosporium thermophilum* (Apinis) *comb. nov.* und *Chrysosporium fergusii sp. nov. = status conidialis* von *Corynascus thermophilus* (Fergus und Sinden) *comb. nov. Archives of Microbiology* 98:366–369.

van Klopotek, A. 1974b. *Thielavia heterothallica spec. nov.*, die perfekte form von *Chrysosporium thermophilum. Archives of Microbiology* 107:223–224.

van Oorschot, C.A.N. 1977. The genus *Myceliophthora. Persoonia* 9:404–408.

van Oorschot, C.A.N. 1980. A revision of *Chrysosporium* and allied genera. *Studies in Mycology* 20:1–89.

van Oorschot, C.A.N., and de Hoog, G.S. 1984. Some hyphomycetes with thallic conidia. *Mycotaxon* 20:129–132.

von Arx, J.A. 1975. *On Thielavia and Some Similar Genera of Ascomycetes.* Netherlands.

von Arx, J.A. 1981a. *The Genera of Fungi Sporulating in Pure Culture.* Lubrecht & Cramer Ltd.

von Arx, J.A. 1981b. On *Monilia sitophila* and some families of Ascomycetes. *Sydowia* 34:12–29.

von Arx, J.A., Figueras, M.J., and Guarro, J. 1988. Sordariaceous ascomycetes without ascospore ejaculation. *Beihefte zur Nova Hedwigia* 94:1–104.

Zhou, P., Zhang, G., Chen, S., Jiang, Z., Tang, Y., Henrissat, B., Yan, Q., Yang, S., Chen, C.F., Zhang, B., and Du, Z. 2014. Genome sequence and transcriptome analyses of the thermophilic zygomycete fungus *Rhizomucor miehei. BMC Genomics* 15:29.

8 The Conflict of Name Change and Synonymies

8.1 INTRODUCTION

For a long time, fungal taxonomists have been faced with the daunting task of understanding the dual nature of certain pleomorphic fungi to produce spores and conidia as sexual and asexual reproductive structures, respectively, within the same species. This led to the dual nomenclature system for the anamorph (asexual forms) and the teleomorph (sexual forms) of the same species, despite the fact that it is illogical to assign multiple names to one species. Dual nomenclature refers to the use of more than one name for a single taxon and was established in the International Code of Botanical Nomenclature (ICBN) in 1910 to accommodate the problem of naming fungi that exhibit pleomorphic life cycles (Cline et al., 2005). Article 59 of the ICBN governs the naming of such fungi. The article thus allows the use, for any taxon, of either the teleomorph or anamorph name as valid. The Fungi kingdom is hyperdiverse and morphologically enigmatic. Approximately 100,000 species of fungi are accepted in the current taxonomy (Kirk et al., 2008); however, more than 400,000 fungal species names—including numerous synonyms—are recorded in the literature, and it is expected that millions of new species still anticipate description (Bass and Richards, 2011).

The binomial nomenclature of Carolus Linnaeus envisaged a unique nomenclature for every organism containing a generic and a species name. It was learned that Linnaeus based his plant taxonomy on floral morphology on the demonstrated notion that each plant had but one type of flower. The taxonomic concept of Linnaeus was extended to fungi, as they were considered to be plants. Due to his strong taxonomic influence, Linnaeus's mycological contemporaries (Fries, the front runner), despite seeing more than one type of "seed" through their lenses, could not argue about pleomorphy in fungi. In due course, this rule deviated and dual nomenclature became rampant in some groups of fungi. Two excellent articles on the history of fungi that produce both meiotic and mitotic spores, that is, pleomorphic fungi, that could help in understanding the origin of dual nomenclature in fungi are Weresub and Pirozynski (1979) and Tulasne and Tulasne's work published in *Selecta Fungorum Carpologia* in 1861.

Taylor (2011) reproduced excerpts of the article by the Tulasne brothers:

> In the Mucedinei [Fries] sees the conidia … but everywhere he categorically denies that there occur "two kinds of sporidia on the same plant," exactly as if he had heard, sounding in his ears, the loud voice of Linnaeus, crying "It would be a remarkable doctrine—that there could exist races differing in fructification, but possessing one and

the same nature and power; that one and the same race could have different fructifica-
tions; for the basis of fructification, which is also the basis of all botanical science, would
thereby be destroyed, and the natural classes of plants would be broken up." (Tulasne
and Tulasne 1861)

They further go on to chastise Linnaeus, adding, "But since the illustrious author
always completely abjured the use of magnifying glasses, and therefore scarcely ever
tried to describe accurately either conidia or spores, we fear (may he pardon the
statement) that he really knew very few seeds of either kind" (Tulasne and Tulasne,
1861). Clearly, they favored pleomorphy in fungi when they wrote, "The fungus upon
which we are now touching (*Pleospora*) is not only almost the commonest of all
belonging to its order, but also affords a wonderful proof of our doctrine concerning
the multiple nature of the seeds of species of fungi" (Tulasne and Tulasne 1861: 248).
One cannot help but speculate whether they presumed not only that their work was
controversial, but also that the mycological world was heading toward dual nomenclature, when they wrote, "As today we have seen the various members of the same
species now unwisely torn from one another against the laws of nature" (Tulasne and
Tulasne 1861: 189). Later, in 1882 Saccardo, in his *Sylloge Fungorum*, propounded
the use of mature anamorph morphology for the classification and identification of
mitosporic fungi. However, his systematics was not based on evolutionary relationships. A separate systematics involving studies on mitosporic development based on
evolutionary relationships began with the work of Vuillemin (1910), Mason (1933,
1937), Hughes (1953), Tubaki (1958), Barron (1968), and Cole and Samson (1979).
These studies were followed by new approaches in fungal systematics, that is, cladistic analysis and access to nucleic acid variation, which are now great paraphernalia
for removing anomalies relating to fungal taxonomy and weeding out redundant
published names while accelerating the naming of newly discovered species.

In the last decade, strategies for maintaining a stable nomenclature have been in
the forefront of fungal taxonomy, including proposals for the correct anamorph–
teleomorph nomenclature, for subgeneric taxa, and for a list of accepted names (Pitt
and Samson, 2007). As for other members of the fungal kingdom, thermophilic fungi
also face misleading taxonomic decisions. Mouchacca (2000) made an attempt to
resolve taxonomic decisions and name changes for a number of thermophilic fungi.
Thermophilic fungi have received considerable attention due to their ability to
produce thermostable enzymes with varied biotechnological applications. Therefore,
the accurate naming of such fungi becomes a prerequisite in their industrial application, as their misidentification could lead to a chaotic state in the binomials of some
thermophiles in published work. Further, it is also known that a few taxa encountered
in ecological studies could have been misidentified, while some names reported in
biotechnological work are obscured with uncertainties.

8.2   THE CONFLICT OVER NAME CHANGE

Names act as the communication tools about different organisms among basic and
applied scientists, students, and the general public. In recent times, a debate has spurred

among mycologists regarding the adoption of a unified name for fungal species exhibiting both teleomorphic and anomorphic stages. Although thermophilic fungi constitute a small group spreading into almost all classes of fungi, many of them have been identified and named based on their sexual (Ascomycetes) and asexual (Deuteromycetes) stage. Names can be based on distinct characteristics, host species association, the locality, personal names (as an honor to another mycologist), or other preferences of individuals (e.g., *Remersonia thermophila* reflects the renowned mycologist Ralph Emerson, and *Chaetomium thermophile* refers to the organism's ability to grow at high temperatures). However, fungal nomenclature had an unusual situation for several decades, creating great confusion among fungal taxonomists and biotechnologists.

Accurate identification of fungi is of utmost practical importance in various fields such as clinical microbiology, plant pathology, biodeterioration of materials, biotechnology, and environmental studies. The dual system of nomenclature for pleomorphic fungi has persisted for more than a century. Although the dual system had many advantages, particularly for the morphological identification of a fungus, in the molecular biology era, this anomaly (dual nomenclature) appears to be inappropriate. The system of dual nomenclature was in conflict with the time-honored principle of one organism, one name. Thus, a change toward unification was postulated mainly by molecular phylogeneticists, while other taxonomists were afraid of a flood of name changes inevitably following such a change. At the same time, mycologists around the world have suggested the need to create a separate MycoCode for the kingdom of fungi (Taylor, 2011). Pedro Crous, a long-term proponent of the dual system, converted to the unifying camp under the influence of David Hawksworth, John Taylor, Keith Seifert, and others and convened a "One Fungus—One Name" symposium in Amsterdam on April 19–20, 2011. On this occasion, the Amsterdam Declaration (Hawksworth et al., 2011) was signed by more than 80 mycologists from around the world, which voted to abolish the dual system of fungal nomenclature. In July 2011, during the 18th session of the International Botanical Congress (IBC) held in Melbourne, the Nomenclature Section of the IBC accepted Hawksworth's proposal (Hawksworth et al., 2009) and the code was renamed: International Code of Nomenclature for algae, fungi, and plants (ICN) (McNeill et al., 2012). Moreover, the Nomenclature Section also voted to abolish Latin descriptions (English will now suffice), to allow the publication of new names in online-only journals (earlier, print was required); however, registration of new fungal names in a publically accessible database, such as Index Fungorum or MycoBank, was made mandatory (Norvell, 2011). Barring some fungal taxonomists, the unified system was welcomed by the majority of scientists around the world, who foresee better communication regarding correct names for a large number of fungi, including many economically important pathogens and industrial organisms.

According to Article 59.1 of the ICN, pleomorphic fungi should have one nomenclaturally correct name in a particular classification system. This further created great confusion as to which name should be legitimate. Walter Gams (2016) reviewed the recent changes in fungal nomenclature and their impact on the naming of microfungi. The author elaborated on Article 59.1 and its implications in the light of names introduced independently for different morphs of a pleomorphic fungus remain legitimate; however, a choice of one of them must be made.

59.1. A name published prior to 1 January 2013 for a taxon of non-lichen-forming *Ascomycota* and *Basidiomycota*, with the intent or implied intent of applying to or being typified by one particular morph (e.g. anamorph or teleomorph), may be legitimate even if it otherwise would be illegitimate under Art. 52 on account of the protologue including a type (as defined in Art. 52.2) referable to a different morph. If the name is otherwise legitimate, it competes for priority (Arts. 11.3 and 11.4; see also Art. 57.2).

Accordingly, introduction of a name on or after January 1, 2013, for a morph different from one previously named for the same species makes it illegitimate. Dayarathne et al. (2016), while reviewing the taxonomic utility of old names in current fungal classification and nomenclature, presented two major consequences of Article 59.1 of the ICN: First, the correct name is now the earliest published legitimate name; that is, the principle of priority applies regardless of the teleomorph or anamorph stage represented by the name-bearing type. Second, many names had been introduced for different morphs of a single taxon; those names would, strictly speaking, be either alternative names (not validly published, if proposed at the same time) or nomenclaturally redundant and illegitimate (if proposed for a taxon where one morph already had a legitimate name). Thus, in view of the potential disruption this would cause, names in those two categories are ruled as validly published and legitimate (Article 59.1). The new rules have started to bring about drastic changes in the nomenclature of fungi, including thermophilic fungi. Now the onus is on mycologists of the world to implement these changes in their own publications, and also update and compile a list of name changes within this important group of fungi. A list of name changes is presented in Table 8.1. Because this may not be a complete inventory, readers must consult the latest literature and the ICN for updating the changed names of taxa. Thermophilic fungi are described and distributed among the Fungi kingdom in different taxonomic groups, with the majority of them belonging to Ascomycetes and Deuteromycetes (Salar and Aneja, 2007). However, in the light of recent changes in fungal taxonomy, the one-fungus, one-name movement, and the new species described, thermophilic fungi may comprise 44 species belonging to a group of 19 genera (*Acremonium*, *Arthrinium*, *Canariomyces*, *Chaetomium*, *Humicola*, *Malbranchea*, *Melanocarpus*, *Myceliophthora*, *Myriococcum*, *Rasamsonia*, *Remersonia*, *Rhizomucor*, *Scytalidium*, *Sordaria*, *Thermoascus*, *Thermomyces*, *Thermomucor*, *Thermophymatospora*, and *Thielavia*) (de Oliveira et al., 2015).

8.3 THE CONFLICT OF ONE FUNGUS, WHICH NAME?

In the last two decades, thermophilic fungi have been prospected for various enzymes of biotechnological relevance. Several of these fungi exhibit pleomorphism, and thus cause a taxonomical problem for industrial biotechnologists, creating great confusion over their names. Following the Amsterdam Declaration, which resulted in the one-fungus, one-name regime, the criteria for the naming of fungi have changed entirely.

TABLE 8.1

Taxonomic Terms Frequently Used for Specimens and Cultures of Fungi and Their Definitions

Term	Taxonomic Definition
Authentic	One named by the author of the name, generally after it was published, or, if the name is a combination, the author of the basionym.
Voucher	One used in a particular study, either for experimentation or to support an identification, enabling the same material to be used by or verified by later researchers.
Representative	One or more from a large set or specimens or cultures considered to serve as vouchers where it is impractical to preserve all those used or cited in a particular study.
Sibling species	Species that share the same, most recent common ancestor.
Cryptic species	Species recognized by nucleic acid variation that had not been recognized as distinct by morphological phenotypes. Once recognized, phenotypic characteristics useful for identification may be discovered in the future.
Clade/subclade	Phylogenetic group consisting of an ancestral species and all its descendants. Clades and subclades can be recognized at any given taxonomic level.
Lineage	Series of species connected by evolutionary descent, not necessarily representing all known descendants.
Cluster/group	Terminal series of phylogenetically related species, used when precise relationships are uncertain.
Protolog	Original description and any other representation of a taxonomic entity.
Type	Entity defining a taxonomic name and indicated as such in the protolog. Species and below are defined by a specimen, whereas higher taxonomic entities are defined by the first lower category.
Neotype	New specimen in accordance with the protolog in case the original type material is lost.
Epitype	Reference specimen in accordance with the protolog when the original material is not interpretable.

However, the immediate problem that was experienced by fungal taxonomists was one fungus, which name? The General Committee of the ICN is empowered to conserve or reject names of fungi, as well as to modify the ICN itself. Although it was made clear that the principle of priority will follow, in some cases it was difficult to implement the coveted rule. For instance, take *Fusarium* (anamorph) and *Gibberella* (teleomorph) at the genus level—the teleomorph name may be chosen without General Committee approval, whereas selection of the anamorph name, even if it is the older name, requires approval.

During the follow-up "One Fungus, Which Name?" symposium held in the Royal Netherlands Academy of Arts and Sciences in Amsterdam on April 12–13, 2012, it was decided that the term *conserved* would be used for names included in the accepted list of names of fungi. Recognizing the fact that there are several cases in some groups of fungi where there could be many names that might merit formal retention or rejection, a new provision was introduced in the code. Now lists of names can be submitted to the General Committee, and after due scrutiny, names accepted on those lists are to be treated as conserved over competing synonyms, which can be listed as *appendices* to the code. Further, there is no restriction on the ranks of names or of taxonomic groupings. A list may well be confined to all names in a particular rank, such as orders, families, genera, or species, within a particular taxon. While preparing the list of correct names of fungi, various terminologies may be used for specimens and cultures other than name-bearing types (Table 8.1). The dual naming system for fungi, although confusing for novices, had been useful during the age of the microscope. Now the criterion of classification and nomenclature has shifted from phenotype to genotype and is largely guided by the analysis of nucleic acid sequence variations, which eventually replaced phenotype with the history of phylogenetic relationships. This is now further expanded, following the introduction of next-generation sequencing, which is discovering genetically distinct populations that deserve species status. As a consequence, such changes will hopefully contribute to nomenclatural stability in the future.

While describing and comparing new species with old names, taxonomists are relying on DNA sequence data, besides the morphological characteristics of the new fungus, and are governed by the ICN. The application of one species, one name is still in its infancy in mycology, primarily because many fungi are still only known in either their sexual (telomorph) or asexual (anamorph) state. In such circumstances, while describing new species, mycologists mostly recover and describe only either state. This means that all legitimate names proposed for a species, regardless of what stage they are typified by, can serve as the correct name for that species. In addition, when publishing new species or revisiting the classification system of heterogeneous genera, phylogenetic data help a great deal. To complement this, the DNA barcode (the internal transcribed spacer [ITS] rDNA gene sequence) serves as the key basis for assigning a taxon to a taxonomic rank and ultimately provides evidence of species novelty. However, for the majority of valid thermophilic fungi, barcode sequences do not exist (de Oliveira et al., 2015), meaning that no intraspecific variation data on the ITS marker exist to help the discovery of new molecular species. Thus, the metagenome forms a resource pool for new sequences to assess the community of fungi from thermal niches and will certainly lead to the discovery of novel thermophilic fungi with potent industrial attributes. The researchers, academicians, and industrial scientists working on thermophilic fungi should work in tandem to achieve a reasonable degree of nomenclatural stability in the times to come when large changes in the nomenclature of these fungi are unavoidable. The rules enacted in the ICN must be followed in letter and spirit while addressing the problem of one fungus, which name?

8.4 TAXONOMIES AND THE NAME CHANGES

As stated elsewhere, the history of thermophilic fungi is more than a century old. Many members of these fungi and their synonyms were described a long time ago, and type material is often lacking or uninterpretable. Ambiguities pertaining to the taxonomy of these fungi have resulted in conflicts and misidentification. Nomenclatural changes usually take a long time to gain wide acceptance among taxonomists and the general public, particularly for fungi, which have immense economic importance.

Thus, taxonomy is a dynamic science that cannot be stifled by nomenclatural protocols. Even strict nomenclatural rules provide a certain degree of liberty. In the process of naming each fungus, teleomorph and anamorph names, as well as their synonyms, are being considered in a way that would increase acceptance and stability in applying one name for one fungus.

Relatively few studies have been carried out on the taxonomy and diversity of thermophilic fungi. Furthermore, on the applied side of the heat-tolerant fungi, a number of thermotolerant fungi have been classified as thermophilic, mainly because there are different systems for differentiating thermophilic and thermotolerant fungi (Mouchacca, 2000). Consequently, many biotechnologists have encountered great confusion with regard to using the term *thermophilic* or *thermotolerant* while working with these fungi. The use of cardinal temperatures as a basis to distinguish thermophilic from thermotolerant fungi is controversial, as many of these fungi do not fit well in this criterion. An alternative way to distinguish thermophilic and thermotolerant strains is to classify fungi with an optimum temperature for growth between 40°C and 50°C but an inability to grow below 20°C as thermophilic, whereas those that can grow below 20°C should be classified as thermotolerant strains. Thus, correct classification of these fungi is of great academic and economic interest, as many of them are involved in biotechnological processes and patent acquisition (Mouchacca, 2000). Owing to the dual nomenclature system in operation since the mid-nineteenth century, various synonyms are often used to describe the same fungus (Table 8.2). In this context, the present state of thermophilic fungi creates complexities when one wants to compare the results of different studies, as authors use different nomenclature systems. To overcome these difficulties, the use of a one-fungus, one-name system is of utmost importance, and the authors must consult the following nomenclatural databases for information on the correct names of fungi:

- Index Fungorum (http://www.indexfungorum.org)
- MycoBank (http://www.mycobank.org)
- Fungal Names (http://www.mycolab.org.cn/en/index.aspx)
- CABI databases (http://www.speciesfungorum.org)

The above databases are important tools where new names are recorded, along with illustrations and descriptions of each taxon. In addition, these databases are also the current repositories of fungal names, such as those for genera and species, including the names of novel fungi (de Oliveira et al., 2015). MycoBank is owned and governed by the International Mycological Association and has servers in Belgium and the Netherlands; similarly, Index Fungorum is run by Landcare Research, New Zealand and Royal Botanical Gardens (RBG), Kew—Mycology, whereas Fungal Names is an initiative of the Institute of Microbiology, Chinese Academy of Sciences (IM-CAS), having servers in Beijing. These three repositories represented by the respective institutions have entered into a memorandum of cooperation (MOC) with the Nomenclature Committee for Fungi (NCF). The MOC was valid until August 2017, which also coincided with the next IBC, held in Shenzhen, China in July 2017 (Redhead and Norvell, 2012). In this congress, the most significant decision was to transfer

TABLE 8.2

List of Name Changes of Thermophilic and Thermotolerant Fungi and Their Synonyms

Sr. No.	New Name	Authority/Reference	Old Name/Synonyms	Authority/Reference	Comments
Zygomycetes					
1	*Rhizomucor miehei*	(Cooney & Emerson) Schipper, 1978	*Mucor miehei*	Cooney & Emerson, 1964	Original description by Cooney and Emerson
			Mucor miehei var. minor	(Cooney & Emerson) Subrahmanyam & Gopalkrihnan, 1984	Unwarranted taxonomic decision
			Rhizomucor nainitalensis	Joshi, 1982	Original description by M.C. Joshi
			Rhizopus nainitalensis	Joshi, 1982	Typographical error of *Rhizomucor nainitalensis*
2	*Rhizomucor pusillus*	(Lindt) Schipper, 1978	*Mucor pusillus*	Lindt, 1886	First-ever thermophilic fungus reported by W. Lindt
			Mucor thermohyalospora	Subrahmanyam, 1981	Resembles *Mucor tauricus* but homothallic
			Rhizomucor pakistanicus	Qureshi & Mirza et al., 1979	Invalid name, superfluous publication, synonym of *Rhizomucor pusillus*
Ascomycetes					
3	*Coonemeria crustacea*	(Apinis & Chesters) Mouchacca, 1997	*Dactylomyces crustaceus*	Apinis & Chesters, 1964	Basionym and synonym of *Coonemeria crustacea*
4	*Coonemeria aegyptiaca*	(Ueda & Udagawa) Mouchacca, 1997	*Thermoascus aegyptiacus*	Ueda & Udagawa, 1983	Basionym of *Coonemeria aegyptiaca*

(Continued)

TABLE 8.2 (CONTINUED)
List of Name Changes of Thermophilic and Thermotolerant Fungi and Their Synonyms

Sr. No.	New Name	Authority/Reference	Old Name/Synonyms	Authority/Reference	Comments
5	*Coonemeria verrucosa*	Yaguchi, Someya, & Udagawa, 1997	*Thermoascus crustaceus* var. *verrucosus*	Yaguchi, Someya, & Udagawa, 1995	Basionym of *Coonemeria verrucosa*
			Thermoascus taitungiacus	Chen & Chen, 1996	Synonym of *Coonemeria verrucosa*
			Paecilomyces taitungiacus	Chen & Chen, 1996	Anamorph of *Coonemeria verrucosa*
6	*Dactylomyces thermophilus*	(Sopp) Apinis, 1967	*Thermoascus thermophilus*	(Sopp) von Arx, 1970	Type species of *Dactylomyces* and a superfluous combination
7	*Melanocarpus thermophilus*	(Abdullah & Al-Bader, 1990) Guarro et al., 1996	*Thielavia minuta* var. *thermophila*	(Cain) Malloch & Cain, 1973	Basionym of *Melanocarpus thermophilus*
8	*Chaetomidium pingtungium*	(Chen & Chen) Mouchacca, 1999	*Thielavia pingtungia*	Chen & Chen, 1996	Basionym of *Chaetomidium pingtungium*
			Anamorphic Fungi		
9	*Myceliophthora fergusii*	(Klopotek) van Oorschot, 1977	*Chrysosporium fermentortritci*	Matsushima & Matushima, 1996	Synonym *fide* Sigler et al., 1998
10	*Myceliophthora thermophila*	(Apinis) van Oorschot, 1977	*Chrysosporium thermophilum*	(Apinis) von Klopotek, 1974	Synonym of *Myceliophthora thermophila*
			Myceliophthora indica	Basu, 1984	Invalid nomenclature, synonym of *Myceliophthora thermophila*

(Continued)

TABLE 8.2 (CONTINUED)
List of Name Changes of Thermophilic and Thermotolerant Fungi and Their Synonyms

Sr. No.	New Name	Authority/Reference	Old Name/Synonyms	Authority/Reference	Comments
11	*Arthrinium pterospermum*	(Cooke & Massee) von Arx, 1981	*Sporotrichum cellulophilum*	Awao, 1986	Not validly published, synonym of *Myceliophthora thermophila*
			Sporotrichum thermophilum	Apinis, 1963	Basionym of *Myceliophthora thermophila*
			Coniosporium pterospermum	von Arx, 1981	Basionym of *Arthrinium pterospermum*
			Humicola stellata var. *gigantea*	Khanna, 1963	Invalid nomenclature
			Pteroconium pterospermum	(Cooke & Massee) Grove, 1971	Synonym of *Arthrinium pterospermum*
12	*Thermomyces lanuginosus*	Tsiklinskaya, 1899	*Humicola brevis* var. *thermoidea*	Straatsma & Samson, 1993	Synonym of *Thermomyces lanuginosus*
			Humicola brevispora	Subrahmanyam, 1975	No valid publication
			Humicola grisea var. *indica*	Subrahmanyam, 1980	Invalid nomenclature
			Humicola lanuginosa	(Griffon & Maublanc) Cooney & Emerson, 1964	Synonym of *Thermomyces lanuginosus*
			Humicola lanuginosa var. *catenulata*	(Griffon & Maublanc) Morinaga et al. 1986	Deviant strain of *Thermomyces lanuginosus*
13	*Scytalidium thermophilum*	(Cooney & Emerson) Austwick, 1976	*Humicola fuscoatra* var. *longispora*	(Cooney & Emerson) Fassatiová, 1967	Superfluous name change

(Continued)

TABLE 8.2 (CONTINUED)
List of Name Changes of Thermophilic and Thermotolerant Fungi and Their Synonyms

Sr. No.	New Name	Authority/Reference	Old Name/Synonyms	Authority/Reference	Comments
			Humicola fuscoatra var. *nigra*	(Traaen) Subrahmanyam, 1982	Invalid nomenclature
			Humicola grisea var. *thermoidea*	(Traaen) Cooney & Emerson, 1964	Synonym of *Torula thermophila*
			Humicola insolens	Cooney & Emerson, 1964	A superfluous name change
			Humicola insolens var. *thermoidea*	(Cooney & Emerson) Ellis, 1982	A superfluous name change
			Humicola nigrescens var. *thermorongeura*	(Omvik) Subrahmanyam, 1982	Invalid nomenclature
			Scytalidium allahabadum	Narain, Srivastava, & Mehrotra, 1983	Synonym of *Scytalidium thermophilum*
			Torula thermophila	Cooney & Emerson, 1964	Basionym of *Scytalidium thermophilum*
14	*Malbranchea cinnamomea*	(Libert) van Oorschot & de Hoog, 1984	*Malbranchea pulchella* var. *sufurea*	(Sacc. & Penzig) Cooney & Emerson, 1964	Synonym of *Malbranchea cinnamomea*
			Malbranchea sufurea	(Miehe) Sigler & Carmichael, 1976	Synonym of *Malbranchea cinnamomea*
			Thermoideum sulfureum	Miehe, 1907	Synonym of *Malbranchea cinnamomea*
			Trichothecium cinnamomeum	Libert, 1830	Basionym of *Malbranchea cinnamomea*
15	*Remersonia thermophila*	(Fergus) Seifert et al., 1997	*Stilbella thermophila*	Fergus, 1964	Basionym of *Remersonia thermophila*

(Continued)

TABLE 8.2 (CONTINUED)
List of Name Changes of Thermophilic and Thermotolerant Fungi and Their Synonyms

Sr. No.	New Name	Authority/Reference	Old Name/Synonyms	Authority/Reference	Comments
16	*Myceliophthora guttulata*	Zhang et al., 2014	–	–	A new thermophilic species of *Myceliophthora*
17	*Myceliophthora heterothallica*	van den Brink et al., 2012	*Corynascus heterothallicus*	(van Klopotek) von Arx, 1981	Teleomorph of *M. heterothallica*
18	*Rasamsonia emersonii*	Houbraken et al., 2011	*Talaromyces emersonii*	Stolk, 1965	Reclassified
19	*Rasamsonia byssochlamydoides*	Houbraken et al., 2011	*Talaromyces byssochlamydoides*		Reclassified
20	*Thermomyces dupontii*	Houbraken et al., 2014	*Talaromyces thermophilus*	Stolk, 1965	Reclassified

Source: Modified from Mouchacca, J., *World J. Microbiol. Biotechnol.*, 16, 881, 2000.

decision making on matters related to the naming of fungi from the International Botanical Congress to the International Mycological Congress (IMC). Beginning on January 1, 2019, it will be necessary to deposit the details of lecto-, neo-, and epitypifications in one of the recognized repositories of fungal names in order for them to be validly published and to establish their priority. A list of name changes and current names of thermophilic and thermotolerant fungi is provided in Table 8.2 for reference.

In an effort to provide a natural system for the fungi, similar to that of plants and animals, unification of fungal nomenclature has been pushed through the Amsterdam Declaration in 2012. Fungal taxonomists around the world are making enormous efforts to minimize the chaos resulting from the enforcement of one fungus, one name by generating fresh lists of valid fungal names that will become established in due course. In this context, Walter Gams (2016) is of a different opinion, stating that the protected names are not the last words in fungal taxonomy, and mycologists cannot be forced to adopt a particular taxonomic system when they do not agree with it. The author further emphasizes that it is presently impossible to effectively squeeze all known species into recognized, available, and strictly monophyletic genera. A complete shift to unified nomenclature will entail the recombination of all included species into a single recognized genus for a particular group.

Taxonomic decisions and name changes for a number of thermophilic fungi have been reviewed by Mouchacca (2000). His synopsis provides the latest legal valid names for several thermophilic fungi (Table 8.2) and aims to suppress the use of ghost binomials having no taxonomic status of any kind and favors the continuous use of the latest legal valid name of a taxon to avoid cases of redundancies by citing binomials of known synonymies. For example, *Mucor miehei*, *Mucor miehei* var. *minor*, and *Rhizomucor nainitalensis* have been renamed *Rhizomucor miehei*. Similarly, *Rhizomucor pusillus* is the final legitimate name for *Mucor pusillus*, *Mucor thermohyalospora*, and *Rhizomucor pakistanicus* in zygomycetes (Table 8.2). In Ascomycetes, several name changes of truly thermophilic fungi have been affected in the past several decades, as presented in Table 8.2. Some of the new names include *Chaetomium pingtungium*, *Coonemeria aegyptiaca*, *Coonemeria crustacea*, *Coonemeria verrucosa*, *Dactylomyces thermophilus*, and *Melanocarpus thermophilus*. Recently, Houbraken et al. (2012) reviewed the genus *Talaromyces* and observed that no thermophiles are found in this genus. Therefore, they proposed a new genus, *Rasamsonia*, to accommodate *Geosmithia argillacea*, *Talaromyces emersonii*, and *Talaromyces byssochlamydoides*. The authors further observed that within the order Eurotiales, the genera *Talaromyces*, *Thermoascus*, *Thermomyces*, *Coonemeria*, and *Dactylomyces* contain thermophilic species (Stolk, 1965; Apinis, 1967; Stolk and Samson, 1972; Mouchacca, 1997). Several studies on the taxonomy of the thermophilic genus *Thermoascus* have been conducted (Stolk, 1965; Apinis, 1967; Mouchacca, 1997; Pitt, 2000). Apinis (1967), however, divided this genus into two: *Thermoascus* was retained for its type species *T. aurantiacus*, whereas *T. thermophilus* and *T. crustaceus* were shifted to *Dactylomyces*. Later, Mouchacca (1997) further divided *Dactylomyces* into two, creating the genus *Coonemeria* for *T. crustaceus*, while *T. aurantiacus* was retained. Houbraken et al. (2012) finally concluded that the thermophilic fungi are restricted to the genera *Thermomyces*, *Thermoascus*, and *Rasamsonia* within the order Eurotiales.

Earlier, we (Sigler et al., 1998) redescribed *Corynascus thermophilus* and its anamorph *Myceliophthora fergusii* and reviewed its taxonomic history. While the species of *Corynascus* is recognized by some to be thermophilic, it was reclassified as belonging to *Myceliophthora* (van den Brink et al., 2012). Now the genus *Myceliophthora* is reported to comprise seven thermophilic species (de Oliveira et al., 2015): *M. fergusii*, *M. fusca*, *M. heterothallica*, *M. hinnulea*, *M. sulfurea*, *M. thermophila*, and the recently discovered new species from China, *M. guttulata* (Zhang et al., 2014).

Among the anamorphic fungi, the names of *Myceliophthora fergusii*, *Myceliophthora thermophila*, *Arthrinium pterospermum*, *Thermomyces lanuginosus*, *Scytalidium thermophilum*, *Malbranchea cinnamomea*, *Remersonia thermophila*, *Myceliophthora guttulata*, *Myceliophthora heterothallica*, *Rasamsonia emersonii*, *Rasamsonia byssochlamydoides*, and *Thermomyces dupontii* have been revised to accommodate their synonyms or invalid names of these fungi. Following the changes in the nomenclature of thermophilic fungi, the recent challenge is how to treat the new diversity seen in the taxonomy of these industrially important fungi. At the genus level, these are reallocations, rank changes, and generic disarticulations or reunifications that stem from studies of more materials and from having to choose one name from two or more current names (de Hoog et al., 2015). Where the names of biotechnologically important fungi do change—many will change in the near future as the unification code is applied, it may reflect the fact that the small group of thermophilic fungi is just one of many industrially important organisms that are involved in production of important metabolites. Further, it is advisable that good taxonomic studies do not always need new names immediately. For biotechnologically important strains, the change in names should be carried out with some delay and after due consultation and validation by a governing body, until a consensus on its name is reached and sufficient stability is achieved.

As suggested by Mouchacca (1997), strict restriction to nomenclatural rules governing citations of fungal binomials is fundamental. Many researchers use their own strains for the production of metabolites; this is, of course, welcome, as some strains of related isolates are highly productive. Therefore, the authors of applied research dealing with thermophiles should necessarily follow such regulations in order to stabilize the names of strains used for applied research or in produced goods. This will definitely and biotechnology.

8.5   CLASSIFICATION OF UNCULTURED SPECIES

One of the most controversial issues within any taxonomy and classification bring an end to the chaotic state prevailing, especially in publications relating to fungal taxonomy system is to come to an agreement about what a species is. To the majority of taxonomists, species is a pragmatic unit, conceived to identify those patterns of recurrence that occur in nature and that are thought to be units, and the basis for the construction of any classification. Post-Darwinism, species appeared as a scientific need for defining the basic unit of the biological order. Therefore, the species concept is an artifact of the mind, that is, an artificial concept that tries to embrace units of biological diversity. The question that is hampering taxonomy even today is whether this category reflects real discrete units in nature. Clearly, the current definition of species does not allow the classification of uncultured species. At most, one can classify something with the category *candidatus*. This category, with a provisional status in taxonomy, was created to accommodate such organisms that could be recognized by 16 SrRNA sequences and a few other parameters. However, after the Amsterdam Declaration, the practice of encouraging classification based on environmental sequences was widely accepted. Such data are usually obtained from metagenomic studies, which mostly include next-generation sequencing data (Hibbett and Taylor, 2013). However, developing community standards for sequence-based classification is the most difficult job. Presently, we have no reliable information regarding the extant of uncultivable fungi present in hot environments in particular. Similarly, we do not have information on their ability to produce metabolites of biotechnological interest in comparison with their culturable counterparts.

After the polymerase chain reaction (PCR) generally became available, mycologists debated whether DNA or even a DNA sequence could act as a type element in describing a species (Reynolds and Taylor, 1991). Their observation now has become reality, as the dawn of environmental sequencing has allowed mycologists to use PCR primers for rDNA to amplify variable regions from DNA isolated from an environmental source. Ecologists believe that most of the fungal species dwelling in their favorite natural environment can be neither cultivated nor collected; consequently, they have to rely on environmental nucleic acid sequences (ENAS) to assess the true fungal diversity. Each of these ecological studies may add hundreds or thousands of ENAS to GenBank. Now there is a hoard of rDNA sequences in various fungal databases that document the existence of fungi for which there is neither a specimen nor a culture. Rather, it has been contemplated that these DNA-only fungi or ENAS can exceed the number of culturable fungi (Taylor, 2011). Furthermore, this imbalance confronts fungal classification and nomenclature that may reduce the challenge of integrating teleomorphic and anamorphic fungi.

For thermophilic fungi, it has been observed that DNA barcodes do not exist or are scanty; thus, we have very little information on intraspecific variation data on the ITS marker to help in the recovery of molecular species from an environmental source. Therefore, to gradually move from culture-based classification to environmental sequence–based classification, mycologists must deposit sequences of thermophilic fungi in a public database, such as the National Center for Biotechnology Information (NCBI) or GenBank. Further, such efforts should also be coupled with the deposition of specimens into well-known and easily accessible culture collections (de Oliveira et al., 2015). The choice of the ITS region as the universal barcode marker for fungi is an important tool in achieving this goal (Schoch et al., 2012). Thus, an important strategy for the identification and classification of thermophilic fungi would be the use of the ITS marker in taxonomic and phylogenetic studies, coupled with morphological characteristics.

The use of metagenomic studies for the discovery of novel putative thermophilic fungi is receiving considerable attention (Chávez et al., 2015). Recently, culture-independent approaches, such as phospholipid fatty acid analysis (PLFA), denaturing gradient gel electrophoresis (DGGE) of PCR-amplified DNA fragments combined with the sequencing of relevant bands, terminal restriction fragment length polymorphism (T-RFLP) analysis, and more recently, next-generation sequencing, have been used to identify fungal organisms (Anderson and Cairney, 2004; Hultman et al., 2010; Lindahl et al., 2013). In an effort to investigate the biodiversity of thermophilic and thermotolerant fungi in two compost samples, Langarica-Fuentes et al. (2014) used culture-based and molecular approaches. They observed that the diversity of thermophilic fungi recovered using 454 pyrosequencing of the ITS region was greater (175 operational taxonomic units [OTUs]) than that observed using culture-based methods (8 morphospecies). Out of these 175 OTUs, 107 OTUs from both compost samples were classified to the rank of family or above. They further observed that molecular studies demonstrated the frequent presence of several thermophilic (*Scytalidium thermophilum* and *Myriococcum thermophilum*) and thermotolerant (*Pseudallescheria boydii*, *Corynascus verrucosus*, and *Coprinopsis* sp.) fungi in the composts, despite the absence of these species from the culture-based analysis. More recently, Holman et al. (2016) observed that fungal diversity and richness were not altered by either compost type or sampling time. They recovered 11 OTUs that formed the core fungal microbiota of the compost against culturable *Remersonia thermophila*, *Talaromyces thermophiles*, and *Thermomyces lanuginosus*. Thus, the environmental sequences from many of these uncultivated thermophilic fungi may have important applications in the biotechnology industry. Additionally, some of the sequences recovered using molecular approaches having no taxonomic affiliation may belong to previously described species for which barcode data do not exist.

8.6   UNWARRANTED TAXONOMIES

Cooney and Emerson (1964), in their monograph, introduced some new thermophilic taxa. However, some of the taxonomic decisions adopted by them proved to be misleading, and their descriptions of novel taxa did not support critical analysis. These limitations prompted subsequent studies, which mainly focused on pending problems while expanding this fascinating group of thermophilic fungi. Apinis and Chester (1964) described *Dactylomyces crustaceaus*. Pugh et al. (1964) reintroduced *Thermomyces*. Stolk (1965) assessed the taxonomic status of *Penicillium dupontii* and *Thermoascus aurantiacus*. Again, Apinis (1967) clarified the generic concepts of *Dactylomyces* and *Thermoascus*. However, in spite of the above and later contributions, not all standing problems received attention. Similarly, *Mucor miehei* was reported by Cooney and Emerson (1964); the introduction of *Mucor miehei* var. *minor* was an unwarranted taxonomic decision by Subrahmanyam and Gopalkrihnan (1984). In 1973, Tansey recovered isolates of a fungus closely resembling *Papulospora thermophila* (now *Myriococcum thermophilum*) from alligator nesting material. He named this isolate BURGOA-PAPULOSPORA sp., based on the names of two generic entities developing "bulbil-like" structures of the Agonomycetes group. In subsequent publications, this was incorrectly reported as *Burgoa papulospora*, causing it to be inadvertently reported as a binomial.

Taxonomic knowledge of several fungal groups is still quite inadequate and often does not yet allow decisions about the delimitation of natural taxa. The fungal world remains alive in its native environment, awaiting discovery. Due to some metabolic attributes of thermophilic fungi, several warranted nomenclatural and taxonomic decisions were not adopted by some mycologists, creating a chaotic state in the cited binomials of some thermophiles in published work (Mouchacca, 2000). Contrary to this, the adoption of unwarranted taxonomic decisions led to misunderstanding and misidentification of some thermophilic fungi. To overcome these uncertainties, mycologists working on thermophilic fungi must fervently adopt a one-name, one-fungus concept, which may facilitate the study of fungal systematics by students and researchers.

REFERENCES

Abdullah, S.K., and Al-Bader, S.M. 1990. On the thermophilic and thermotolerant mycoflora of Iraq. *Sydowia* 42:1–7.

Anderson, I.C., and Cairney, J.W. 2004. Diversity and ecology of soil fungal communities: Increased understanding through the application of molecular techniques. *Environmental Microbiology* 6:769–779.

Apinis, A.E. 1963. Occurrence of thermophilous microfungi in certain alluvial soils near Nottingham. *Nova Hedwigia* 5:57–78.

Apinis, A.E. 1967. *Dactylomyces* and *Thermoascus*. *Transactions of the British Mycological Society* 50:573–582.

Apinis, A.E., and Chester, C.G.C. 1964. Ascomycetes of some salt marshes and sanddunes. *Transactions of the British Mycological Society* 47:419–435.

Austwick, P.K.C. 1976. Environmental aspects of *Mortierella wolfii* infection in cattle. *New Zealand Journal of Agricultural Research* 19:25–33.

Awao, T. 1986. *Sporotrichum cellulophilum*. *CBS Newsletter* 7:9.

Barron, G.L. 1968. *The Genera of Hyphomycetes from Soil*. Baltimore: Williams & Wilkins.

Bass, D., and Richards, T.A. 2011. Three reasons to re-evaluate fungal diversity 'on Earth and in the ocean'. *Fungal Biology Reviews* 25:159–164.

Basu, M. 1984. *Myceliophthora indica* in a new thermophilic species from India. *Nova Hedwigia* 40:85–90.

Chávez, R., Fierro, F., Ramón, O., Rico, G., and Vaca, I. 2015. Filamentous fungi from extreme environments as a promising source of novel bioactive secondary metabolites. *Frontiers in Microbiology* 6:903. doi: 10.3389/fmicb.2015.00903.

Chen, K.Y., and Chen, Z.C. 1996. A new species of *Thermoascus* with a *Paecilomyces* anamorph and other thermophilic species from Taiwan. *Mycotaxon* 50:225–240.

Cline, E.T., Ammirati, J.F., and Edmonds, R.L. 2005. Does proximity to mature trees influence ectomycorrhizal fungus communities of Douglas-fir seedlings? *New Phytologist* 166: 993–1009.

Cole, G.T., and Samson, R.A. 1979. *Patterns of Development in Conidial Fungi*. London: Abb. Pitman Publishing Ltd.

Cooney, D.G., and Emerson, R. 1964. *Thermophilic Fungi: An Account of Their Biology, Activities, and Classification*. San Francisco: W.H. Freeman and Company.

Dayarathne, M.C., Boonmee, S., Braun, U., Crous, P.W., Daranagama, D.A., Dissanayake, A.J., Ekanayaka, H., et al. 2016. Taxonomic utility of old names in current fungal classification and nomenclature: Conflicts, confusion & clarifications. *Mycosphere* 7(11): 1622–1648.

de Hoog, G.S., Chaturvedi, V., Denning, D.W., Dyer, P.S., Frisvad, J.C., Geiser, D., Graser, Y., et al. 2015. Name changes in medically important fungi and their implications for clinical practice. *Journal of Clinical Microbiology* 53:1056–1061.

de Oliveira, T.B., Gomes, E., and Rodrigues, A. 2015. Thermophilic fungi in the new age of fungal taxonomy. *Extremophiles* 19:31–37.

Ellis, D.H. 1982. Ultrastructure of thermophilic fungi. V. Conidial ontogeny in *Humicola grisea* var. *thermozidea* and *H. insolens*. *Transactions of the British Mycological Society* 78:129–133.

Fassatiová, O. 1967. Notes on the genus *Humicola* Traaen. II. *Ceská Mykologie* 21:78–89.

Fergus, C.L. 1964. Thermophilic and thermotolerant molds and actinomycetes of mushroom compost during peak heating. *Mycologia* 56:267–284.

Gams, W. 2016. Recent changes in fungal nomenclature and their impact on naming of microfungi. In *Biology of Microfungi, Fungal Biology, ed.* D.W. Li. Cham, Switzerland: Springer International Publishing.

Guarro, J., Abdullah, S.K., Al-Bader, S.M., Figueras, M.J., and Gene, J. 1996. The genus *Melanocarpus*. *Mycological Research* 100:75–78.

Hawksworth, D.L., Crous, P.W., Dianese, J.C., Gryzenhout, M., Norvell, L.L., and Seifert, K.A. 2009. Proposals to amend the code to make it clear that it covers the nomenclature of fungi, and to modify the governance with respect to names of organisms treated as fungi. *Taxon* 58:658–659.

Hawksworth, D.L., Crous, P.W., Redhead, S.A., Reynolds, D.R., Samson, R.A., and Seifert, K.A. 2011. The Amsterdam Declaration on fungal nomenclature. *IMA Fungus* 2:105–112.

Hibbett, D.S., and Taylor, J.W. 2013. Fungal systematics: Is a new age of enlightenment at hand? *Nature Reviews Microbiology* 11:129–133.

Holman, D.B., Hao, X., Topp, E., Yang, H.E., and Alexander, T.W. 2016. Effect of co-composting cattle manure with construction and demolition waste on the archaeal, bacterial, and fungal microbiota, and on antimicrobial resistance determinants. *PloS One* 11(6):e0157539.

Houbraken, J., de Vries, R.P., and Samson, R.A. 2014. Modern taxonomy of bio-technologically important *Aspergillus* and *Penicillium* species. *Advances in Applied Microbiology* 86:199–249.

Houbraken, J., Dijksterhuis, J., and Samson, R.A. 2012. Diversity and biology of heat-resistant fungi. In *Stress Responses of Foodborne Microorganisms*, ed. H.-C. Wong, 331–353.

Houbraken, J., Frisvad, J.C., and Samson, R.A. 2011. Fleming's penicillin producing strain is not *Penicillium chrysogenum* but *P. rubens*. *IMA Fungus* 2:87–95.

Hughes, S.J. 1953. Fungi from the Gold Coast. II. *Mycological Papers* 50:1–104.

Hultman, J., Vasara, T., Partanen, P., Kurola, J., Kontro, M.H., Paulin, L., Auvinen, P., and Romantschuk, M. 2010. Determination of fungal succession during municipal solid waste composting using a cloning-based analysis. *Journal of Applied Microbiology* 108:472–487.

Joshi, M.C. 1982. A new species of *Rhizomucor* from India. *Sydowia* 35:100–103.

Khanna, P.K. 1963. *Humicola stellata* var. *gigantean* [*Asstellatus* var. *giganteus*]. *Current Science* 32:175.

Kirk, P.M., Cannon, P.F., Minter, D.W., and Stalpers, J.A. 2008. *Dictionary of the Fungi.*

Libert, M.A. 1830. *Trichothecium cinnamomeum. Plantae Cryptogamae quas in Arduenna Coll,* Nr. 1013.

Langarica-Fuentas, A., Zafar, U., Heyworth, A., Brown, T., Fox, G., and Robson, G.D. 2014. Fungal succession in an in-vessel composting system characterized using 454 pyrosequencing. *FEMS Microbiology Ecology* 88:296–308.

Lindahl, B.D., Nilsson, R.H., Tedersoo, L., Abarenkov, K., Carlsen, T., Kjøller, R., Kõljalg, U., Pennanen, T., Rosendahl, S., Stenlid, J., and Kauserud, H. 2013. Fungal community analysis by high-throughput sequencing of amplified markers—A user's guide. *New Phytologist* 199:288–299.

Lindt, W. 1886. Mitteilungen ber einige neue pathogene Schimmelpilze. *Archives in Experimental Pathology and Pharmacology* 21:269–298.

Malloch, D., and Cain, R.F. 1973. The genus *Thielavia. Mycologia* 65:1055–1077.

Mason, E.W. 1933. Annotated account of fungi received at the Imperial Mycological Institute, List II (Fasc. 2). *Mycological Papers* 3:1–67.

Mason, E.W. 1937. Annotated account of fungi received at the Imperial Mycological Institute, List II, Fasc. 3 gen. part. *Mycological Papers* 4:69–99.

Matsushima, K., and Matsushima, T. 1996. Fragmenta mycologica II. *Matsushima Mycological Memoirs* 9:31–40.

McNeill, J., Barrie, F.R., Buck, W.R., Demoulin, V., Greuter, W., Hawksworth, D.L., Herendeen, P.S., et al. 2012. International Code of Nomenclature for algae, fungi, and plants (Melbourne Code) adopted by the Eighteenth International Botanical Congress, Melbourne, Australia, July 2011. *Regnum Vegetabile* 154:1–140.

Miehe, H. 1907. *Die Selbsterhitzung des Heus. Eine Biologische Studie.* Jena, Germany: Gustav Fischer.

Mirza, J.H., Khan, S.M., Begum, S., and Shagufta, S. 1979. *Mucorales of Pakistan.* Faisalabad, Pakistan: University of Agriculture.

Morinaga, T., Kanda, S., and Nomi, R. 1986. Lipase production of a new thermophilic fungus, *Humicola lanuginosa var. catenulata. Journal of Fermentation Technology* 64(5):451-453.

Mouchacca J. 1997. Thermophilic fungi: Biodiversity and taxonomic status. *Cryptogamie Mycologie* 18:19–69.

Mouchacca, J. 1999. Thermophilic fungi: Present taxonomic concepts. In *Thermophilic Moulds in Biotechnology*, eds. Johri, B.N., Satyanarayana, T., and Olsen, J., 43–83. Dordrecht: Kluwer Academic Publishers.

Mouchacca, J. 2000. Thermophilic fungi and applied research: A synopsis of name changes and synonymies. *World Journal of Microbiology and Biotechnology* 16:881.

Narain, R., Srivastava, R.B., and Mehrotra, B.S. 1983. *Scytalidium allahabadum. Zentralblatt für Mikrobiologie* 138:570.

Norvell, L.L. 2011. Report of the Nomenclature Committee for Fungi: 17. *Taxon* 60:610–613.

Pitt, J.I. 2000. Toxigenic fungi: Which are important? *Medical Mycology* 1:17–22.

Pitt, J.I., and Samson, R.A. 2007. Nomenclatural considerations in naming species of *Aspergillus* and its teleomorphs. *Studies in Mycology* 59:67–70.

Pugh, G.J.F., Blakeman, J.P., and Morgan-Jones, G. 1964. *Thermomyces verrucosus* sp. nov. and *T. languginosus. Transactions of the British Mycological Society.* 47(1):115–121.

Redhead, S.A., and Norvell, L.L. 2012. MycoBank, Index Fungorum, and Fungal Names recommended as official nomenclatural epositories for 2013. *IMA Fungus* 3:44–45.

Reynolds, D.R., and Taylor, J.W. 1991. Nucleic acids and nomenclature: Name stability under Article 59. In *Improving the Stability of Names: Needs and Options*, ed. D.L. Hawksworth, 171–177. [Regnum Vegetabile no. 123.] Königstein, Germany: Koeltz Scientific Books.

Saccardo, P.A. 1882. Fungi Veneti novi vel critici v. mycologiae Veneti addendi (adjectis nonnullis extra-Venetis). Series XIII. *Michelia* 2(8):528–563.

Salar, R.K., and Aneja, K.R. 2007. Thermophilic fungi: Taxonomy and biogeography. *Journal of Agricultural Technology* 3:77–107.

Schipper, M.A.A. 1978. On the genera *Rhizomucor* and *Parasitella. Studies in Mycology* 17:53–71.

Schoch, C.L., Seifert, K.A., Huhndorf, S., Robert, V., Spouge, J.L., André Levesque, C., Chen, W., and Fungal Barcoding Consortium. 2012. Nuclear ribosomal internal transcribed spacer (ITS) region as a universal DNA barcode marker for fungi. *Proceedings of the National Academy of Sciences of the United States of America* 109:6241–6246.

Seifert, K.A., Samson, R.A., Boekhout, T., and Louis-Seize, G. 1997. *Remersonia*, a new genus for *Stilbella thermophila*, a thermophilic mould from compost. *Canadian Journal of Botany* 75:1158–1165.

Sigler, L., Aneja, K.R., Kumar, R., Maheshwari, R., and Shukla, R.V. 1998. New records from India and redescription of *Corynascus thermophilus* and its anamorph *Myceliophthora fergusii*. *Mycotaxon* 68:185–192.

Sigler, L., and Carmichael, J.W. 1976. Taxonomy of *Malbranchea* and some other Hyphomycetes with arthroconidia. *Mycotaxon* 4:349–488.

Stolk, A.C. 1965. Thermophilic species of *Talaromyces* Benjamin and *Thermoascus* Miehe. *Antonie van Leeuwenhoek* 31:262–276.

Stolk, A.C., and Samson, R.A. 1972. The genus *Talaromyces*—Studies on *Talaromyces* and related genera II. *Studies in Mycology* 2:1–65.

Straatsma, G., and Samson, R.A. 1993. Taxonomy of *Scytalidium thermophilum*, an important thermophilic fungus in mushroom compost. *Mycological Research* 97:321–328.

Subrahmanyam, A. 1975. *Studies on morphology*, culture characters and biological activities of some thermophilic fungi. PhD thesis, Poona University, Poona, India.

Subrahmanyam, A. 1980. A new thermophilic variety of *Humicola grisea* var. indica. *Current Science* 49:30–31.

Subrahmanyam, A. 1981. Studies on thermomycology. *Mucor thermohyalospora* sp. nov. *Bibliotheca Mycologica* 91:421–423.

Subrahmanyam, A. 1982. Studies on thermomycology—Notes on the genus *Humicola traaen*. *Hindustan Antibiotics Bulletin* 24:41–46.

Subrahmanyam, A., and Gopalkrihnan, K.S. 1984. Notes on thermophilic fungi. *Indian Botanical Reporter. Prof. K.B. Deshpande Commemoration* 33–36.

Tansey, M.R. 1973. Isolation of thermophilic fungi from alligator nesting material. *Mycologia* 65:594–601.

Taylor, J.W. 2011. One fungus = one name: DNA and fungal nomenclature twenty years after PCR. *IMA Fungus* 2:113–120.

Tsiklinskaya, P. 1899. Sur les mucedinees thermophiles. *Annales de l'Institut Pasteur, Paris* 13:500–505.

Tubaki, K. 1958. Studies on the Japanese Hyphomycetes, 5. Leaf and stem group with a discussion of the classification of Hyphomycetes and their perfect stages. *Journal of the Hattori Botanical Laboratory* 20:142–244.

Tulasne, L.R., and Tulasne, C. 1861. *Selecta Fungroum Carpologia*. Vol. 1. Paris: Imperial Typographer. (Reprint 1931, Clarendon Press, Oxford.)

Ueda, S., and Udagawa, S.I. 1983. *Thermoascus aegyptiacus*, a new thermophilic ascomycete. *Transactions of the Mycological Society of Japan* 24:135–142.

van den Brink, J., Samson, R.A., Hagen, F., Boekhout, T., and de Vries, R.P. 2012. Phylogeny of the industrial relevant, thermophilic genera *Myceliophthora* and *Corynascus*. *Fungal Diversity* 52:197–207.

van Oorschot, C.A.N. 1977. The genus *Myceliophthora*. *Persoonia* 9:401–408.

van Oorschot, C.A.N., and de Hoog, G.S. 1984. Some hyphomycetes with thallic conidia. *Mycotaxon* 20:129–132.

von Arx, J.A. 1970. *The Genera of Fungi Sporulating in Pure Culture*. Lehre, Germany: Cramer.

von Arx, J.A. 1981. *The Genera of Fungi Sporulating in Pure Culture*. Port Jervis, New York: Lubrecht & Cramer Ltd.

von Klopotek, A. 1974. Revision of thermophilic *Sporotrichum* species: *Chrysosporium thermophilum* (Apinis) comb. nov. and *Chrysosporium fergusii* spec. nov. equal status conidialis of *Corynascus thermophilus* Fergus and (Sinden) comb. nov. *Archives of Microbiology* 98:365–369.

Vuillemin, P. 1910. Description d'un type de chaque ordre de Conidiosporés. *Bulletin des Séances du Société des Sciences de Nancy, Séries* 11:138–143.

Weresub, L.K., and Pirozynski, K.A. 1979. Pleomorphism of fungi as treated in the history of mycology and nomenclature. In *The Whole Fungus*, ed. B. Kendrick, 17–25. Ottawa: National Museums of Canada.

Yaguchi, T., Someya, A., and Udagawa, S.I. 1995. *Thermoascus crustaceus* var. *verrucosus*. *Mycoscience* 36:151.

Yaguchi, T., Someya, A., and Udagawa, S.I. 1997. *Coonemeria verrucosa*. *Cryptogamie Mycologie* 18:32.

Zhang, Y., Wu, W.P., Hu, D.M., Su, Y.Y., and Cai, L. 2014. A new thermophilic species of *Myceliophthora* from China. *Mycological Progress* 13:165–170.

Section III

Biotechnological Applications

Bacteriological Applications

9 Role of Thermophilic Fungi in Composting

9.1 INTRODUCTION

To a layman, fungi are notorious organisms destined to cause disease in living beings or deteriorate materials such as wood, bread, and cloth. However, historically they have been exploited by man in many ways, and their uses are expanding rapidly. The Egyptians and Romans prized mushrooms, and there are records of their being eaten in China between 25 BC and 220 AD (Wang, 1985). The use of large fungi or mushrooms as food is quite commonplace and has traditionally been associated with meats. In Malawi, they are regarded as a meat analogue (Morris, 1984). This thought was also upheld by Francis Bacon, who in his *Sylva Sylvarum* of 1927 described mushrooms as yielding "so delicious a meat." More recently, however, filamentous fungi have found their way into the industrial development of various kinds of mycoproteins and/or supplements for various substrates for human and animal consumption. The development of "Quorn," a mycoprotein produced from *Fusarium graminearum*, is a success story, and it now festoons the shelves of British supermarkets (Trinci, 1992).

Throughout the globe, agricultural operations generate huge quantities of manure, which must be eliminated in a manner that is consistent with public health guidelines. During the last three decades, several researchers have tried to utilize filamentous fungi, particularly thermophilic strains, for the development of environmentally friendly compost that does not emit odors into the environment. Composting refers to the controlled aerobic conversion of mixed organic materials into a form that is suitable for addition to soil or for growing plants, including mushroom cultivation. Traditionally, composting was carried out for piling agricultural residues or organic materials and allowing them to stay in the agricultural field until the next planting season, by which time the dark brown material would be ready for soil application. The modern practice of composting started in 1921 with the setting up of the first industrial station in Austria for the bioconversion of urban organic waste into compost. Biologically, composting represents an amazing example of solid-state fermentation wherein crude wastes, such as sewage, sludge, refuse, animal dung, kitchen wastes, agro-residues, and industrial wastes, are treated through microbial routes for obtaining composts that are nutritionally enriched and suitable for various uses (Rawat et al., 2005). According to Miller (1994), the overall objectives of composting and anticipated benefits are

- Achievement of a suitable bulk density. Compost makes a more physically stable landfill and can be easily stored, transported, and disposed of than the original material, as the bulk density of the former is higher.

- Modification of complex polysaccharides and plant materials.
- Biological removal of readily available nutrients to avoid overheating.
- Building up of an appropriate biomass and a variety of microbial products.
- Establishment of selectivity.
- Conversion of nitrogen into a stable organic form.
- Sanitation, that is, the killing of pathogenic microorganisms, larvae, and weeds.

In addition to the above, the primary goals of composting include the safe handling of the organic wastes and enhancement of the soil's fertility. Furthermore, the advantages of composting include the cost-effective, sustainable procurement of a substitute for synthetic fertilizers (Beck, 1997; Moss et al., 2002); the establishment of a bacterial environment that supports healthy root systems (Tiquia et al., 2002; Danon et al., 2008); and the conversion of objectionable organic wastes into a stable, benign substance having less volume (Dickson et al., 1991; Dougherty, 1998). Dougherty (1998) further expanded the list of benefits of composting to include the reduction of odors, especially in the case of feces composting; the improvement of the bulking and water-holding capacities of soils; the enhancement of the biodegradation process compared with what would happen in nature; the destruction of weed seeds, pathogens, and insect eggs; the incorporation of nitrogen, phosphorus, and potassium into more suitable chemical compounds; and the inhibition of molds and fungi in soils.

With the expansion of suburban populations encroaching upon formerly rural areas, more people are being exposed to mushroom compost preparation facilities. Recently, the general public has been intolerant of odors emanating from substrate preparation, and these odors are becoming a serious problem for composting operations (Beyer et al., 1997). Presently, the mushroom composting process is typically based on Sinden's work, who shortened phase I down to about a week (exclusive of the collecting and wetting of horse manure), followed by a short phase II of about 6 days in trays or tunnels (Sinden and Hauser, 1953). This is well known as a short method of composting and is used by most mushroom growers in Europe, and developing countries like India and China are also nurturing their mushroom industry with the short method of composting.

Mushroom compost production is based on a mixture of biological, chemical, physical, and ecological parameters that are not exactly defined. Traditionally, compost is produced from wheat straw, straw-bedded horse manure, chicken manure, and gypsum. After mixing and moistening these ingredients for about 2–3 days, the mixture is subjected to a phase I composting process. The mixed ingredients are stacked in windrows in the open air for exposure to an uncontrolled self-heating process for up to 1 week. During self-heating, temperatures in the windrows range from ambient to 80°C and ammonia and foul-smelling compounds are emitted, causing environmental problems. Phase II is essentially an aerobic process carried out by maintaining the compost temperature at 45°C for 6 days in shallow layers in mushroom houses or tunnels. This is preceded by pasteurization for 8 h at 70°C to prevent an outbreak of pests and molds during phase II (Straatsma et al., 1995b). The biotechnological applications of thermophilic fungi in composting stem from (1) their ability to hydrolyze plant polymers, thereby hastening the process of decomposition; (2) their role in the production of nutritionally rich compost to increase the quality and yield of mushrooms; and (3) their suitability as agents of experimental systems for genetic manipulation for use in recombinant DNA technology.

In this chapter, the bioconversion of lignocellulosic materials, physicochemical aspects of composts, ecology of thermophilic fungi in mushroom compost, methods of composting, growth promotion of *Agaricus bisporus* (*Agaricus brunnescens*) by thermophilic fungi, and physiological interactions of thermophilic fungi and their role in co-composting are discussed, and the prospects for future study are identified.

9.2   BIOCONVERSION OF LIGNOCELLULOSIC MATERIALS

The most important basic material for the production of compost is various kinds of cellulosic materials, such as straw of wheat, paddy, rye, barley, and oats. However, wheat and rye are preferred because they are firmer and give better texture (Gerrits, 1985). The main function of straw is to provide a reservoir of cellulose, hemicellulose, and lignin, which is utilized by the mushroom mycelium as a carbon source during its growth. During composting, a great deal of organic matter is broken down into simpler compounds. Readily accessible compounds are degraded first, and recalcitrant materials, such as inorganic compounds, cellulose, and lignin, accumulate (Derikx et al., 1990). Cellulose is a linear molecule composed of repeating cellobiose units held together by β-glycosidic linkages. Lignin is a complex phenylpropanoid polymer that surrounds and strengthens the cellulose–hemicellulose framework (Kringstad and Lindstrom, 1984). It is believed that hemicellulose acts as a molecular bonding agent between the cellulose and lignin fractions (Torrie, 1991). The optimal temperature of 50°C for the mineralization rate during phase I coincides with cellulase production by bacteria and fungi (Gilbert and Hazelwood, 1993). Cellulases are mixtures of several enzymes that act in concert to hydrolyze crystalline cellulose to its monomeric component, glucose.

The exact mechanism of cellulose hydrolysis is quite obscure (Duff and Murray, 1996), although a number of possible models have been proposed. The classic, and perhaps the simplest, one is presented in Figure 9.1. The mechanism of cellulose hydrolysis is envisioned as an initial attack by endocellulases that cleave internal β-1,4 linkages in amorphous sections of cellulose. Exocellulases (CBH-I and CBH-U in *Trichoderma*) then cleave cellobiose units from the nonreducing ends of the cellulose chain. Finally, β-glucosidase converts cellobiose to glucose monomers (Goyal et al., 1991; Walker and Wilson, 1991). These enzymes act in a synergistic or cooperative manner.

In general, composting is considered to be an aerobic process (Finstein and Morris, 1975), suggesting high biological activity. The oxygen consumption rates of compost during phase I increase logarithmically with temperature from 20°C to 70°C (Schulze, 1962). Derikx et al. (1990) observed a sharp decline in the oxygen consumption rates above 60°C, indicating the inhibitory effect of elevated temperatures on biological activity. They also observed that mineralization dropped to zero at 70°C, and the oxygen consumption rate decreased to 25% of its initial value. High biological activity is characterized by the production of stench, hydrogen sulfide, methane thiol, dimethyl sulfide, and dimethyl trisulfide. Anaerobic conditions at elevated temperatures favor

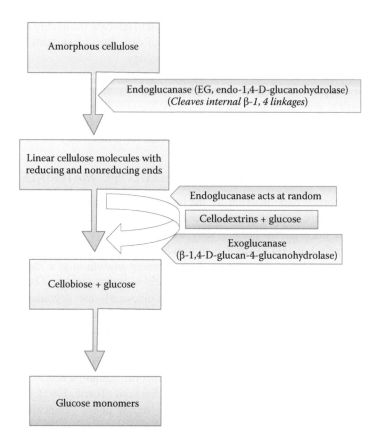

FIGURE 9.1 Scheme of breakdown cellulose in lignocellulosic materials. (Modified from Chahal, D.S., and Overend, R.P., Ethanol fuel from biomass, in *Advances in Agricultural Microbiology*, ed. N.S. Subba Rao, 581–641, New Delhi, Oxford and IBH Publishing Co, 1982; Duff, S.J.B., and Murray, W.D., *Bioresource Technology*, 55:1–33, 1996.)

the formation of carbon disulfide and dimethyl disulfide, indicating that their nonbiological formation causes much of the environmental problems. Their formation can be decreased to 75% if phase I is performed in tunnels, because of the aerobic environment provided by the forced ventilation (Derikx et al., 1991).

During the composting process, thermophilic and thermotolerant fungi show a whole spectrum of physiological behavior with regard to their ability to decompose plant polymers. Evidently, some quite complex interspecific relationships exist in this respect (Chang, 1967). Some thermophilic fungi, that is, *Chaetomium thermophile*, *Humicola insolens*, *Aspergillus fumigatus*, and *Torula thermophila*, can hydrolyze cellulose present in straw (Flannigan and Sellars, 1972). A few fungi, such as *Talaromyces thermophilus*, may be weakly lignolytic (Tansey et al., 1977). Sharma (1991) found no lignin degradation during phases I and II; however, he reported a decrease in the lignin content of the spent compost. Bonnen et al. (1994) convincingly demonstrated the ability of *Agaricus bisporus* to produce lignin-degrading peroxidases. In contrast, Iiyama et al. (1994) reported a relative increase in the lignin content of compost during the composting and cultivation of mushrooms. They observed that the structure of lignin was altered rather than degraded. Till (1962) showed that good yields of mushrooms can be obtained on autoclaved straw supplemented with organic nitrogen. Extensive degradation of straw is not a prerequisite for a high yield of mushrooms. Tunnel phase I compost produces the same quantity of mushrooms as traditional outdoor compost that has been degraded much more intensively (Straatsma et al., 1995b).

9.3 PHYSICOCHEMICAL ASPECTS OF COMPOSTS

All types of composts, such as municipal compost, industrial compost, mushroom compost, vermicomposts, and garden compost, are characteristically unique with regard to the raw materials used, their method of preparation, their physicochemical properties, and the end product formed. The most important parameters governing the entire composting process include the initial C:N ratio of substrates, moisture content, temperature, bed size, initial pH value, ammonia, and carbon dioxide. The early phase of composting is the most dynamic part of the process and is characterized by a rapid surge in temperature, large swings in pH, and the degradation of simple organic compounds (Schloss et al., 2003).

9.3.1 INITIAL C:N AND C:P RATIO

The C:N ratio is one of the important parameters most often considered in studies pertaining to the composition of composted mixtures (Brito et al., 2008; Vernoux et al., 2009). The rate of decomposition is markedly influenced by the size of the compost pile, as well as the C:N ratio of the composting material. Beck (1997) observed that if the C:N ratio is too high, the temperature in the compost pile will fail to rise, whereas if C:N is too low, the mixture may emit unpleasant odors. The C:N ratios of some common compost materials are given in Table 9.1. The initial C:N values, ranging from 25:1 to 40:1, have been recommended for preparing good-quality compost mixtures. A C:N ratio of 30:1 is considered ideal for the activity of most microbes. However, it is difficult to define the upper end of this range owing to the wide range of density, particle size, and lignin contents of different lignocellulosic materials. Further, compared with the C:N ratio, the ratio of carbon to phosphorus, that is, C:P, is also crucial in composting; however, it has received little research attention. Brown et al. (1998) observed that phosphorus can be a limiting factor when the composting mixture contains a high proportion of paper. The authors found that the initial C:P ratio should be between 120:1 and 240:1 when the C:N ratio is 30:1.

A major challenge in composting is to minimize the loss of gaseous nitrogen stored in the tissues of living or dead organisms. Due to enzymatic activity, organic substances in the compost raw materials are broken down, resulting in the production of ammonium (NH_4^+) in a process known as ammonification. Under alkaline conditions, ammonium ion is converted into ammonia gas, which can escape to the atmosphere if not recaptured (Hubbe et al., 2010). To retain the ammonia gas, moistening of the outer surface layer of the compost pile is generally recommended. Ammonium is further converted to nitrite and nitrate by *Nitrosomonas* and *Nitrobacter*, respectively, under aerobic conditions. On the contrary, under anaerobic

TABLE 9.1
C:N Ratio of Some Common Compost Materials

Sr. No.	Compost Material	C:N Ratio	Comments
1	Sawdust	500:1	
2	Paper	200:1	
3	Straw	40–80:1	Too high in C if used alone
4	Cornstalks	60:1	
5	Used mulch	60:1	
6	Dry leaves	60:1	
7	Old hay	30:1	
8	Wilted greens	20:1	Ideal range by itself
9	Vegetable scraps	20:1	
10	Grass clippings	15:1	
11	Legume hay	15:1	
12	Manure	15:1	Too high in N if used alone
13	Kitchen scraps	10:1	
14	Blood meal	5:1	

Source: Storl, W.E., *Culture and Horticulture: A Philosophy of Gardening*, Bio-Dynamic Literature, Wyoming, RI, 1979. See also course content created by Robinson, D., Introductory class on-line (subject 4): BD materials and compost, Biodynamic Farming and Gardening Association, Myrtle Point, OR.

conditions nitrite is converted to ammonia or nitrogen gas, both of which eventually escape to the atmosphere. The C:N elemental ratio is often used to evaluate the degree of completion of a composting process.

Similarly, one expects the carbon content to decline as CO_2 is released; in contrast, the retention of organic nitrogen is usually regarded as highly desirable. A good correlation between CO_2 evolution on the surface of the compost pile and microbial activity has been obtained by Johri and Rajni (1999). They also observed that in spite of the low number of CFU during peak heating, CO_2 evolution is high due to a higher rate of respiration of the abundant thermophilic microflora. Weigant (1992) advocated the effect of respiratory CO_2 of *Scytalidium thermophilum* as the most likely reason for the growth-promoting effect of this fungus on mushroom yield. According to Johri and Rajni (1999), CO_2 evolution can be measured by the following equation:

$$CO_2 \text{ evolved} = \frac{\sqrt{\% \text{ carbohydrate}}}{^{C}/_{N} \text{ residues} \times \% \text{ lignin}}$$

The rate of decomposition in compost depends on the carbon, nitrogen, lignin, and carbohydrate composition of the composting organic material.

9.3.2 MOISTURE CONTENT

An appropriate level of moisture content is important in the composting process, as it facilitates the substrate decomposition through mobilizing microbial activities and provides better conditions for nitrogen fixation in the compost. A moisture content of 50%–60% (dry weight basis) is ideal for composting and microbial activity (Horng, 2003). Liang et al. (2003) observed that 50% moisture content in the composting mixture is the minimum requirement for maintaining high microbial activity. Further, under low moisture conditions the ammonium and ammonia present generate a high vapor pressure, thus creating conditions for the loss of nitrogen. As ammonia is readily soluble in water, a higher moisture content prevents ammonia escape from the compost and promotes nitrogen fixation. In addition, the ratio between dry matter, water, and air is crucial in composting, as diverse raw materials have different moisture contents. Typically, in mushroom compost the recommended optimum moisture is 75% at the time of filling, 69% at the time of spawning, and 66% after the spawn run (Johri and Rajni, 1999; Rawat, 2004). Different moisture contents have been recommended for a variety of composting mixtures. For example, initial moisture contents of 69% for composting poultry manure and wheat straw (Petric et al., 2009) and 46% for composting kitchen garbage (Hwang et al., 2002) have been found to be optimum. The change in moisture contents also affects the change in temperature during the entire process of composting.

9.3.3 TEMPERATURE

Temperature is one of the key variables of the composting process, as it affects the kind of microorganisms that are likely to thrive within the compost pile. Several authors have focused on changes in temperature during composting and observed that it is an important parameter to monitor composting efficiency, because it affects not only the biological reaction rates and the dynamic population of microbes but also the physicochemical characteristics of composts (Dougherty, 1998; Namkoong et al., 2002; Antizar-Ladislao et al., 2005; Tang et al., 2007; Vernoux et al., 2009). A gradient of temperature is obtained in the compost pile, generally within 48 h of piling the moistened raw materials (Figure 9.2). In a compost pile, temperature is directly proportional to the biological activity. With the acceleration of the metabolic rate of microbes, the temperature within the pile increases; in contrast, low metabolic rates of microorganisms result in a lower temperature in the compost pile. As depicted in Figure 9.2, a temperature of up to 80°C can be attained in the center of the pile. The maximum temperature that can be attained in the composting system varies with the type of composting raw materials. For example, a maximum temperature of 70°C can be attained within 4 days in a metro waste composting system meant for processing 150 t of general municipal waste daily (Rawat, 2004). Similarly, a peak temperature of 67°C was obtained after 8 days while composting wheat straw that remained above 50°C for about 3 weeks (Chang and Hudson, 1967). Likewise, Pedro et al. (2001) observed that composting of garden waste occurs at 60°C–70°C during the initial phase by self-heating, whereas the latter phase operates at 45°C.

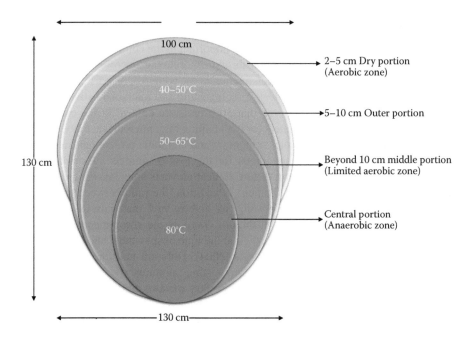

FIGURE 9.2 A vertical section of mushroom compost pile showing different zones of temperature.

9.3.4 PILE SIZE

The optimum size of a compost pile depends on whether the pile is aerated, whether it is turned, and whether its extremities are partly contained in insulating materials (Hubbe et al., 2010). Dickson et al. (1991) recommended a minimum volume of approximately 1 m^3 to ensure sufficient self-insulation, so that composting material will heat up. In addition, the size of compost pile is important not only in temperature buildup and maintaining an appropriate microbial equilibrium and successional pattern, but also to induce Maillard reactions. A Maillard reaction refers to the fixing of free ammonia through reactions with carbohydrates and lignatious polymers. Maillard reactions transform carbohydrates into a chemical form that is accessible to the mushroom mycelium but not to the competitors, and hence allows for compost selectivity.

9.3.5 INITIAL pH VALUE OF COMPOST

The initial pH value of composting raw materials is critical for allowing the onset of microbial decomposition. In general, an initial pH in the range of 4.2–7.2 (Dickson et al., 1991) or 7–7.5 (Tchobanoglous et al., 1993) has been suggested. Hultman (2009) reported that production of lactic and acetic acids during initial decomposition of biomass leads to an acidic pH of 4.2–5.5. As the biodegradation proceeds and thermogenesis sets in, the pH can rise up to 9, resulting in the release of obnoxious ammonia from the compost heap, and thereafter the pH usually sets around neutral as the compost matures. A rise in pH is sometimes considered evidence for the colonization of microorganisms and successful composting. Golueke (1972) reported the optimum pH values for bacterial (6–7.5) and fungal (5.5–8.9) activities in composts. In the case where the initial pH of the composting materials is highly acidic, the addition of lime or fly ash is recommended (Beck, 1997). In contrast, during excessive release of ammonia (pH 8.5–9), the addition of ammonium sulfate could counter such a release of ammonia (Ekinci et al., 2000). A pH of 7.5 is ideal for the optimum activity of cellulose, as cellulose is the key enzyme involved in the decomposition of lignocellulosic materials, which are the major component of composts.

9.4   ECOLOGY OF THERMOPHILIC FUNGI
IN MUSHROOM COMPOST

The role of thermophilic fungi, bacteria, and actinomycetes in the self-heating of the mushroom compost was first investigated by Waksman and his coworkers in the 1930s (Waksman and McGrath, 1931; Waksman and Nissen, 1932; Waksman and Cordon, 1939). It is now well established that the rise in temperature and the decomposition of composting plant materials are brought about by thermophilic microorganisms, including fungi. Successions of fungi in wheat straw compost have been studied by Chang and Hudson (1967). Their results are probably typical for any compost. Thermophilic microorganisms in compost have received extensive investigation (Fermor et al., 1985). A number of thermophilic fungi frequently isolated from mushroom compost are enumerated in Table 9.2. Fergus (1964) isolated and described eight species of thermophilic and thermotolerant fungi from mushroom compost collected from the surface and interior of composts during phase II of peak heating. The isolated fungi were *Aspergillus fumigatus*, *Chaetomium thermophile*, *Humicola grisea* var. *thermoidea*, *Humicola insolens*, *Humicola lanuginosa*, *Mucor pusillus*, *Talaromyces dupontii*, and *Stilbella thermophila*. Satyanarayana (1978) reported the significant presence of thermophilous fungi, such as *Acremonium albamensis*, *Achaetomium macrosporum*, *Rhizopus microsporus*, *Absidia corymbifera*, *Theilavia minor*, and *Torula thermophila*, during the composting of paddy straw. Similarly, Sandhu and Sidhu (1980) isolated two thermophilic fungi, *Rhizopus microsporus* and *Mucor pusillus*, from decomposing sugarcane bagasse, whereas Lacey (1974) isolated *Humicola*, *Allescharia*, *Sporotrichum*, and others from the sugarcane bagasse. Ogbonna and Pugh (1982) isolated *Mucor meihei*, *Melanocarpus albomyces*, *Talaromyces thermophilus*, *Thermoascus aurantiacus*, *Thermoascus thermophilus*, and *Theilavia terrestris* from composts in Nigeria. They also observed that thermophilic fungi such as *Chaetomium thermophile* var. *coprophile*, *Myceliophthora thermophila*, *Aspergillus fumigatus*, and *Sporotrichum pulverulentum* isolated from municipal compost were able to convert tree bark into compost.

The course of fungal succession in mushroom composts may be partially explained by the ecophysiological data available (Evans, 1971; Fergus and Amelung, 1971; Chapman, 1974; Rosenberg, 1975, 1978). The ability to use complex carbon sources and thrive at high temperatures is the most important characteristic of the successful

TABLE 9.2
Thermophilic and Thermotolerant Fungi Found in Mushroom Compost

Fungus	Detected in Our Study[a]	Log$_{10}$ (CFU/g)	References
		Zygomycetes	
Absidia corymbifera	+	2.9	Chang and Hudson (1967), Basuki (1981), Straatsma et al. (1994a)
Rhizomucor miehei	+	3.1	Eicker (1977), Straatsma et al. (1994a)
Rhizomucor pusillus	+	3.4	Fergus (1964), Chang and Hudson (1967), Hayes (1969), Cailleux (1973), Seal and Eggins (1976), Eicker (1977), Fermor et al. (1979), Basuki (1981), Bilai (1984), Straatsma et al. (1994a)
		Ascomycetes	
Chaetomium thermophile	+	3.7	Fergus (1964), Chang and Hudson (1967), Cailleux (1973), Seal and Eggins (1976), Eicker (1977), Fermor et al. (1979), Basuki (1981), Bilai (1984), Zhang et al. (2017)
Corynascus thermophilus	–		Fergus and Sinden (1969), Straatsma et al. (1994a)
Emericella nidulans	+	3.0	Cailleux (1973), Basuki (1981), Straatsma et al. (1994a)
Talaromyces emersonii	+	2.9	Straatsma et al. (1994a)
Talaromyces thermophilus	+	2.9	Fergus (1964), Chang and Hudson (1967), Eicker (1977), Straatsma et al. (1994a)
Thermoascus aurantiacus	+	3.6	Bilai (1984), Straatsma et al. (1994a)
Thermoascus aurantiacus var. *levispora*	–		Straatsma et al. (1994a)
Thermoascus crustaceus	–		Straatsma et al. (1994a)
Thielavia terrestris	–		Straatsma et al. (1994a)
Myriococcum albomyces	+	3.5	Fergus (1971), Seal and Eggins (1976), Straatsma et al. (1994a)
Myriococcum thermophilum	–		Straatsma et al. (1994a)
Chaetomium sp.	–		Straatsma et al. (1994a), Zhang et al. (2017)
Rasamsonia composticola	–		Su and Cai (2013)
Rasamsonia emersonii	–		Sebok et al. (2015)
		Basidiomycetes	
Coprinus cinereus	–		Straatsma et al. (1994a)
Basidiomycetes 1	+		
Basidiomycetes 2	+		

(Continued)

TABLE 9.2 (CONTINUED)
Thermophilic and Thermotolerant Fungi Found in Mushroom Compost

Fungus	Detected in Our Study[a]	Log₁₀ (CFU/g)	References
		Deuteromycetes	
Aspergillus fumigatus	+	3.9	Fergus (1964), Hayes (1969), Chang and Hudson (1967), Cailleux (1973), Olivier and Guillaumes (1976), Seal and Eggins (1976), Eicker (1977), Fermor et al. (1979), Basuki (1981), Bilai (1984), Straatsma et al. (1994a), Zafar et al. (2013)
Hormographiella aspergillata	–		Straatsma et al. (1994a)
Humicola insolens	+	3.9	Vijay and Pathak (2014)
Malbranchea sulfurea	+	3.1	Chang and Hudson (1967), Eicker (1977), Straatsma et al. (1994a)
Paecilomyces variotii	–		Straatsma et al. (1994a)
Torula thermophila (= Scytalidium thermophilum)	+	4.2	Fergus (1964), Hayes (1969), Chang and Hudson (1967), Cailleux (1973), Olivier and Guillaumes (1976), Seal and Eggins (1976), Eicker (1977), Fermor et al. (1979), Basuki (1981), Bilai (1984), Straatsma et al. (1994a), Zafar et al. (2013), Souza et al. (2014), Vijay and Pathak (2014)
Stilbella thermophila	+	2.8	Fergus (1964), Hayes (1969), Cailleux (1973), Seal and Eggins (1976), Straatsma et al. (1994a)
Thermomyces lanuginosus	+	3.7	Fergus (1964), Hayes (1969), Chang and Hudson (1967), Cailleux (1973), Seal and Eggins (1976), Eicker (1977), Fermor et al. (1979), Basuki (1981), Bilai (1984), Straatsma et al. (1994a), Zafar et al. (2013), Souza et al. (2014), Sebok et al. (2015)
Thermomyces ibadanensis	–		Souza et al. (2014)
Unidentified taxon	+	2.8	Salar and Aneja (2007)

Source: Salar, R.K., and Aneja, K.R., *Agric. Technol. Bangkok* 3, 241–253, 2007.

[a] –, not detected; +, isolated from compost.

colonizers of composts. On wheat straw composts, *Rhizomucor pusillus* disappears early in the succession and seems to be a primary sugar fungus. *Thermomyces lanuginosus* persists as a secondary sugar fungus in mutualistic relationships with some of the true cellulose decomposers of composts (Chang, 1967; Hedger and Hudson, 1974; Deacon, 1985). *Chaetomium thermophile, Malbranchea sulfurea,* and *Scytalidium thermophilum* grow fast and are known to produce cellulose, which is essential for the breakdown of cellulolytic materials. *Scytalidium thermophilum* is reported to be the climax species in composts (Straatsma and Samson, 1993). The only time that fungi are not active in the compost is during the peak heating phase of outdoor composting. The maximum temperature phase kills off all fungi in the center (Figure 9.2) and allows recolonization as the temperature falls to below each fungus's upper temperature limit for growth (Chang and Hudson, 1967). In general, one might suspect that the higher the maximum temperature for growth, the more rapid the recolonization. From the data of Cooney and Emerson (1964), Kumar and Aneja (1999), Rosenberg (1975), Singh and Sandhu (1982), and Tansey and Brock (1972), it is observed that there are only a few degrees of difference in the upper temperature limits (mostly between 55°C and 60°C) of the various thermophilic fungi.

Willenborg and Hindorf (1985) studied the fungal flora in mushroom culture substrate from the beginning of composting to the final stage of mushroom picking. They observed that 60.7% of the total microflora comprised thermophilic fungi. In our study (Salar and Aneja, 2007) on the role of thermophilic fungi in the growth promotion of button mushroom, we isolated 18 thermophilic fungi, including 2 basidiomycetes and 1 sterile form from phase I and phase II composts, and these represented most of the known thermophilic taxa (Table 9.2). During phase I isolation, it was observed that the plates were frequently overcrowded with the ubiquitous *Aspergillus fumigatus* and *Rhizomucor* spp. The isolation of a few isolates producing only sterile mycelium proved problematic taxonomically. Macroscopically, young cultures of *Torula thermophila* and *Chaetomium thermophile* resembled closely and were fast growing. *Rhizomucor pusillus, Rhizomucor miehei,* and *Absidia corymbifera* were also very similar in gross morphology and were fast growing. Seven thermophilic and thermotolerant fungi were isolated from phase I compost; the rest of the species was isolated from phase II compost (Table 9.2). The zygomycetous fungi (Table 9.2) are very common in wheat straw compost (Fergus, 1964; Chang and Hudson, 1967; Chahal et al., 1976; Straatsma et al., 1994a). During phase I, the total count was quite low, as revealed by low colony-forming units (CFU) per gram (Table 9.3). The fungi isolated from phase II compost were *C. thermophile, Emericella nidulans, Thermoascus aurantiacus, Myriococcum albomyces, Humicola insolens, Malbranchea sulfurea, T. thermophila, Stilbella thermophila,* and *Thermomyces lanuginosus.* Two basidiomycetous species were also visually observed in phase II compost. A species producing only sterile mycelium was also isolated from phase II. Furthermore, most species nearly disappeared after phase II composting. Fungi recovered from compost at the end of phase II were almost exclusively *T. thermophila, Humicola insolens,* and *C. thermophile.* The population density of *T. thermophila* was highest (15849 CFU/g) compared with the other two fungi (Table 9.2). Our survey (Salar and Aneja, 2007) of thermophilic fungi in mushroom composts provided valuable isolates of *T. thermophila, C. thermophile, Malbranchea sulfurea,* and

TABLE 9.3

Growth Rates of *Agaricus bisporus* on Sterilized Compost Inoculated with Different Thermophilic Fungi Singly and in Combinations

Species	Mycelial Extension Rate in Tubes (mm/day)[a]	Radial Growth Rate in Petri Dishes (mm/day)
Control	6.1e	4.9 ± 0.5
Chaetomium thermophile	6.3d	ND
Malbranchea sulfurea	7.1b	6.1 ± 0.9
Thermomyces lanuginosus	5.4g	ND
Torula thermophila	6.6c	5.3 ± 0.8
C. thermophile + *M. sulfurea*	7.1b	ND
C. thermophile + *T. lanuginosus*	6.0e	ND
C. thermophile + *T. thermophila*	6.6c	ND
M. sulfurea + *T. lanuginosus*	5.8f	ND
M. sulfurea + *T. thermophila*	7.7a	ND
T. lanuginosus + *T. thermophila*	6.0e	ND
C. thermophile + *M. sulfurea* + *T. lanuginosus* + *T. thermophila*	6.5c	9.0 ± 1.0

Note: CV = 1.5%. ND = Not determined. Radial growth rate values are given as ± standard deviation.
[a] Values with the same letter are not significantly different (DMRT; $p < 0.05$).

T. lanuginosus. The first three fungi are known to adapt and colonize the compost frequently (Straatsma et al., 1994a, 1994b), and *T. lanuginosus* is known to promote the decomposition rate when grown in combination with cellulolytic fungi (Deacon, 1985). The composition and genetic diversity of fungal populations during phase II of compost production for the cultivation of *Agaricus subrufescens* was determined by Souza et al. (2014) using culture-dependent and culture-independent methods. They identified *Scytalidium thermophilum*, *T. lanuginosus*, and *Thermomyces ibadanensis* as the most abundant species. The latter species was identified for the first time from mushroom compost.

Recently, Vos et al. (2017) used an alternate technique to quantify bacterial and fungal biomass in compost during colonization of *Agaricus bisporus*. They used chitin content in the compost as an indicator of the total fungal biomass in the compost. They observed that the chitin content increased during a 26-day period from 576 to 779 nmol/g *N*-acetylglucosamine compost in the absence of *A. bisporus* (negative control). They also observed a similar increase in the presence of this mushroom-forming fungus. They further reported that the fungal phospholipid-derived fatty acid (PLFA) marker C18:2ω6, indicative of the living fraction of the fungal biomass, decreased from 575 to 280 nmol/g compost in the negative control, whereas it increased to 1200 nmol/g compost in the presence of *A. bisporus*. They suggested that fungal biomass can make up 6.8% of the compost after *A. bisporus* colonization, 57% of which is dead. Furthermore, their results showed that *A. bisporus* impacts biomass and composition of bacteria in compost. Molecular techniques, like sequencing of rDNA genes and internal transcribed spacer (ITS) sequences, have also been employed in monitoring the microbial diversity in composts (Ivors et al., 2000; Souza et al., 2014; Holman et al., 2016).

9.5   ROLE OF HYDROLYTIC ENZYMES OF THERMOPHILES IN COMPOSTING

Composts harbor structurally diverse and functionally active microbial communities that are important to the composting process both in nature and under controlled environmental conditions. During the composting process, various organic materials are converted into simpler units of organic carbon and nitrogen. The overall efficiency of organic material breakdown depends on the microbes and their activities (Raut et al., 2008). Thermophilic fungi play a great role in the degradation of organic materials by way of secreting various types of hydrolytic enzymes, including the most important cellulolytic and xylanolytic enzymes. Considering the fact that these fungi produce enzymes that maintain their activities at high temperatures, detailed studies of their enzymes are likely to provide a better understanding of their biodegradative potential. Furthermore, enzymes from thermophilic fungi are often more stable at higher temperatures than the enzymes from their mesophilic counterparts, and some even show stability at up to 80°C (Margaritis and Merchant, 1983; Margaritis et al., 1986). It has also been observed that biomass-degrading enzymes from thermophilic fungi consistently display higher hydrolytic ability, despite the fact that the enzyme titers are lower than those of more conventionally used species (Wojtczak et al., 1987; Lee et al., 2014). A detailed discussion on the enzymes produced by these fungi is presented in Chapter 11.

During composting, the pioneer microflora, such as *Mucor pusillus* and *Aspergillus fumigatus*, having pH optima just below 7 and an optimum temperature around 40°C, is able to utilize simple sugars. However, when self-heating and ammonification start and the pH reaches 9, such microflora disappear from the compost. *Thermomyces lanuginosus* and *Talaromyces thermophilus* have relatively high pH and temperature optima. Their high thermal death points offer a selective advantage during periods of peak heating. They do not degrade cellulose and have a moderate growth rate. The thermophilic *Scytalidium thermophilum* and *Chaetomium thermophilum* grow fast and degrade cellulose and hemicellulose strongly. Microorganisms degrade about 40% of dry matter of compost, the dry matter being of potential value for the nutrition of mushroom mycelium (Straatsma et al., 1994b). During phase II, when compost temperature drops significantly, the thermophilic *Sporotrichum thermophile* and *Scytalidium thermophilum* and the mesophilic *Coprinus cinereus* and *Clitopilus pinsitus* appear and can utilize cellulose and hemicellulose in compost at a comparatively slower rate (Chang and Hudson, 1967; Straatsma et al., 1994a). Vijay and Pathak (2014) investigated the degradation of hemicellulose, cellulose, and lignin at various pH levels by measuring xylanase, exocellulase, endocellulase, β-glucosidase, and laccase activity on different agricultural residues and on different compost formulations. They observed that all these enzymes play an important role in degrading the polymeric components of agricultural residues in monomeric form or sugars that later can easily be utilized by different thermophilic fungi, *Humicola insolens* in particular. The highest activity of all the cellulolytic and lignolytic enzymes was recorded at pH 8.0 in all agricultural residues and formulations by all strains.

The wide enzymatic potential displayed by thermophilic fungi with different physicochemical characteristics has been exploited by preinoculation of compost by several workers. Straatsma et al. (1994b) reported the growth promotion of *Agaricus bisporus* on sterilized compost by nine thermophilic fungi: *Chaetomium thermophilum*, an unidentified *Chaetomium* sp., *Malbranchea sulfurea*, *Myriococcum thermophilum*, *Scytalidium thermophilum*, *Stilbella thermophila*, *Thielavia terrestris*, and two unidentified basidiomycetes. The growth-promoting species are not pioneers of compost, and they are probably all cellulolytic. In our study (Salar and Aneja, 2007), precolonization of compost with four thermophilic fungi, *Chaetomium thermophile*, *Malbranchea sulfurea*, *Thermomyces lanuginosus*, and *Torula thermophila*, either singly or in different combinations, proved beneficial for the growth promotion of *Agaricus bisporus*. All these species are able to produce a wide array of extracellular enzymes (Maheshwari et al., 2000). In contrast, Savoie and Libmond (1994) have stimulated the composting process by addition of commercial polysaccharolytic enzymes in the substrate. They observed that compost treated with polysaccharidases resulted in increased enzymatic activities, number of bacteria, and solubilization of carbon and nitrogen. However, no effect on mushroom yield was observed by the authors.

9.6 METHODS OF MUSHROOM COMPOSTING

9.6.1 LONG METHOD OF COMPOSTING

For preparing mushroom compost, the traditional long method of composting is employed by most seasonal growers. However, this method exists only in a few pockets of the world due to its low mushroom yield and proneness to attack by pathogens, as well as it being a more time-consuming and labor-intensive process. It is generally prepared outdoors or under a shed on a cemented floor so that mite infestation does not occur. The raw materials used for composting have been given by different organizations, institutions, and workers and can be selected for making compost. Wheat straw compost is generally used for cultivation of button mushroom. Wheat straw is spread in a layer of 8–10 in. thickness over the floor of the composting yard. Sprinkle water on the straw for wetting two or three times a day for 2 consecutive days. Urea, calcium ammonium nitrate, and wheat bran are thoroughly mixed separately and covered with damp gunny bags for 14–16 h. These ingredients are now mixed with a prewetted straw on the floor and heaped into a pile with a stack mold. The entire pile is opened and spread over the composting yard on the third or fourth day for 45–60 min, and this process is called turning, which is repeated on every third day (Figure 9.3). At each turning, water is sprinkled to make up the loss of water due to evaporation. At the third turning, half of the gypsum amount is added, and the remaining gypsum is added on the fourth turning. At the fifth turning,

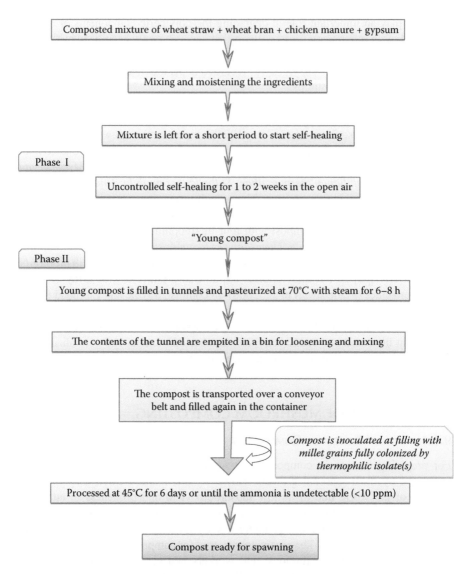

FIGURE 9.3 Protocol for the environmentally controlled production of compost inoculated with thermophilic isolate(s).

the insecticide Nemagon is added and thorough mixing of the straw is done; later, the pile is opened and left for 3 days until the ammonia smell dissipates. The compost is ready to use in about 25–28 days, in which a lignoprotein complex is formed that favors the growth of white button mushroom, and also decreases the carbon/nitrogen ratio with the addition of nitrogen sources. However, the microbiological and chemical reactions operating inside the substrate to convert it into a suitable medium for mushroom cultivation remain unsystematic under uncontrolled environmental conditions of composting.

Improvements in the yield and shortening the duration of the long method of composting were attempted by inoculating potent strains of *Scytalidium thermophilium* and *Humicola insolens* and their consortium in compost at 0 days of composting (Vijay and Pathak, 2014). The composting period was reduced to 20 days, compared with the present 28 days normally taken when long method of composting (LMC) is employed. Moreover, a significantly higher yield was also obtained in compost inoculated with thermophilic molds. Straatsma et al. (1989) hypothesized that *S. thermophilum* increases the growth of *Agaricus bisporus* by an unknown mechanism.

9.6.2   SHORT METHOD OF COMPOSTING

Owing to several limitations associated with the long method of composting and with the realization of the role of microbes and temperature in the composting process, improved methods of composting came into existence in the early 1950s. In practice, two phases are distinguished in the short method of composting. Phase I is performed in the open or under a roof where the moistened ingredients are stacked in long heaps or piles. A transverse section of a compost pile is shown in Figure 9.2. The material heats up and is turned several times (Figure 9.3). Processing of phase I in tunnels is presently under investigation, and its acceptability is debated (Straatsma et al., 1994b; Gerrits et al., 1995). Phase II composting is essentially performed in tunnels or mushroom houses that allow satisfactory control of environmental conditions, that is, temperature and air supply. The tunnels are used for large-scale operations. Before filling the "young compost" (any substrate prior to phase II) in tunnels, it is subjected to phase I (Figure 9.4) to soften the straw, which allows an optimal amount of compost to be filled into the cropping rooms. This is also essential for high biological efficiency (BE) of the substrate (Gerrits et al., 1994). BE refers to the percent fresh

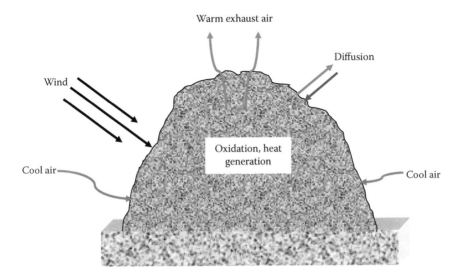

FIGURE 9.4   Convection currents and other factors affecting the aeration of a static mushroom compost pile. (Modified from Hubbe, M.A. et al., Bioresources, 5:2808–2854, 2010.)

weight of mushrooms produced from a given dry weight of compost ingredients; it is an indication of the efficiency underlying the bioconversion process that transforms straw and supplements into mushrooms (Leonard and Volk, 1992).

Young compost is filled in tunnels and pasteurized at 70°C for 8 h before processing at 45°C for 5–6 days. The dimension of a typical mushroom tunnel is 100 m^2 to fill compost at 2 m height or 1 t/m^2 (Van Lier et al., 1994). For inoculating the compost with a thermophilic isolate, the compost is cooled to 40°C and inoculated at filling in the second tunnel. Ventilative heat management is applied to the compost in the tunnel to keep the temperature within the range of 45°C–50°C. Typically, the total circulation is 200 m^3/t/h and ventilation with fresh ambient air is about 20 m^3/t/h (Van Lier et al., 1994). For oxidation of the substrate, only 2 m^3/t/h is required; therefore, most of the ventilation air is used to remove excess water to control temperature and not to supply oxygen. The loss of water and the high process temperature are important for good-quality compost. Air circulation rates below 100 m^3/t/h result in lower water loss from the substrate, thus making it vulnerable to desiccation, especially in the bottom section of the tunnel (Van Lier et al., 1994). Processing of the compost is continued until volatile NH$_3$ is undetectable (<10 ppm). After processing, the compost is cooled and inoculated with *A. bisporus* spawn.

9.6.3 ANGLO-DUTCH METHOD

The Anglo-Dutch method was developed by continuous research in several European countries (England, Holland, and Belgium) and Australia (Gerrits and Van Griensven, 1990; Nair and Price, 1991; Harper et al., 1992; Noble and Gaze, 1994). It is also known as the cold-process or low-temperature composting technique. This method involves the maintenance of an optimum temperature for the development of thermophilic fungi, namely, *Scytalidium thermophilum* and *Humicola* species. This method operates in two phases, a short pasteurization phase of 4–6 h at 60°C, followed by the conditioning phase at 41°C for a week. The whole process takes about 6–9 days and provides excellent selectivity with excellent outdoor control and significant savings of raw materials.

9.6.4 INRA METHOD

This method of indoor composting is used in countries like France, Italy, Belgium, and Austria, where large quantities of mushroom compost are prepared by it (Laborde, 1994). INRA is named after the L'Institut National de la Recherché Agronomique (French National Institute for Agricultural Research). This method is also known as a double-phase high-temperature process and involves two phases that are carried out indoors. In the first phase, compost ingredients are processed at a constant temperature of 80°C for 2–3 days under indoor conditions. In the second phase, the temperature is kept at 50°C for 5–7 days. Most microorganisms, including the desirable microbes, are killed in phase I due to a very high operating temperature. Therefore, reinoculation of the compost with the desirable thermophilic fungi becomes necessary. However, reinoculation can be avoided in situations where proper ventilation in the tunnel during phase I is maintained or, alternatively, phase I is being carried out in the bunker system. Straatsma et al. (1995a) have worked out a suitable formula for the inoculation of *Scytalidium thermophile*, an important thermophilic fungus for obtaining compost of the desired quality. The INRA method is also known as environmentally controlled composting (ECC), rapid indoor composting, aerated rapid composting, or express preparation of substrate (Laborde et al., 1993).

The processing of compost has been rationalized during the last three decades. Further, a number of innovations and modifications have been introduced in the composting process that involve an outdoor phase I, as it has been increasingly subjected to environmental audit. In general, tunnel processing (indoor composting) is advantageous for the bulk treatment of compost because it is cost-effective; it also provides the opportunity to clean the exhaust air of ammonia and stench. Ammonia can easily be removed by acid washing, and stench caused by sulfur-containing organics may be removed by biofiltration or sodium hypochlorite washing (Op den Camp et al., 1992). Some advantages of the indoor composting technology over other methods of composting include (1) less composting time than the long or short methods of composting; (2) higher mushroom yield (30–35 kg/100 kg of compost); (3) excellent control and management of environmental pollutants, conforming to civic laws; (4) less loss of raw materials, and thus an increased end product; and (5) improved selectivity of mature compost for *Agaricus bisporus*.

9.7 GROWTH PROMOTION OF *AGARICUS BISPORUS* BY THERMOPHILIC FUNGI

In the last decade, a new driving force has entered into the field of composting that is associated with the growth promotion of *A. bisporus* mycelium by thermophilic fungi, in particular *Scytalidium thermophilum* (Straatsma et al., 1995a). The inoculation of thermophilic fungi has shown that compost colonization by selected isolates is successful, and that microbial manipulation of phase II composting is feasible (Straatsma et al., 1994a; Kumar, 1996). The radial growth rate of mushroom mycelium never exceeds 3 mm/day (Last et al., 1974). Unfortunately, a growth-promoting effect of *S. thermophilum* on the mycelium of *A. bisporus* is not found on agar media (Renard and Cailleux, 1973). The high hyphal extension rates of *A. bisporus* on compost in the presence of thermophilic fungi may have an ecological significance: it may be able to grow as fast as possible, thereby colonizing as much substrate as possible. Once the substrate has been occupied, the mushroom mycelium seems to be able to prevent occupation by other microorganisms, either by consuming them (Fermor and Wood, 1981; Fermor and Grant, 1985) or by excretion of carbon monoxide (Stoller, 1978), which effectively inhibits the growth of most competing organisms but inhibits the growth of the mushroom mycelium itself only partly (Derikx et al., 1990).

Our studies (Salar and Aneja, 2007) concerning the effect of thermophilic fungi on the mycelial extension rate of *A. bisporus* when inoculated in sterilized compost singly and in combinations (Table 9.3) have indicated that thermophilic fungi prompted the growth of *A. bisporus* by an unknown mechanism. The linear growth rate of the mycelium of *A. bisporus* on sterilized compost in tubes (Figure 9.5a) was 6.1 mm/day. After an initial lag phase of 1 day, the growth rates were constant. The effects of *T. lanuginosus* when inoculated singly or in combination with another

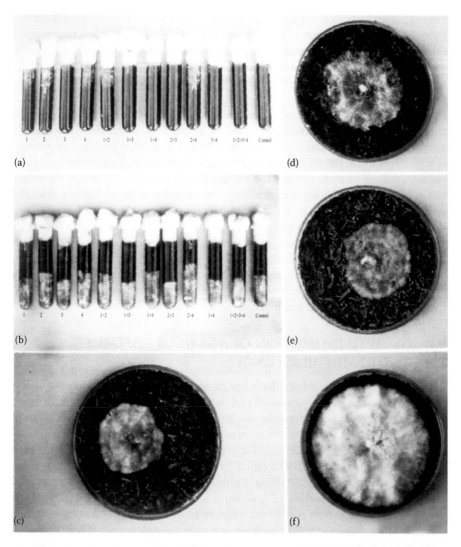

FIGURE 9.5 (a) Young compost colonized by four thermophilic fungi: (1) *Chaetomium thermophile*, (2) *Torula thermophila*, (3) *Thermomyces lanuginosus*, and (4) *Malbranchea sulfurea*; (b) Colonization of *Agaricus bisporus* mycelium on compost treated with various thermophilic fungi as in (a); (c–f): Mycelial growth of *Agaricus bisporus* 8 days after inoculation on (c) sterile compost, (d) compost treated with *M. sulfurea*, (e) compost treated with *T. thermophila* and (f) compost treated with *M. sulfurea* + *T. thermophila*.

thermophilic fungus on compost have been insignificant, resulting in lower growth rates than those of the control (Table 9.3). This could be due to the competition between organisms for products of hydrolytic enzymes. Therefore, it is difficult to establish the initial growth rates of thermophilic fungi. The other three species tested, *C. thermophile*, *Malbranchea sulfurea*, and *T. thermophila*, enhanced the growth of

A. bisporus to rates of 6.3, 7.1, and 6.6 mm/day, respectively, when inoculated singly. A mixed inoculum consisting of *M. sulfurea* and *T. thermophila* proved best and promoted the growth of *A. bisporus* mycelium to the plateau of 7.7 mm/day (Table 9.3), which was significantly higher than that from all other treatments. This finding indicates some specificity of the growth-promoting factors. An analysis of growth rate data using Duncan's Multiple Range Test (DMRT) is given in Table 9.3. The treatments *M. sulfurea* and *M. sulfurea* + *C. thermophile* were insignificant. Similarly, *T. thermophila*, *T. thermophila* + *C. thermophile*, and *C. thermophile* + *M. sulfurea* + *T. lanuginosus* + *T. thermophila* were also insignificant, as indicated by the DMRT alphabet in Table 9.3. The coefficient of variation (CV) of all the treatments was 1.5%.

A. *bisporus*, when grown in petri dishes on compost inoculated with *M. sulfurea* (Figure 9.5d), *T. thermophila* (Figure 9.5e), and *M. sulfurea* + *T. thermophila* (Figure 9.5f), produced radial growth rates (mm/day) of 6.1, 5.3, and 9, respectively (Figure 9.5). The growth rate for the control was low (4.9 mm/day). These two species of thermophilic fungi appeared most promising and were used for more controlled preparation of the substrate for *A. bisporus* cultivation. The growth of *A. bisporus* on control dishes was dense compared with growth on compost treated with either *M. sulfurea* or *T. thermophila*. Fluffy growth occurred when compost treated with both fungi was used as a substrate *for A. bisporus* growth (Figure 9.5d and e).

After inoculation with *A. bisporus* spawn at the rate of 0.5% (w/w) and incubation for 15 days, the inoculated composts were fully colonized by *A. bisporus* mycelium. The pH of inoculated compost before spawning was 6.5, and that of the control was 6.9, indicating weaker colonization (Gerrits, 1988). Mushroom yields from compost inoculated with *M. sulfurea* + *T. thermophila* were high (1990 g/5 kg of compost), almost twice that from pasteurized control. The yield from control compost was clearly below (1020 g/5 kg of compost) that from inoculated compost and was significantly lower, as shown by pair comparison using DMRT (Table 9.4). Low yields

TABLE 9.4

Compost Inoculation with Thermophilic Fungi and Cropping of *Agaricus bisporus*

Treatment	pH of Compost before Spawning (g/5 kg of compost)[a]	Yield of Mushrooms (%)	Biological Efficiency
Control	6.9	1020	20.4
Malbranchea sulfurea	6.5	1910	38.2
Torula thermophila	5.3	1355	27.1
M. sulfurea + *T. thermophila*	6.0	1990	39.8

Note: CV = 2.18%.

[a]   Average of five replications; the differences among treatments are significant (DMRT; $p < 0.05$).

of *A. bisporus* linked to the absence of thermophilic fungi might be explained by nonselective and/or toxic properties of experimental composts.

The effect of thermophilic fungi on the growth rate of mushroom mycelium in sterilized compost is quite remarkable. The radial growth rate of mushroom mycelium on any laboratory medium never exceeds 3 mm/day (Last et al., 1974). Based on our experimental data, we were convinced that thermophilic fungi, in particular *T. thermophila* and *M. sulfurea*, provide a trigger for the enhanced growth of *A. bisporus* acting by an unknown mechanism. This may be taken as an indication that the results of this study could be extrapolated to what actually happens during the production of mushroom compost. Unfortunately, a growth-promoting effect of *S. thermophilum* (syn. = *Torula thermophila*) on *A. bisporus* is not found on agar media (Renard and Cailleux, 1973). Actinomycetes and other bacteria might play a role in successful colonization of *S. thermophilum* during composting (Straatsma et al., 1989). Till (1962) showed that a good yield of mushrooms can be obtained on a noncomposted sterile mixture containing mainly straw and organic nitrogen. Carbon dioxide concentrations in the range of 0.3%–1.0% generate a higher extension rate of mushroom mycelium (Wiegant, 1992). The lower growth rates observed in petri dishes than culture tubes remain unexplained. This has also been reported for other fungi (Dickson, 1935; Trinci, 1973). The probable reason for a higher growth rate in culture tubes may be the ventilation caused by the ventilation channel in tubes, Salar and Aneja (2007). But it could not be explained unless otherwise demonstrated.

Wiegant (1992) suggested that CO_2 produced by *S. thermophilum* at 0.4%–0.5% CO_2 (v/v) explained the growth-promoting effect. Heterotrophic CO_2 fixation in the primary metabolism of *A. bisporus* is well known (Bachofen and Rast, 1968; Le Roux and Couvy, 1972), and established mycelia could maintain growth by the production of their own respiratory CO_2. Straatsma et al. (1995a) concluded that the CO_2 level influences the duration of the adaptation period rather than the extension rate. They showed that at an optimal CO_2 level, the mycelial extension rate of *A. bisporus* on compost fully grown with *S. thermophilum* was twice that of *A. bisporus* on compost without *S. thermophilum*. Therefore, the effects of thermophilic fungi and CO_2 seemed to be distinct.

9.8 CO-COMPOSTING

Agricultural operations often generate large amounts of manure, which if applied directly to agricultural land can have an adverse effect on soil, water, and air quality through contamination, odor and gas emissions, and nutrient leaching (Larney and Hao, 2007). Composting of agricultural wastes offers several advantages, including decreasing the load of microbial pathogens and phytotoxins, making the product more nutrient stable (Larney et al., 2006). Moreover, composting may also decrease the concentration of excreted antimicrobials, resistance determinants, and resistant bacteria (Arikan et al., 2007; Sharma et al., 2009; Xu et al., 2015), along with decreasing transportation costs, as composting significantly reduces the volume and mass of the manure (Larney et al., 2000).

Recent trends in the co-composting of manure or agricultural wastes with construction and demolition waste have attracted the attention of researchers (Holman et al., 2016). In developing countries and also in developed countries, construction and demolition waste makes up about 25% of the total solid municipal waste. Co-composting of manure with construction and demolition waste offers a potential alternative to make manure safe for soil amendment and also divert construction and demolition waste from municipal landfills. Recently, Holman et al. (2016) investigated the dynamics of the archaeal, bacterial, and fungal microbiota in two different types of composted feedlot cattle manure and determined the effects of the addition of construction and demolition waste on these microbiota, over a 99-day period. They also assessed the effect that these compost mixtures have on the concentration of antimicrobial resistance determinants in the cattle manure. It was hypothesized that the addition of construction and demolition waste would not affect the microbial parameters of composting. Therefore, the addition of construction and demolition waste as a cosubstrate in composting does not alter the compost microbiota; its inclusion in composted manure offers a safe, viable option for diverting construction and demolition waste from landfills.

9.9 FUTURE PROSPECTS

Since conventional composting is coupled with emission of ammonia and annoying odors, the search for environmentally friendly composting techniques is of great interest. Indoor composting research is now being targeted toward the microbial manipulation of phase II compost. Thermophilic fungi are believed to contribute significantly to the quality of compost. Thermophilic fungi in phase II compost are believed to contribute to a good crop of mushrooms in the following ways: (1) by decreasing the concentration of ammonia in the compost, which would otherwise counteract the growth of the mushroom mycelium; (2) by immobilizing nutrients in a form apparently available to the mushroom mycelium; and (3) possibly by having a growth-promoting effect on the mushroom mycelium, as has been demonstrated for *Scytalidium thermophilum* and several other thermophilic fungi (Wiegant, 1992). However, mechanisms behind the growth stimulation of *Agaricus bisporus* by these fungi remain unresolved. It is advisable to perform phase I as a natural semi-controlled composting process for a good crop yield of mushrooms (Overstijns, 1995). However, because of the growing concern about the environmental impact of the production of compost and pressure from environmentalists, it will be necessary to reduce the length of phase I. By using tunnels, it is now already possible to drop phase I outdoors completely (Gerrits et al., 1995). There are prospects for using spent compost for bioremediation purposes (Buswell, 1994) and as an organic soil stabilizer (Gerrits, 1994). Further research on indoor phase I composting and biodegradation of lignocellulosics is needed. Although *A. bisporus* is capable of degrading the lignin in compost, the specific components of the enzyme system utilized by this fungus during the biodegradation of lignin are not known with certainty. Bonnen et al. (1994) found that the activities of laccase and manganese peroxidase of *A. bisporus* in compost are related to the degradation of compost lignin.

What is presently needed in the mushroom industry is to render the compost preparation procedure more rational and controllable by identifying the important factors that affect yield and quality. It is hoped that recombinant DNA technologies may provide an impetus to the study of enzymes involved in bioconversion and carbon metabolism. Earlier, the cellulase and laccase genes of *A. bisporus* were cloned (Wood and Thurston, 1991; Perry et al., 1993), and Schaap et al. (1994) cloned some housekeeping genes of carbon metabolism. Our knowledge of mechanisms underlying straw and compost degradation before cropping will be increased rapidly if *in situ* microscopic studies in straw and compost can be performed using gene probes for *S. thermophilum*, including other thermophilic fungi and *A. bisporus*. Such work requires a multidisciplinary approach with input from disciplines such as mycology, mushroom science, molecular biology, biotechnology, and biochemistry.

REFERENCES

Antizar-Ladislao, B., Lopez-Real, J., and Beck, A.J. 2005. Laboratory studies of the remediation of polycyclic aromatic hydrocarbon contaminated soil by in-vessel composting. *Waste Management* 25:281–289.

Arikan, O.A., Sikora, L.J., Mulbry, W., Khan, S.U., and Foster, G.D. 2007. Composting rapidly reduces levels of extractable oxytetracycline in manure from therapeutically treated beef calves. *Bioresource Technology* 98:169–176.

Bachofen, R., and Rast, D. 1968. Carboxylierungsreaktionen in *Agaricus bisporus* III. Pyruvat und Phosphoenolpyruvat als CO_2-Acceptoren. *Archives of Microbiology* 60:217–234.

Basuki, T. 1981. *Ecology and productivity of the paddy straw mushroom [Volvariella volvacea (Bull ex Fr.) Sing]*. PhD thesis, University of Wales, Cardiff.

Beck, M. 1997. *The Secret Life of Compost: A Guide to Static-Pile Composting—Lawn, Garden, Feedlot or Farm*. Austin, TX: Acres USA.

Beyer, D.M., Heinemann, P., Labance, S., and Rhoades, T. 1997. The effect of covering compost piles with microporous membrane on mushroom substrate preparation process and fresh mushroom yield. *Mushroom News* 45:14–20.

Bilai, V.T. 1984. Thermophilic micromycete species from mushroom composts. *Mikrobiology Zhurnal (Kiev)* 46:35–38.

Bonnen, A.M., Anton, L.H., and Orth, A.B. 1994. Lignin degrading enzymes of the commercial button mushroom, *Agaricus bisporus*. *Applied and Environmental Microbiology* 60:960–965.

Brito, L.M., Coutinho, J., and Smith, S.R. 2008. Methods to improve the composting process of the solid fraction of dairy cattle slurry. *Bioresource Technology* 99:8955–8960.

Brown, K.H., Bouwkamp, J.C., and Gouin, F.R. 1998. The influence of C:P ratio on the biological degradation of municipal solid waste. *Compost Science & Utilization* 6:53–58.

Buswell, J.A. 1994. Potential of spent mushroom substrate for bioremediation purposes. *Compost Science & Utilization* 2:31–36.

Cailleux, R. 1973. Mycoflore du compost destine a la culture du champignon de couche. *Revue de Mycology* 37:14–35.

Chahal, D.S., and Overend, R.P. 1982. Ethanol fuel from biomass. In *Advances in Agricultural Microbiology*, ed. N.S. Subba Rao, 581–641. New Delhi: Oxford and IBH Publishing Co.

Chahal, D.S., Sekhon, A., and Dhaliwal, B.S. 1976. Degradation of wheat straw by the fungi isolated from synthetic mushroom compost. In *Proceedings of the 3rd International Biodegradation Symposium*, ed. J.M. Shamplay and A.M. Kaplan. London: Applied Science Publishers.

Chang, Y. 1967. The fungi of wheat straw compost. II. Biochemical and physiological studies. *Transactions of the British Mycological Society* 50:667–677.

Chang, Y., and Hudson, H.J. 1967. The fungi of wheat straw compost I. Ecological studies. *Transactions of the British Mycological Society* 50:649–666.

Chapman, E.S. 1974. Effect of temperature on growth rate of seven thermophilic fungi. *Mycologia* 66:542–546.

Cooney, D.G., and Emerson, R. 1964. *Thermophilic Fungi: An Account of Their Biology, Activities, and Classification.* San Francisco: W.H. Freeman and Co.

Danon, M., Franke-Whittle, I.H., Insam, H., Chen, Y., and Hadar, Y. 2008. Molecular analysis of bacterial community succession during prolonged compost curing. *FEMS Microbiology Ecology* 65:133–144.

Deacon, J.W. 1985. Decomposition of filter paper cellulose by thermophilic fungi acting singly, in combination, and in sequence. *Transactions of the British Mycological Society* 85:663–669.

Derikx, P.J.L., Op den Camp, H.J.M., Van der Drift, C., Van Griensven L.J.L.D., and Vogels, G.D. 1990. Biomass and biological activity during the production of compost used as a substrate in mushroom cultivation. *Applied and Environmental Microbiology* 56:3029–3034.

Derikx, P.J.L., Simons, F.H.M., Op den Camp, HJ.M., Van der Drift, C., Van Griensven L.J.L.D., and Vogels, G.D. 1991. Evolution of volatile sulfur compounds during laboratory-scale incubations and indoor preparation of compost used as a substrate in mushroom cultivation. *Applied and Environmental Microbiology* 57:563–567.

Dickson, H. 1935. Studies on *Coprinus sphaerosporus* II. The influence of various morphological and physiological characters. *Annals of Botany* 49:181–204.

Dickson, N., Richard, T., and Kozlowski, R. 1991. *Composting to Reduce the Waste Stream—A Guide to Small Scale Food and Yard Waste Composting.* NRAES-43. Ithaca, NY: Natural Resource, Agricultural, and Engineering Service, Cooperative Extension.

Dougherty, M., ed. 1998. *Composting for Municipalities: Planning and Design Considerations.* Ithaca, NY: Natural Resource, Agricultural, and Engineering Service, Cooperative Extension.

Duff, S.J.B., and Murray, W.D. 1996. Bioconversion of forest products industry waste cellulosics to fuel ethanol: A review. *Bioresource Technology* 55:1–33.

Eicker, A. 1977. Thermophilic fungi associated with the cultivation of *Agaricus bisporus. Journal of South African Botany* 43:193–207.

Ekinci, K., Keener, H.M., and Elwell, D.L. 2000. Composting short paper fiber with broiler litter and additives part I: Effects of initial pH and carbon/nitrogen ratio on ammonia emission. *Compost Science & Utilization* 8:160–172.

Evans, H.C. 1971. Thermophilous fungi of coal spoil tips. II. Occurrence, distribution and temperature relationships. *Transactions of the British Mycological Society* 57:255–266.

Fergus, C.L. 1964. Thermophilic and thermotolerant molds and actinomycetes of mushroom compost during peak heating. *Mycologia* 56:267–283.

Fergus, C.L. 1971. The temperature relationships and thermal resistance of a new thermophile *Papulaspora* from mushroom compost. *Mycologia* 63:426–431.

Fergus, C.L., and Amelung, R.M. 1971. The heat resistance of some thermophilic fungi on mushroom compost. *Mycologia* 63:675–679.

Fergus, C.L., and Sinden, J.W. 1969. A new thermophilic fungus from mushroom compost: *Thielavia thermophila* spec. nov. *Canadian Journal of Botany* 47:1635–1637.

Fermor, T.R., and Grant, W.D. 1985. Degradation of fungal and actinomycete mycelia by *Agaricus bisporus. Journal of General Microbiology* 131:1729–1734.

Fermor, T.R., Randle, P.E., and Smith J.F. 1985. Compost as a substrate and its preparation. In *The Biology and Technology of the Cultivated Mushroom*, ed. P.B. Flegg, D.M. Spencer, and D.A. Wood, 81–109. Chichester, UK: John Wiley & Sons.

Fermor, T.R., Smith J.F., and Spencer, D.M. 1979. The microflora of experimental mushroom composts. *Journal of Horticulture Science* 54:137–147.

Fermor, T.R., and Wood, D.A. 1981. Degradation of bacteria by *Agaricus bisporus* and other fungi. *Journal of General Microbiology* 126:377–387.

Finstein, M.S., and Morris, M.L. 1975. Microbiology of municipal solid waste composting. *Advances in Applied Microbiology* 19:113–151.

Flannigan, B., and Sellars, P.N. 1972. Activities of thermophilous fungi from barley kernels against arabinoxylan and carboxymethylcellulose. *Transactions of the British Mycological Society* 58:338–341.

Gerrits, J.P.G. 1985. Developments in composting in the Netherlands. *Mushroom Journal* 146:45–53.

Gerrits, J.P.G. 1988. Compost treatment in bulk for mushroom growing. *Mushroom Journal* 182:471–475.

Gerrits, J.P.G. 1994. Composition, use and legislation of spent mushroom substrate in the Netherlands. *Compost Science & Utilization* 2:24–30.

Gerrits, J.P.G., Amsing, J.G.M., Straatsma, G., and Van Greinsven, L.J.L.D. 1994. Indoor compost: Fase I processen van 5 dagen in tunnels. *Champignoncultuur* 38:338–345.

Gerrits, J.P.G., Amsing, J.G.M., Straatsma, G., and Van Greinsven, L.J.L.D. 1995. Phase I process in tunnels for the production of *Agaricus bisporus* compost with special reference to the importance of water. In *Science and Cultivation of Fungi*, ed. T.J. Elliot, 203–211. Rotterdam, the Netherlands: Balkema.

Gerrits, J.P.G., and Van Griensven, L.J.L.D. 1990. New developments in indoor composting (tunnel process). *Mushroom Journal* 205:21–29.

Gilbert, H.J., and Hazelwood, G.P. 1993. Bacterial cellulases and xylanases. *Journal of General Microbiology* 139:187–194.

Golueke, C.G. 1972. *Composting. A Study of the Process and Its Principles*. Emmaus, PA: Rodale Press.

Goyal, A., Ghosh, B., and Eveleigh, D. 1991. Characteristics of fungal cellulases. *Bioresource Technology* 36:37–50.

Harper, E., Miller, F.C., and Macauley, B.J. 1992. Physical management and interpretation of an environmentally controlled composting ecosystem. *Australian Journal of Experimental Agriculture* 32:657–667.

Hayes, W.A. 1969. Microbiological changes in composting wheat straw/horse manure mixtures. *Mushroom Science* 7:173–186.

Hedger, J.N., and Hudson, H.J. 1974. Nutritional studies of *Thermomyces lanuginosus* from wheat straw compost. *Transactions of the British Mycological Society* 62:129–143.

Holman, D.B., Hao, X., Topp, E., Yang, H.E., and Alexander, T.W. 2016. Effect of co-composting cattle manure with construction and demolition waste on the archaeal, bacterial, and fungal microbiota, and on antimicrobial resistance determinants. *PLoS One* 11:e0157539. doi: 10.1371/journal.pone.0157539.

Horng, J.M. 2003. *Food Wastes Utilized Effectively*. Taiwan: Environmental Protection Union of Taiwan.

Hubbe, M.A., Nazhad, M., and Sánchez, C. 2010. Composting of lignocellulosics. *Bioresources* 5:2808–2854.

Hultman, J. 2009. Microbial diversity in the municipal composting process and development of detection methods. PhD thesis, Department of Ecological and Environmental Sciences, Faculty of Biosciences and Institute of Biotechnology and Vikki Graduate School in Biosciences, University of Helsinki, Finland.

Hwang, E.J., Shin, H.S., and Tay, J.H. 2002. Continuous feed, on-site composting of kitchen garbage. *Waste Management & Research* 20:119–126.

Iiyama, K., Stone, B.A., and Macauley, B.J. 1994. Compositional changes in compost during composting and growth of *Agaricus bisporus*. *Applied and Environmental Microbiology* 60:1538–1546.

Ivors, K.L., Collopy, P.D., Beyer, D.M., and Kang, S. 2000. Identification of bacteria in mushroom compost using ribosomal RNA sequence. *Compost Science & Utilization* 8:247–253.

Johri, B.N., and Rajni. 1999. Mushroom compost: Microbiology and application. In *Modern Approaches and Innovations in Soil Management*, ed. D.J. Bagyaraj, A. Verma, K.K. Khanna, and H.K. Kheri, 345–358. Meerut: Rastogi Publications.

Kringstad, K.P., and Lindstrom, K. 1984. Spent liquors from pulp bleaching. *Environmental Science & Technology* 18:236A–248A.

Kumar, R. 1996. Taxophysiological studies on thermophilous fungi from northern Indian soils. PhD thesis. Kurukshetra University, Kurukshetra, India.

Kumar, R., and Aneja, K.R. 1999. Biotechnological applications of thermophilic fungi in mushroom compost preparation. In *From Ethnomycology to Fungal Biotechnology*, ed. J. Singh and K.R. Aneja, 115–126. New York: Plenum Publishers.

Laborde, I.J. 1994. Controlled composting indoors (indoor composting): An overview of the current technique. *Mushroom Information* 9:5.

Laborde, J., Lanzi, B., Francescutti B., and Giordani, E. 1993. Indoor composting: General principles and large scale developments in Italy. In *Mushroom Biology and Mushroom Products*, ed. S. Chang, J.A. Buswell, and S. Chiu, 93–113. Hong Kong: Chinese University Press.

Lacey, J. 1974. Allergy in mushroom workers. *Lancet* 366. doi: 10.1016/S0140-6736(74)93133-X.

Larney, F.J., and Hao, X. 2007. A review of composting as a management alternative for beef cattle feedlot manure in southern Alberta, Canada. *Bioresource Technology* 98:3221–3227.

Larney, F.J., Buckley K.E., Hao, X., and McCaughey, W.P.. 2006. Fresh, stockpiled, and composted beef cattle feedlot manure: Nutrient levels and mass balance estimates in Alberta and Manitoba. *Journal of Environmental Quality* 35:1844–1854.

Larney, F.J., Olson A.F., Carcamo, A.A., and Chang, C. 2000. Physical changes during active and passive composting of beef feedlot manure in winter and summer. *Bioresource Technology* 75:139–148.

Last, F.T., Holdings, M., and Stone, O.M. 1974. Effects on cultural conditions on the mycelial growth of healthy and virus infected cultivated mushroom, *Agaricus bisporus*. *Annals of Applied Biology* 76:99–111.

Lee, H., Lee, Y.M., Jang, Y., Lee, S., Lee, H., Ahn, B.J., Kim, G.H., and Kim, J.-J. 2014. Isolation and analysis of the enzymatic properties of thermophilic fungi from compost. *Mycobiology* 42:181–184.

Leonard, T.J., and Volk, T.J. 1992. Production of specialty mushrooms in North America: Shitake and morels. In *Frontiers in Industrial Mycology*, ed. G.F. Leatham, 1–23. New York: Chapman & Hall.

Le Roux, P., and Couvy, J. 1972. Fixation et metabolisme du gaz carbonique dans le mycelium et le cordon mycelien d'*Agaricus bisporus*. *Mushroom Science* 8:641–646.

Liang, C., Das, K.C., and McClendon, R.W. 2003. The influence of temperature and moisture contents regimes on the aerobic microbial activity of a biosolids composting blend. *Bioresource Technology* 86:131–137.

Maheshwari, R., Bharadwaj, G., and Bhat, M.K. 2000. Thermophilic fungi: Their physiology and enzymes. *Microbiology and Molecular Biology Reviews* 64:461–488.

Margaritis, A., and Merchant R.F. 1983. Production and thermal characteristics of cellulose and xylanase enzymes from *Thielavia terrestris*. *Biotechnology and Bioengineering Symposium* 13:299–314.

Margaritis, A., Merchant R.F.J., and Yaguchi, M. 1986. Thermostable cellulases from thermophilic microorganisms. *Critical Reviews in Biotechnology* 4:327–367.

Miller, F.C. 1994. Conventional composting system. In *Agaricus Compost*, ed. N.G. Nair, 1–18. Windsor: Australian Mushroom Growers Association.

Morris, B. 1984. Macrofungi of Malawi: Some ethno-botanical notes. *Bulletin of the British Mycological Society* 49–57.

Moss, L.H., Epstein, E., and Logan, T. 2002. *Evaluating Risks and Benefits of Soil Amendments Used in Agriculture*. 99-PUM-1. Alexandria, VA: Water Environment Research Foundation.

Nair, N.G., and Price, C. 1991. A composting process to minimize odour pollution. In *Science and Cultivation of Edible Fungi*, ed. M.J. Maher, 205–206. Rotterdam, the Netherlands: Balkema.

Namkoong, W., Hwang, E.Y., Park, J.S., and Choi, J.-Y. 2002. Bioremediation of diesel-contaminated soil with composting. *Environmental Pollution* 119:23–31.

Noble, R., and Gaze, R.H. 1994. Controlled environment composting for mushroom cultivation: Substrates based on wheat and barley straw and deep litter poultry manure. *Journal of Agricultural Sciences* 123:71–79.

Ogbonna, C.I.C., and Pugh, G.J.F. 1982. Nigerian soil fungi. *Nova Hedwegia* 36:795–808.

Olivier, J.M., and Guillaumes, J. 1976. Etude ecologique des composts de champignonnieres. I. Evolution de la microflore pendant I' incubation. *Annals of Phytopathology* 8: 283–301.

Op den Camp, H.J.M., Pot, A., Van Griensven, L.J.L.D., and Gerrits, J.P.G. 1992. Stankproduktie tijdens 'Indoor Verse Compostbereiding' (IVC) en het effect van luchtbehandeling met een luchtwasser. *Champignoncultuur* 36:319–325.

Overstijns, A. 1995. Indoor composting. *Mushroom News* 43:16–26.

Pedro, M., Haruta, S., Hazaka, M., Shimada, R., Yoshida, C., Hiura, K, Ishii, M., and Igarashi, Y. 2001. Denaturing gradient gel electrophoresis analysis of microbial community from field-scale composter. *Journal of Bioscience and Bioengineering* 91:159–165.

Perry, C.R., Matcham, S.E., Wood, D.A., and Thurston, C.F. 1993. The structure of laccase protein and its synthesis by the commercial mushroom *Agaricus bisporus*. *Journal of General Microbiology* 139:171–178.

Petric, I., Sestan, A., and Sestan, I. 2009. Influence of initial moisture content on the composting of poultry manure with wheat straw. *Biosystems Engineering* 104:125–134.

Raut, M.P., Prince William S.P.M., Bhattacharyya, J.K., Chakrabarti, T., and Devotta, S. 2008. Microbial dynamics and enzyme activities during rapid composting of municipal solid waste—A compost maturity analysis perspective. *Bioresource Technology* 99:6512–6519.

Rawat, S. 2004. Microbial diversity of mushroom compost and xylanase of Scytalidium thermophilum. PhD thesis, G.B. Pant University of Agriculture and Technology, Pantnagar, India.

Rawat, S., Agarwal, P.K., Chaudhary, D.K., and Johri, B.N. 2005. In *Microbial Diversity: Current Perspectives and Applications*, ed. T. Satyanarayana and B.N. Johri, 181–206. New Delhi: I.K. International Pvt. Ltd.

Renard, Y., and Cailleux, R. 1973. Contribution a l'etude des micro-organismes du compost destine a la culture du champignon de couche. *Revue de Mycologie* 37:36–47.

Rosenberg, S.L. 1975. Temperature and pH optima for 21 species of thermophilic and thermotolerant fungi. *Canadian Journal of Microbiology* 21:1535–1540.

Rosenberg, S.L. 1978. Cellulose and lignocellulose degradation by thermophilic and thermotolerant fungi. *Mycologia* 70:1–13.

Salar, R.K., and Aneja, K.R. 2007. Significance of thermophilic fungi in mushroom compost preparation: Effect on growth and yield of *Agaricus bisporus* (Lange) Sing. *Agricultural Technology, Bangkok* 3:241–253.

Sandhu, D.K., and Sidhu, M.S. 1980. The fungus succession on decomposing sugarcane bagasse. *Transactions of the British Mycological Society* 75:281–286.

Satyanarayana, T. 1978. Thermophilic microorganisms and their role in composting process. PhD thesis, Sagar University, Sagar, India.

Savoie, J.-M., and Libmond, S. 1994. Stimulation of environmentally controlled mushroom composting by polysaccharidases. *World Journal of Microbiology and Biotechnology* 10:313–319.

Schaap, P.J., Van der Vlugt, R.A.A., De Greet, P.W.J., Mueller, Y., Van Griensven, L.J.L.D., and Visser, J. 1994. Strategies for cloning of housekeeping genes of *Agaricus bisporus* [abstract]. In *Fifth International Mycological Congress*, Vancouver, BC, August 14–21, 190.

Schloss, P.D., Hay, A.G., Wilson, D.B., and Walker, L.P. 2003. Tracking temporal changes of bacterial community fingerprints during the initial stages of composting. *FEMS Microbiology Ecology* 46:1–9.

Schulze, K.L. 1962. Continuous thermophilic composting. *Applied Microbiology* 10 (2): 108–122.

Seal, K.J., and Eggins, H.O.W. 1976. The upgrading of agricultural wastes by thermophilic fungi. In *Food from Wastes*, ed. G.G. Birch, K.J. Parker, and J.T. Wargan, 58–78. London: Applied Science Publishers.

Sebok, F., Dobolyi, C., Bobvos, J., Szoboszlay, S., Kriszt, B., and Magyar, D. 2015. Thermophilic fungi in air samples in surroundings of compost piles of municipal, agricultural and horticultural origin. *Aerobiologia* 32:255–263.

Sharma, H.S.S. 1991. Biochemical and thermal analyses of mushroom compost during preparation. *Mushroom Science* 13:169–179.

Sharma, R., Larney, F.J., Chen, J., Yanke, L.J., Morrison, M., Topp, E., McAllister, T.A., and Yu, Z. 2009. Selected antimicrobial resistance during composting of manure from cattle administered sub-therapeutic antimicrobials. *Journal of Environmental Quality* 38:567–575.

Sinden, J.W., and Hauser, E. 1953. The nature of composting process and its relation to short composting. *Mushroom Science* 2:123–130.

Singh, S., and Sandhu, D.K. 1982. Growth response of some thermophilous fungi at different incubation temperatures. *Proceedings of the Indian Academy of Sciences (Plant Science)* 91:153.

Souza, T.P., Marques, S.C., da Silveira e Santos, D.M., and Dias, E.S. 2014. Analysis of thermophilic fungal populations during phase II of composting for the cultivation of *Agaricus subrufescens*. *World Journal of Microbiology and Biotechnology* 30:2419–2425.

Stoller, B.B. 1978. Detection and evaluation of carbon monoxide, ethylene, and oxidants in mushroom beds. *Mushroom Science* 10:445–449.

Storl, W.E. 1979. *Culture and Horticulture: A Philosophy of Gardening*. Wyoming, RI: Bio-Dynamic Literature. See also course content created by Robinson, D. Introductory class on-line (subject 4): BD materials and compost. Myrtle Point, OR: Biodynamic Farming and Gardening Association.

Straatsma, G., Gerrits, J.P.G., Augustijn, M.P.A.M., Op den Camp, H.J.M., Vogels, G.D., and Van Griensven, L.J.L.D. 1989. Population dynamics of *Scytalidium thermophilum* in mushroom compost and stimulatory effects on growth rate and yield of *Agaricus bisporus*. *Journal of General Microbiology* 135:751–759.

Straatsma, G., Olijnsma, T.W., Gerrits, J.P.G., Amsing, J.G.M., Op den Camp, H.J.M., and Van Griensven, L.J.L.D. 1994a. Inoculation of *Scytalidium thermophilum* in button mushroom compost and its effect on yield. *Applied and Environmental Microbiology* 60:3049–3054.

Straatsma, G., Olijnsma, T.W., Van Griensven, L.J.L.D., and Op den Camp, H.J.M. 1995a. Growth promotion of *Agaricus bisporus* mycelium by *Scytalidium thermophilum* and

CO_2. In *Science and Cultivation of Fungi*, ed. T.L. Elliott, 289–291. Rotterdam, the Netherlands: Balkema.

Straatsma, G., and Samson, R.A. 1993. Taxonomy of *Scytalidium thermophilum*, an important thermophilic fungus in mushroom compost. *Mycological Research* 97:321–328.

Straatsma, G., Samson, R.A., Olijnsma, T.W., Gerrits, J.P.G., Op den Camp, H.J.M., and Van Griensven, L.J.L.D. 1995b. Bioconversion of cereal straw into mushroom compost. *Canadian Journal of Botany* 73:1019–1024.

Straatsma, G., Samson, R.A., Olijnsma, T.W., Op den Camp, H.J.M., Gerrits, J.P.G., and Van Griensven, L.J.L.D. 1994b. Ecology of thermophilic fungi in mushroom compost, with emphasis on *Scytalidium thermophilum* and growth stimulation of *Agaricus bisporus* mycelium. *Applied and Environmental Microbiology* 60:454–458.

Su, Y.Y., and Cai, L. 2013. *Rasamsonia composticola*, a new thermophilic species isolated from compost in Yunnan, China. *Mycological Progress* 12:213–221.

Tang, J.C., Shibata, A., Zhou, Q.X., and Katayama, A. 2007. Effect of temperature on reaction rate and microbial community in composting of cattle manure with rice straw. *Journal of Bioscience and Bioengineering* 104:321–328.

Tansey, M.R., and Brock, T.D. 1972. The upper temperature limits for eukaryotic organism. *Proceedings of the National Academy of Sciences of the United States of America* 69:2426–2428.

Tansey, M.R., Murrmann, D.N., Behnke, B.K., and Behnke, E. 1977. Enrichment, isolation, and assay of growth of thermophilic and thermotolerant fungi in lignin containing media, *Mycologia* 69:463–476.

Tchobanoglous, G., Theisen, H., and Vigil, S. 1993. *Integrated Solid Waste Management: Engineering Principles and Management Issues*, McGraw-Hill International Edition, 1–949. Singapore: McGraw-Hill.

Till, O. 1962. Champignonkultur aufsterilisiertem Nahrsubstrat und die Wiederverwendung von abgetragenem Kompost. *Mushroom Science* 5:127–133.

Tiquia, S.M., Wan, J.H.C., and Tam, N.F.Y. 2002. Microbial population dynamics and enzyme activities during composting. *Compost Science & Utilization* 10:150–161.

Torrie, J. 1991. *Extracellular/3-0-mannanase activity from Trichoderma harzianum E58*. PhD thesis, University of Ottawa.

Trinci, A.P.J. 1973. The hyphal growth unit of wild type and spreading colonial mutants of *Neurospora crassa*. *Archiv Fur Mikrobiologie* 91:127–136.

Trinci, A.P.J. 1992. Myco-protein: A twenty-year overnight success story. *Mycological Research* 96:1–13.

Van Lier, J.J.C., Van Ginkel, J.T., Straatsma, G., Gerrits, L.P.G., and Van Griensven, L.J.L.D. 1994. Composting of mushroom substrate in a fermentation tunnel: Compost parameters and a mathematical model. *Netherlands Journal of Agricultural Science* 42:271–292.

Vernoux, A., Guiliano, M., Le Dreau, Y., Kister, J., Dupuy, N., and Doumenq, P. 2009. Monitoring of the evolution of an industrial compost and prediction of some compost properties by NIR spectroscopy. *Science of the Total Environment* 407:2390–2403.

Vijay, B., and Pathak, A. 2014. Exploitation of thermophilic fungi in compost production for white button mushroom (*Agaricus bisporus*) cultivation—A review. In *Proceedings of the 8th International Conference on Mushroom Biology and Mushroom Products (ICMBMP8)*, New Delhi, India, November 19–22, 2014 Volume I and II 292–308.

Vos, A.M., Heijboer, A., Boschker, H.T.S., Bonnet, B., Lugones, L.G., and Wösten, H.A.B. 2017. Microbial biomass in compost during colonization of *Agaricus bisporus*. *AMB Express* 7:12. doi: 10.1186/s13568-016-0304-y.

Waksman, S.A., and Cordon, T.C. 1939. Thermophilic decomposition of plant residues in composts by pure and mixed cultures of microorganisms. *Soil Science* 47:217–224.

Waksman, S.A., and McGrath, J.M. 1931. Preliminary study of chemical processing in the decomposition of manure by *Agaricus compestris*. *American Journal of Botany* 18:573–581.

Waksman, S.A., and Nissen, W. 1932. On the nutrition of the cultivated mushroom, *Agaricus campestris*, and the chemical changes brought about by this organism in the manure compost. *American Journal of Botany* 19:514–537.

Walker, L.P., and Wilson, D.B. 1991. Enzymatic hydrolysis of cellulose: An overview. *Bioresource Technology* 36:3–14.

Wang, Y.C. 1985. Mycology in China with emphasis on review of the ancient literature. *Acta Mycologica Sinica* 4:133–140.

Wiegant, W.M. 1992. Growth characteristics of the thermophilic fungus *Scytalidium thermophilum* in relation to production of mushroom compost. *Applied and Environmental Microbiology* 58:1301–1307.

Willenborg, A., and Hindorf, H. 1985. Investigations on the fungal flora in mushroom culture substrate from various Rhineland installations. *Champignon* 284:26–38.

Wojtczak, G., Breuil, C., Yamada, J., and Saddler, J.N. 1987. A comparison of the thermostability of cellulases from various thermophilic fungi. *Applied Microbiology and Biotechnology* 27:82–87.

Wood, D.A., and Thurston, C.F. 1991. Progress in the molecular analysis of *Agaricus* enzymes. In *Genetics and Breeding of Agaricus*, ed. L.J.L.D. Van Griensven, 81–86. Wageningen, the Netherlands: Pudoc.

Xu, S., Sura, S., Zaheer, R., Wang, G., Smith, A., Cook, S., Olson, A.F., Cessna, A.J., Larney, F.J., and McAllister, T.A. 2015. Dissipation of antimicrobial resistance determinants in composted and stockpiled beef cattle manure. *Journal of Environmental Quality* 45:528–536.

Zafar, U., Houlden, A., and Robson, G.D. 2013. Fungal communities associated with the biodegradation of polyester polyurethane buried under compost at different temperatures. *Applied and Environmental Microbiology* 79(23):7313–7324.

Zhang, Y., Wu, W., and Cai, L. 2017. Polyphasic characterisation of *Chaetomium* species from soil and compost revealed high number of undescribed species. *Fungal Biology* 121 (1):21–43.

10 Bioremediation and Biomineralization

10.1 INTRODUCTION

With the advent of the Industrial Revolution, the ever-increasing environmental pollution caused by toxic and persistent heavy metals has resulted in deleterious ecological effects and poses a serious threat to animals and human health. Several industries, such as electroplating, electronic circuit production, steel and nonferrous processing, and chemical and pharmaceutical industries, discharge a variety of metal-laden wastewater into the environment. Urbanization, modern agricultural practices, and so forth, are equally responsible for such types of pollution. As a result, the earth's surface has become contaminated with natural and xenobiotic toxic compounds. The most important examples include polluted aquifers and other water bodies, mining areas, and industrial sites contaminated with petrochemical products, pesticides, heavy metals, radionuclides, and so forth. With the increased use of chemical fertilizers, pesticides, insecticides, and so forth, in modern agriculture practice, increased concentrations of heavy metals, such as Pb, Cd, Hg, Zn, and Fe, are creating an ecological imbalance. These metal contaminants are dispersed in the biosphere through various types of industrial effluents, municipal wastewaters (Table 10.1), and so on.

Heavy metals released to the environment are increasing continuously as a result of industrial activities and technological development, posing a significant threat to the environment and public health because of their toxicity, accumulation in the food chain, and persistence in nature. It is therefore important to develop new methods for metal removal and recovery from dilute solutions, and for heavy metal ions to be reduced to very low concentrations. The use of conventional technologies, such as ion exchange, chemical precipitation, reverse osmosis, and evaporative recovery, for this purpose is often inefficient and/or very expensive. Metals are directly and/or indirectly involved in all aspects of microbial growth, metabolism, and differentiation. Many metals, such as K, Na, Mg, Ca, Mn, Fe, Cu, Ni, Co, Zn, and Mo, are essential for a variety of biological functions, whereas it is not known whether some others, such as Al, Ag, Cd, Sn, Au, Sr Hg, Ti, and Pb, have any essential biological function. All these elements can interact with microbial cells and be accumulated as a result of physico-chemical mechanisms and transport systems of varying specificity, independent of or directly or indirectly dependent on metabolism. Some of these processes are of biotechnological importance, being relevant to metal removal and recovery from mineral deposits, as well as industrial effluents for industrial use or environmental bioremediation. There are several factors affecting the removal and recovery of these metals from the aqueous phase, like temperature, pH, agitation rate, and metal concentration.

The basic principle of bioremediation entails the use of naturally occurring microorganisms to remove or neutralize pollutants from a contaminated site, such as

TABLE 10.1

Predominant Heavy Metals in Various Industrial Effluents

Metal	Industrial Effluent
As	Fertilizer, paint, textile
Cd	Fertilizer, paint, electroplating, pigment, battery
Cr	Tannery, electroplating, paint, photography
Cu	Electrical, paint, electroplating, rayon
Pb	Electrical, paint, electroplating, battery
Ni	Iron and steel, paint, electroplating
Zn	Galvanizing, iron and steel, electroplating, battery
Hg	Chlor-alkali, cement

soil and waters. In the recent past, microbe-mediated bioremediation has emerged as a potential alternative to conventional treatment methods. In the majority of cases, a consortium of microorganisms is involved in the biodegradation of the contaminants, rather than a single microbial species. In order to understand bioremediation in its real sense, one must consider certain allied terms, such as biodegradation, mineralization, biodeterioration, biotransformation, bioaccumulation, and biosorption. These terms are frequently used with minor differences but often overlappingly. Biodegradation pertains to all biologically mediated breakdowns of chemical compounds, and complete biodegradation leads to mineralization. Biotransformation refers to a step in the biochemical pathway (series of steps) that leads to the conversion of a molecule (precursor) into a product that may be less harmful. Biodeterioration is the breakdown of economically useful compounds, but often the term has been used to refer to the degradation of normally resistant substances, such as metals, plastics, drugs, cosmetics, paint, sculptures, and wood products. Bioaccumulation or biosorption is the accumulation of toxic compounds or metals inside the microbial cell without any degradation of the toxic molecule. This treatment method can be useful in aquatic environments, where the organisms can be removed after being loaded with the toxic compound. A variety of living and dead biomasses of bacteria, algae, fungi, and plants is capable of sequestering toxic metals from wastewaters. A plethora of literature exists on the biosorption of metal ions by bacteria, algae, fungi, and plants. The biosorption or bioaccumulation of metals is the most effective and extensively used approach for bioremediation. The growth of microorganisms in their natural environment can be promoted to attain maximal recovery of pollutants. It can be achieved by the addition of certain key nutrients, such as nitrogen and phosphorus, which are usually present in growth-limiting concentrations in natural substrata. This enables the natural microbial population to develop and metabolize the pollutant.

While the application of mesophilic microorganisms in bioremediation has received considerable attention in recent years, the potential of thermophilic organisms has largely remained unexplored. Although a large number of studies have been conducted on bioremediation using thermophilic bacteria (Sar et al., 2013), studies on thermophilic fungi are poorly investigated. Since the discovery of thermophilic microorganisms, scientists have been exploring the diversity of these organisms for a variety of reasons, including fundamental studies, biotechnological applications, and economic benefits (Hasyim et al., 2011; Spain and Krumholz, 2011; Gugliandolo et al., 2012). Recently, it has been realized that thermophilic microorganisms have developed various adaptation strategies to cope with harsh environmental conditions. One such adaptation exhibited by these organisms is resistance to diverse heavy metals that is associated with their extreme environmental living conditions (Bengtsson et al., 1995; Watkin et al., 2009). Similarly, immobilization of toxic heavy metals using sulfide-producing microorganisms has been reported as an effective means of treating some metal-contaminated sites (Crawford and Crawford, 1996). Thermophilic microorganisms, including thermophilic fungi with higher metal resistance and metabolic attributes at elevated temperatures, may exhibit increased metal remediation, including solubilization and mineralization.

10.2   BIOREMEDIATION

Bioremediation refers to the use of microbes to transform environmental contaminants of natural or anthropogenic origin to benign products. Lately, this pollution treatment technology is considered an effective and eco-friendly alternative to conventional remediation strategies, with significant disadvantages including the cost-intensive process, incomplete remediation, high input of reagents and energy, and generation of large quantities of sludge and other waste products for further disposal (Sar et al., 2013). However, the process of bioaccumulation and biosorption of metals by microorganisms is not new. The accumulation and biosorption of heavy metals by fungi has received more attention due to their extraordinary biosorption of metal ions on the surface of their mycelia, intracellular uptake of metal ions, and chemical transformation of metal ions by fungi (Singh, 2006). Biosorption is a pseudo-ion-exchange process in which metal ion is exchanged for a counterion in the biomass or resin. In general, the filamentous fungi possess higher adsorption capacities for heavy metal removal. Similarly, a thermophilic fungal biomass could be utilized for the biosorption of heavy metals, but unfortunately, this group of fungi is little explored compared with the mesophiles. In order to understand how bioremediation takes place at the level of the ecosystem, we first must consider the various roles and biological activities of metals and the biochemistry of biodegradation.

10.2.1   Heavy Metals as Environmental Pollutants

The term *heavy metals* described the presence of As, Cd, Sb, Be, Cu, Co, Cr, Mo, Ag, V, W, Ni, Pb, Fe, Hg, Zn, and so forth. It has been felt that the impact of these contaminants to aquatic resources is highly disastrous, even if they are present in traces. Some of the metal ions, however, like Cu, Mn, Zn, and Fe, also serve as micronutrients and are required for plants and animals, including humans, while some metal ions, like Cr, Pb, and Ni, are hazardous and create complications in the ecology of the receiving system, and some of the metals, for example, Ag, Au, Pt, W, and Th, are highly precious in nature. When present in excess (not in the form of ore, minerals, etc.) in any ecosystem, they exert a variety of adverse effects on biota and humankind. These effects may range from general symptoms, like irritation, discomfort, and excessive sensation in sensory organs, to chronic diseases, like chronic bronchitis, asthma, and in severe cases, death (Table 10.2). In addition to municipal wastewater, effluents from industries such as electroplating, automobile, airplane manufacturing, chemical and petrochemical, petroleum refinery, tannery, textile dyeing, battery manufacturing, metallurgical, mining, iron and steel, and pesticide and insecticide manufacturing (Table 10.1), contribute large quantities of metallic contaminants to aqueous environments (Choi et al., 2009). Further, the illegal dumping of garbage, the disposal of chemical effluents, geological weathering, and natural calamities like volcanoes and forest fires are some of the other important activities that cause metal contamination of the environment.

During the last few decades, several synthetic chemicals have found their way into the ecosystem as a result of their large-scale production and usage causing environmental hazards. Plastics, pesticides, lignins, chloro-aromatics, dyes, and pigments are posing a great disposal problem. Since the number and diversity of compounds known to be degraded by microbes are truly vast, it is optimistically viewed that these man-made pollutants can be removed with the help of suitably manipulated or

TABLE 10.2
Effect of Heavy Metals on Plants, Animals, and Humans

Heavy Metal	Effects of Industrial Effluent
Al	Shortening of roots of plants
Ag	Irreversible skin graying in humans
As	Severe poisoning in humans
B	Stunting, prostrate forms, deforming and browning of leaves in plants
Ba	Acts as muscle stimulant, affects heart function in animals
Co	Chlorosis in plants
Cu	Chlorosis, reduction of size and seeds of chlorophyll, corollas
Cr	Carcinogenic to humans, causes chlorosis in plants
Fe	Darkening of leaves
Mn	Chlorosis, blotching in plants
Mo	Abnormally colored shoot, chlorosis, prone to insect attack
Ni	Chlorosis and necrosis in plants
Pb	Serious cumulative and acute body poisoning in humans
Se	Causes loss of hair, dermal changes in humans and animals
U	Variation in color of flowers, abnormal fruits, stimulated growth, more chromosomes
Zn	Chlorosis, blotching, etc.

structured microbes. This optimism is not based on theoretical proposition; in the last couple of decades, such manipulation has been carried out in a number of places and resulted in the evolution of specialized microbes (superbugs) to degrade specific environmental pollutants.

10.2.2 METALS AS A PRECIOUS COMPONENT OF LIFE

Certain metals, like Na, K, Ca, Cu, Co, Fe, Ni, Mg, Mn, and Zn, are considered essential micronutrients and are required in trace quantities for the growth and survival of several organisms, yet they all can exert toxicity when present in high concentrations. Metals like Fe, Cu, Zn, and Mn are required for normal metabolism in plants and animals. There are also some minor elements, known as trace elements, or other metalloids that play important roles in the functioning of living organisms, including those of microorganisms. Additionally, these elements could participate in other activities: (1) forming the structure of proteins and pigment, (2) redox processes, (3) regulating osmotic pressure, (4) maintaining the ionic balance, and (5) acting as an enzyme component of the cells (Pilon-Smits et al., 2009; Zaidi et al., 2012). Similarly, some other elements, such as Al, Co, Se, and Si, play a significant role in promoting plant growth and may be essential for particular species (Pilon-Smits et al., 2009). Likewise, Zn plays an important role in cellular division and amplification, and protein synthesis, and contributes to carbohydrate, lipid, and nucleic acid metabolism (Collins, 1981). In contrast, the structure and composition of microbiota (Wani and Khan, 2010) and plant growth are reported to be significantly affected when the concentrations of such trace elements exceed the normal level (Oves et al., 2016).

The uptake of heavy metals from soil has both a direct and indirect effect on microbial composition, metabolism, and differentiation. The interaction of metals and their compounds with soil microbes depends on several factors, including metal species, interacting organisms and their habitat, structure and composition, and microbial functions. Some metals, such as Al, Cd, Hg, and Pb, are highly toxic and their biological functions remain unknown. Metal toxicity results from the displacement of essential ions from their active binding sites; they accumulate within cells and may affect enzyme selectivity, deactivate cellular functions, and damage the DNA structure, occasionally resulting in cell death. Similarly, plants typically need a continuous nutritional supply in order to remain healthy. Two criteria are generally used to define metals as essential nutrients for healthy plants: (1) the metal should be required by the plant to complete its life cycle, and (2) it should be part of an essential plant constituent or metabolite. Inadequate supply of a certain nutrient results in the development of symptoms of nutritional deficiency, which when severe may lead to early mortality of plants. On the other hand, when the nutrient supply exceeds the required quantities, it may cause toxicity, and in some cases may even lead to the death of plants at very high levels of nutrient enriched with heavy metals.

Heavy metals also have great impact on human health. They often enter the human body through different food chains, inhalation, and ingestion. They have long been used by humans for making metal alloys and pigments for paints, cement, paper, rubber, and other materials. Personnel working in these industries are frequently exposed to high doses of heavy metals. In spite of their known toxic effects, the use of heavy metals in some countries is increasing. However, once the heavy metal enters the human body by any means, it is likely to stimulate the immune system and may cause nausea, anorexia, vomiting, gastrointestinal abnormalities, and dermatitis. The toxicity of heavy metals can also disrupt or damage the mental and central nervous systems; change blood composition; and damage the lungs, kidneys, liver, and other important organs (Oves et al., 2016). Besides their toxic effects, some metals, like Cu, Se, and Zn, are known to play many significant and indispensable roles in human metabolism. For example, Cu acts as a cofactor for various enzymes of redox cycling at lower concentrations; conversely, at higher concentrations it disrupts human metabolism, leading to anemia and damage to vital organs, such as the liver and kidneys, in addition to stomach and intestinal irritation.

10.2.3 STRATEGIES TO CONTROL HEAVY METAL CONTAMINATION

Humans have strived to live in an environment that is free from pollutants and diseases. Although the biodegradation of wastes is a centuries-old technology, it is only in last few decades that serious attempts have been made to harness the natural biodegradative potential, with the aim of large-scale technological applications for effective and affordable environmental restoration. Without the use of the tangible biotechnological approaches, our planet would undoubtedly be an unpleasant place to live. In this context, bioremediation has become a widely used technology all over the world, particularly in the developed world.

Control of heavy metal contamination seems to be most economical at the source of its generation (*in situ* bioremediation). Since at this point the concentration of metals remains very high, remediation at such a site would be more effective and also helpful in preventing further dispersal of pollutants to natural sources, like soil, water, and air. In general, the adopted remediation system must adhere to the following specifications:

- Be able to achieve regulatory standards
- Be economically viable and acceptable for the industry
- Be environmentally acceptable and not cause any type of secondary pollution

Strategies to control metal pollution include conventional techniques, such as chemical precipitation, ultrafiltration, solvent extraction, electrodeposition, biological treatment, ion exchange, and adsorption using agro-industrial wastes as biosorbents. Wastewater from industries like electroplating, dye, textiles, metal, engineering, and pesticides contains higher concentrations of heavy metal. Similarly, bioremediation using live microorganisms and immobilized microbes and spent microbial biomasses from fermentation industries is yet another approach for removing heavy metals. Although a detailed discussion on pollution control strategies is beyond the scope of this book, a brief overview of conventional technologies is presented here for the convenience of readers.

10.2.3.1 Conventional Treatment Techniques

Physicochemical methods, such as chemical precipitation, chemical oxidation or reduction, electrochemical treatment, evaporative recovery, filtration, ion exchange, and membrane technologies, have been extensively used to remove heavy metal ions from industrial wastewater. These processes may be ineffective or expensive, especially when the heavy metal ions are in a solution containing in the order of 1–100 mg/L dissolved heavy metal ions (Volesky, 1990). Several inexpensive adsorbents, such as fly ash, bituminous coal, blast furnace slag, coconut shell, mango seed and shell, bagasse, used waste tea leaves, and wood bark, have been investigated by researchers to remove heavy metals from wastewaters (Gunatilake, 2015). However, most of these methods have some limitations (Table 10.3), making their use practically uneconomical from an environmental perspective.

Physical separation involves the use of mechanical screening, hydrodynamic classification, gravity concentration, floatation, magnetic separation, electrostatic separation, and attrition scrubbing. The efficiency of physical separation methods depends on various characteristics of the adsorbent, such as particle size distribution, particulate shape, clay content, moisture content, humic content, heterogeneity of the matrix, density between the matrix and metal contaminants, magnetic properties, and hydrophobic properties of the particle surface.

Chemical precipitation is widely used for the removal of heavy metals from inorganic effluents in industry owing to its simple operation. Chemical precipitation of effluents produces fine insoluble precipitates of heavy metals, such as hydroxide,

TABLE 10.3
Physicochemical Processes for Treatment of Wastewater Containing Heavy Metal

Process	Detrimental Factors
Membrane filtration	Expensive, durability of membranes
Liquid–liquid extraction	Limited application
Carbon adsorption	Expensive adsorbent, requires regeneration
Ion exchange	Expensive adsorbent, requires regeneration
Electrolytic treatment	Limited application
Precipitation	Forms sludge, often very gelatinous
Coagulation–flocculation	Forms hydrated sludge
Chemical reduction	Limited application
Floatation	Forms hydrated materials
Vitrification	Very expensive, landfill required
Evaporation	Energy-intensive process
Crystallization	Time dependent, landfill required

Source: Adapted from Crusberg, T.C. et al., Biomineralization of heavy metals, in *Fungal Biotechnology in Agriculture, Food and Environmental Applications*, ed. D.K. Arora, Marcel Dekker, New York, 2004, 409–418.

sulfide, carbonate, and phosphate, by reacting with dissolved metals in the solution. Further, chemical precipitants, coagulants, and flocculation processes are used to increase the particle size of the precipitates in order to remove them as sludge (Fu and Wang, 2011). As the metals precipitate and form solids, they can easily be removed, and low metal concentrations can be discharged. The commonly used precipitation technique makes use of hydroxide treatment due to its relative simplicity, low cost of precipitant (lime), and ease of pH control.

Coagulation and flocculation is a mechanism of electrostatic interaction between pollutants and coagulant–flocculant agents. Coagulation reduces the net surface charge of the colloidal particles and stabilizes them by electrostatic repulsion (López-Maldonado et al., 2014), whereas through additional collision and interaction with inorganic polymers, flocculation continuously increases the particle size to discrete particles. As the discrete particles are further flocculated into larger particles, they can be easily removed by filtration or floatation. However, the production of large quantities of sludge and the transfer of toxic metals into a solid phase are some of the limitations of this process.

Membrane filtration refers to removing suspended solid contaminants, such as heavy metals, using membrane filters of varying sizes. Depending on the size of contaminant, various filtration techniques, such as ultrafiltration, nanofiltration, and reverse osmosis, can be employed to remove heavy metals from wastewaters. In ultrafiltration, permeable membranes (pore size 5–20 nm) are used to separate heavy metals, macromolecules, and suspended solids from inorganic solution. The efficiency of filtration further depends on the molecular weight of the separation compounds, concentration of metals, pH, and pressure. A similar technique involving complexation–ultrafiltration proves to be a promising alternative to technologies based on precipitation and ion exchange. This is a hybrid approach that involves the selective concentration of heavy metals in solution through metal binding polymers in combination with ultrafiltration. In the complexation–ultrafiltration process, cationic forms of heavy metals are first complexed by a macroligand to facilitate an increase in their molecular weight to a size larger than that of the pores of the selected membrane (Trivunac and Stevanovic, 2006). The main advantages of the complexation–ultrafiltration process include the high separation selectivity due to the use of selective binding and the low energy requirements of the process. In reverse osmosis, pressure is employed to force a solution to pass through a semi-permeable membrane that retains the solute on one side and allows the pure solvent to pass to the other side. With reverse osmosis, many types of molecules and ions from solutions, including bacteria, can be removed, and its separation efficiency depends on solute concentration, pressure, and water flux rate.

Electrodialysis is a membrane separation technique that employs ionized species in the solution and is passed through an ion-exchange membrane by applying an electric potential. The membranes used are composed of thin sheets of plastic materials with either anionic or cationic characteristics. When a solution containing ionic species passes through the cell compartments, the cations migrate toward the cathode and the anions toward the anode, crossing the cation-exchange and anion-exchange membranes (Chen and Chen, 2003). The limitation of this technique is the frequent replacement of membranes and the associated corrosion.

Electrochemical processes use electricity to pass current through an aqueous metal-bearing solution having an insoluble anode and a cathode plate. In this process, heavy metals are precipitated in a weak acidic or neutralized catholyte as hydroxides. Electrochemical treatments of wastewater involve electrodeposition, electrocoagulation, electroflotation, and electrooxidation (Shim et al., 2014). It is the most frequently used precipitation technique, forming coagulants by electrolytic oxidation and destabilizing the contaminants to form flocs. In this process, charged ionic metal species are removed from wastewater by allowing it to react with anion in the effluent. The major advantages of this technique are reduced sludge production, no use of chemicals, and ease of operation.

Ion exchange is a technique that is based on the attraction of soluble ions from the liquid phase to the solid phase. This is a cost-effective technology involving low-cost materials, and is particularly suitable in industries that produce effluents with low concentrations of heavy metals. In this process, cations or anions containing a special ion exchanger are used to remove metal ions from the solution. Ion-exchange resins are nonaqueous solid materials that can absorb positively or negatively charged ions from an electrolyte solution and release other ions with the same charges into the solution in an equivalent amount. The positively charged ions in cationic resins (such as hydrogen and sodium ions) are exchanged with positively charged ions (such as nickel, copper, and zinc ions) in solutions. Likewise, the negatively charged ions in the resins (such as hydroxyl and chloride ions) can be replaced by the negatively charged ions (such as chromate, sulfate, nitrate, cyanide, and dissolved organic carbon) in wastewaters.

10.2.3.2   Bioaccumulation of Heavy Metals

Bioaccumulation refers to intracellular uptake and biochemical transformation of metal ions by living cells. Among various types of microorganisms, fungi have received considerable attention in the recent past. Several fungi, including thermophilic fungi, are known to accumulate heavy metals. In particular, the use of fungal microorganisms to remediate environmental pollutants is known as mycoremediation (Singh, 2006). The bioremediation approach employing viable microorganisms is rather complex and depends on a number of variables. In metabolically active cells, the accumulation of heavy metals is performed in several ways:

1. **Accumulation on external surface:** It is executed by binding the metal ions or their precipitation on the microbial cell wall or membranes. A number of microorganisms have been reported to remediate metal ions, like Fe, Mn, Co, Ni, Pb, and U, either by chelation with cell wall ligands or as sulfides or oxides or as metal precipitants.
2. **Accumulation within the cell:** Some of the microorganisms have shown their ability to accumulate a variety of metal ions intracellularly, although the rate of accumulation of these contaminants varies considerably within a single microbe. It is also known that when metabolically active metals like Fe, Mg, and Co are present in the external environment, their accumulation is rapid. In contrast, nonmetabolic ions, such as Cu, Cd, Ag, Mn, Co, Ni, and Pb, are accumulated at a slower pace.

3. **Accumulation by exopolysaccharides:** Exopolysaccharides secreted by a number of microbes have the affinity to bind various metal ions. These exopolysaccharides can be exclusively used to remediate metal-containing solutions released by various industries.

4. **Extracellular precipitation:** Certain microorganisms produce sulfide anions (S^{2-}) and oxalate anions ($C_2O_4^{2-}$), which can form very insoluble salts with heavy metal ions exhibiting very low-solubility products. Acid phosphatase produced by several species of *Penicillium* and *Aspergillus* has been employed to remove copper from solutions.

5. **Extracellular complexation:** Extracellular complexation refers to the formation of metal complexes as a result of the reaction of an extracellular polymeric substance from microorganisms with metal ions. Metals like Fe, Mo, and V have very high affinity to form metal complexes.

6. **Volatilization of metal ions:** By microbial actions, certain metal complexes become volatilized from solution. Metals like As, Te, Se, and Hg are methylated by microbial metabolism, and the metabolic products of these reactions are volatilized readily.

It is generally observed that metal toxicity results from the displacement of essential ions from their active binding sites, interaction with ligands, and formation of free radicals, leading to protein inactivation and DNA damage. Microorganisms dwelling in contaminated environments have evolved several mechanisms of resistance to heavy metal concentration, which otherwise is lethal to other microbiota. Thermophilic microorganisms, including fungi, often inhabit such environments that are enriched in heavy metals. Owing to their prolonged persistence in such environments, they are able to adapt and restrict the intracellular metals within sublethal levels. This is possibly accomplished through (1) efflux of metals that enter cells by either specific or nonspecific transporters; (2) intracellular compartmentalization within safe sectors of cell, reducing the cytoplasmic availability of metals; (3) intra- or extracellular entrapment of metals by complexation with microbially generated ligands; and (4) enzymatic transformations reducing metal toxicity (Nies, 1999; Bruins et al., 2000; Chatziefthimiou et al., 2007). Thus, using living microbes or microbial consortia for bioremediation of heavy metals is a cost-effective technology that is less intrusive and accumulates toxicants and sequesters them for large-scale removal. In contrast, a nonliving microbial biomass does not depend on requirements for growth, metabolic energy, and transport. Additionally, due to the lack of protons produced during metabolism, a nonliving biomass shows strong affinity for metal ions. Further, such a biomass is unaffected by the problems of toxicity, a major advantage of biosorption. Plenty of fungal biomass is produced as a waste by-product of large-scale industrial fermentation, which after pretreatment by washing with acids and/or bases before final drying and granulation can be used as a biosorbent for metal recovery. All these factors contribute to reducing the final cost of the bioremediation process.

10.2.3.3 Biosorption of Heavy Metals

Biosorption is a metabolism-independent pseudo-ion-exchange process wherein a metal ion is exchanged for a counterion in the biomass or resin. In general, the filamentous fungi possess higher adsorption capacities for heavy metal removal. Factors affecting the efficiency of biosorption include several extrinsic factors, such as type of metal, ionic form in solution, and the functional site. Other factors, such as pH, temperature, biomass concentration, type of biomass preparation, initial metal ion concentration and metal characteristics, and concentration of other interfering ions, are also important in evaluating the degree of biosorption. Biosorption and recovery can be exaggerated during stirring induced by a magnetic field (Gorobets et al., 2004).

In fungi, several mechanisms of metal biosorption are reported, including ion exchange, chelation, adsorption, crystallization, and precipitation, followed by ion entrapment in inter- and intrafibrillar capillaries, spaces of the polysaccharide material, and diffusion through the cell wall and membranes. The applicability and effectiveness of these mechanisms differ according to the fungal species used, the origin of the biomass, and its processing. Fungi are also reported to remove both soluble and insoluble metals from solution and can leach metals from solid waste as well. Solubilization and complexing of metal ions is carried out by protons, organic acids, phosphatases, and other metabolites produced by the majority of fungi. Thermophilic fungi are also known to produce different kinds of acids (Maheshwari et al., 2000) that help in the leaching of metals from self-composting solid waste. Oxalic acid is reported to leach a number of metals, such as Al, Fe, and Li, resulting in the formation of metal oxalate complexes.

The sorptive performance of a microbial biomass is determined on the basis of adsorption isotherms. In this context, the *Langmuir and Freundlich equations* (*Langmuir model*) are commonly employed in evaluation of the adsorption isotherms at a constant temperature. This model is applicable to monolayer sorption onto a surface with a finite number of identical sites and is given by

$$q = q_{max} \frac{bC_f}{1 + bC_f}$$

where q (mg/g) is the uptake of the metal, q_{max} is the maximum uptake, C_f (mg/g) is the equilibrium (final) concentration of metal in the solution, and b is a constant.

The empirical *Freundlich model* based on sorption on a heterogeneous surface is given by

$$q = KC_f^{1/n}$$

where K and n, the Freundlich constants, are indicators of adsorption capacity and adsorption intensity, respectively. The equation can be linearized in logarithmic form, and the Freundlich constants can be determined.

For any biosorbent, rapid uptake of metals is a desirable property. However, it is not always safe to rely on data generated by the above isotherms, particularly for systems operating under variable environmental conditions. Due to variations pertaining to multiple sites, the nature of the sorbent material, pH, and the complex chemistry of metal ions, the process of biosorption generates irregular biosorption isotherms. Therefore, to obtain accurate results, at least different sorption systems must be compared at the same equilibrium concentrations, as the presence of other ions or co-cations also influences the sorption system by superfluous interactions with the metal species.

10.2.3.4 Immobilized Biosorbent for Bioremediation

The efficiency of microorganisms, as well as biosorbents prepared from a nonviable biomass, could be improved by immobilizing the same by some entrapping agents, such as alginates, polyacrylamide, and silica gel. As alginates, the use of sodium alginates, calcium chloride, barium, and strontium chloride is quite common. This is because calcium helps in stabilization, while barium maintains stability and provides necessary rigidity, and strontium makes adequate permeability to immobilisates. When enzymes are used, they offer several advantages, which include efficient regeneration and prolonged activity over a considerable period of exposure. Rome and Gadd (1991) have used immobilized biomasses of *Saccharomyces cerevisiae* and other fungi for the removal of certain heavy metals and radionuclides in batch and continuous mode. They observed that the process is biphasic, with surface biosorption being followed by energy dependence in flux. Lewis and Kiff (1988) have used an immobilized fungal biomass for the removal of heavy metals and observed that entrapped biosorbents efficiently remove metal ions from the aqueous phase.

10.2.3.5 Recovery of Metals and Regeneration of Biomass

Repeated use of biosorbent materials is desirable in order to make the process economically viable. The aim of desorption is to regenerate the biomass for the next step in the biosorption and recovery of a biosorbed metal. A variety of desorbents are used to recover metals from biosorbed materials, such as salts of sodium, potassium, and calcium (chlorides, carbonates, bicarbonates, and hydroxides); mineral acids like HCl, H_2SO_4, $HClO_4$, and HNO_3; ethylenediaminetetraacetic acid (EDTA); nitriloacetic acid (NTAA); and physical treatments, like boiling. Selection of the desorbent and the desorption process is important, so that these can easily integrate into the entire process of biosorption. The regeneration of the biosorbent reduces the cost of the biosorption process and allows for the recovery of metal from the liquid phase. Singh (2006) elegantly reviewed the work on fungal biosorption carried out in the last few decades.

10.2.4 THERMOPHILIC FUNGI IN BIOREMEDIATION

The role of fungi in the bioremediation of heavy metals, industrial effluents, and dye decolorization has gained momentum since the early 1980s. By using fungal organisms, successful technologies have been developed for the bioremediation of environmental pollutants. Thermophilic fungal biomasses could be utilized for the biosorption of heavy metals, but this group is little explored compared with its mesophilic counterpart. Bengtsson et al. (1995) used the biomass of *Talaromyces emersonii* CBS 814.70, a thermophilic fungus, for the biosorption of uranium. Similarly, certain species of *Mucor* and *Rhizopus* have been found to be useful in the accumulation and recovery of heavy metals and radionuclides from wastewater and mining operations.

Synthetic dyes are manufactured and used in large quantities in the textile, food processing, paper and pulp, cosmetics, and pharmaceutical industries. The textile industries account for two-third of the total dye stuff market. Dyes have been used increasingly in the textile and dying industries because of their ease and cost-effectiveness in synthesis, firmness, and variety in color compared with that of natural dyes. The most commonly used dyes in the textile industries are azo dyes, for example, Reactive Blue MR, Orange M2R, Yellow M4G, Black HFGR, and Red M8B. Dyes are designed in such a way that they are resistant to light, water, and oxidizing agents, and therefore most dyes cannot be treated by conventional physical and chemical processes. Dye colors are visible in water concentrations as low as 1 mg/L, whereas textile processing wastewater normally contains more than 10–200 mg/L of dye concentration. Removal of dyes from the effluents or their degradation before discharge is a great environmental challenge for the industries (Baldrian and Gabriel, 2003). In our recent study, the potential of 10 indigenous fungi isolated from soil samples of dye disposal sites was evaluated to decolorize five textile azo dyes, that is, Reactive Blue MR, Orange M2R, Yellow M4G, Black HFGR, and Red M8B. In pure culture, it was observed that the thermophilic *Humicola insolens* and mesophilic *Humicola brevis*, *Aspergillus terrus*, *Aspergillus flavus*, *Aspergillus niger*, and *Rhizopus* sp. were highly efficient in decolorizing (bioaccumulation) textile dyes. The study also depicted that *Rhizopus* sp. was highly efficient in decolorizing (81.01%) a mixture of five dyes. More recently, Taha et al. (2014) investigated the decolorizing capability of a thermophilic fungus, *Thermomucor indicae-seudaticae*, at varying temperatures and dye concentrations using azure B, Congo red, trypan blue, and Remazol Brilliant Blue R. They observed that the inactivated biomass was more effective in decolorization than the living biomass of *T. indicae-seudaticae*, *Aspergillus fumigatus*, and their mixed coculture. They further reported that the inactivated biomass of *T. indicae-seudaticae* rapidly and effectively decolorizes dyes in the temperature range of 30°C–55°C at 100, 500, and 1000 mg/L concentrations compared with either *Aspergillus fumigatus* or the mixed culture. In addition, an acidic pH favored effective adsorption by *T. indicae-seudaticae*. Thus, biological treatment involving living and nonliving microbial biomasses is a cost-effective and eco-friendly process compared with physical and chemical processes.

Enzymes such as alkali-stable xylanase, cellulose, and pectinases from thermophilic fungi are used to reduce organic load, and thereby BOD, in the effluents of the paper and pulp industry. In paper industries, the addition of thermophilic fungal phytase to these enzymes not only reduces the phytic acid content of the pulp but also improves the quality of the paper (Singh and Satyanarayana, 2011; Singh, 2014). The wastewater from fruit processing industries contains the recalcitrant pectic material, which is difficult to degrade by natural microflora during the activated sludge treatment. Some thermophilic fungi, like *Myceliophthora thermophila*, are known to produce alkaline pectinases in solid-state fermentation (SSF) as well as submerged fermentation (SmF) using wheat bran and citrus peel as the substrates (Kaur and Satyanarayana, 2004). The authors observed that the fungus produces 330-fold higher enzyme titers in SSF than in SmF, which can be effectively utilized in the treatment of wastewater containing pectic substances. Similarly, the effluents from oil industries are rich in fatty materials; therefore, lipase-producing thermophilic fungi are ideal in the treatment of such effluents. *Mucor pusillus*, *Humicola lanuginosa*, *Humicola grisea* var. *thermoidea*, and *Talaromyces thermophilus* are reported to produce lipases (Somkuti et al., 1969; Arima et al., 1972). Malaysia is the largest producer of palm oil, and more than 200 industries process palm oil throughout the year, generating enormous effluents every day. The effluent is viscous brown liquid with acidic pH and usually has a temperature of more than 60°C at the point of discharge. The effluent is retained in digestion tanks for a longer duration, allowing the colonization of a range of thermophilic and mesophilic fungi (Kuthubutheen, 1981). During the first week of digestion, several thermophilic fungi, namely, *Aspergillus fumigatus*, *Chaetomium thermophile* var. *dissitum*, *Mucor pusillus*, *Thermoascus aurantiacus*, *Thermomyces lanuginosus*, *Melanocarpus albomyces*, and *Scytalidium thermophilum*, are reported to be present (Kuthubutheen, 1981). As the temperature of the effluents declines, the number and types of thermophilic fungi also decline, paving the way for the growth and colonization of mesophilic fungi. The author also observed that diluted effluents (50%) allow faster growth of the most vigorous thermophiles (two to three times) than of the most vigorous mesophiles. *Rhizopus arrhizus* (Kumar et al., 1993), *T. lanuginosus* strain Y-38, and *R. miehei* and *T. lanuginosus* (Noel and Combes, 2003) have been observed to produce lipases in submerged fermentation. A range of thermostable lipases from thermophilic fungi have found their application in the eco-friendly treatment of the effluents from oil industries.

10.3   BIOMINERALIZATION

Biomineralization refers to a process by which living organisms internally or externally form inorganic minerals. It is an extremely widespread phenomenon exhibited by taxonomically diverse organisms. However, microorganisms, including archaea, bacteria, and fungi, have great potential to produce biominerals and have attracted the attention of biogeochemists world over. Despite the fact that the hallmark of biomineralization is the control that organisms exert over the mineralization process, it has been observed by earth scientists for the last five-plus decades that biologically produced minerals often contain embedded within their compositions signatures that reflect the external environment in which the animal lived (Weiner and Dove, 2003). The process of biomineralization is induced by biological macromolecules and operates at moderate conditions of temperature, pH, and pressure. The resultant biominerals differ from pure inorganic minerals in the sense that they are composed of organic macromolecules and inorganic hybrids. The most common biominerals are phosphate and carbonate salts of calcium; other biominerals include calcium oxalate, oxides of manganese and iron, and silicates. Realizing the importance of fungi in biomineralization, a new branch of science, geomycology, a term coined by Professor Geoffrey M. Gadd and his team, has emerged. Geomycology is defined as the scientific study of the roles of fungi in processes of fundamental importance to geology, which encompasses the bioweathering of rocks and minerals, soil formation, the transformation and accumulation of metals, and the cycling of elements and nutrients.

Based on the extent of biological control over the process, biomineralization is divided into two fundamentally different groups: biologically induced mineralization (BIM) and biologically controlled mineralization (BCM) (Lowenstam, 1981). BIM is the process where an organism modifies the local microenvironment, creating conditions that support extracellular chemical precipitation of mineral phases. In this process, the organism does not appear to control biomineralization. In contrast, in BCM, a great deal of genomic control of an organism over biomineralization is exerted by accumulating mineral precursor ions to form mineral phases in specific shapes and at specific locations. A third category, known as biologically influenced mineralization (or organomineralization sensu strict), involving a biological matrix to initiate or enhance crystal nucleation and growth, with an influence on mineral morphology has also been recognized (Dupraz et al., 2009). It is the specific nature and degree of control that is central to understanding the extent of biological control of the elemental compositions of biominerals. Considering the ubiquity and capacity for production of mineral-transforming metabolites, fungi may be involved in many nonspecific mineral transformations in the environment at varying scales (Hutchens, 2009; Rosling et al., 2009).

Certain urease-producing fungi, such as *Pestalotiopsis* sp. and *Myrothecium gramineum*, isolated from calcareous soil, could precipitate calcite ($CaCO_3$), strontianite ($SrCO_3$), vaterite in different forms, ($CaCO_3$, $[Ca_xSr_{1-x}]CO_3$), and olekminskite ($Sr[Sr, Ca] [CO_3]_2$), suggesting that urease-positive fungi could play an important role in the environmental fate, bioremediation, or biorecovery of Sr or other metals and radionuclides that form insoluble carbonates (Li et al., 2015). Similarly, many fungi, like *Aspergillus niger* and *Serpula himantioides*, can produce several metal oxalates by interacting with a variety of metals and metal-bearing minerals, such as Ca, Cd, Co, Cu, Mg, Mn, Sr, Zn, Ni, and Pb (Sayer and Gadd, 1997; Gadd et al., 2014). These fungi can transform insoluble manganese oxide minerals, including those produced biogenically, into manganese oxalates (Gadd, 2017). Similarly, *Aspergillus niger* is known to produce biominerals of phosphate, such as calcium, aluminum, and iron phosphate (Liang et al., 2015), and silicates, such as phyllosilicate mineral (Wei et al., 2012).

Despite the fact that thermophilic fungi are able to produce an array of extracellular enzymes, such as phytases and laccases, that can be used in the biomineralization of a variety of different metals, studies on their role in this process are scant. Lian et al. (2008) studied the role of thermophilic fungus *Aspergillus fumigatus* in the biomineralization of potassium-bearing minerals to enhance the release of mineralic potassium. They conducted two separate experiments, one with the mineral grains and fungal cells in direct contact, and the other employing a membrane (pore size 0.22 μm) to separate the two. After a period of more than 30 days, they observed that irrespective of the experimental setup, the concentration of free K in the culture was significantly higher than that of the control experiments where no living organism was present. In addition, the occurrence of mineral–cell physical contact enhanced potassium release by an additional factor of 3 to 4 in comparison with the experiment in which the organism and mineral grains were separated by a membrane. In the case of contact experiments, they observed the formation of mycelium–mineral aggregates, as revealed by electron probe microanalysis. Furthermore, atomic force microscopy imaging indicated the possible ingestion of mineral particles by the fungal cells. Their results suggested that *Aspergillus fumigatus* promoted K release by means of at least three possible routes: (1) through the complexation of soluble organic ligands; (2) by engaging the immobile biopolymers, such as the insoluble components of secretion; and (3) via mechanical forces in association with the direct physical contact between cells and mineral particles.

Fungal biomineralization is a rapidly expanding field, primarily because of the rapid development of novel methods to qualitatively and quantitatively analyze the interactions occurring at the organismal level. In addition to this, the role of fungi in biomineralization also stems from the ability of their hyphae to traverse diverse substrates and their organic matter degradation, through either saprotrophy or parasitic and pathogenic interactions. Similar to bacteria, they also play a major role in the mobilization and immobilization of mineral and metal compounds. Numerous examples of induced biomineralization through fungal metabolic activity (i.e., through heterotrophic pathways) have been documented in the literature. However, concerted efforts to highlight the role of fungal processes from a diverse group of fungi, including thermophilic fungi, are needed to harness their true biotechnological potential.

REFERENCES

Arima, K., Wen, H., and Beppu, T. 1972. Lipase from *Bacillus* T. *Agricultural and Biological Chemistry* 11:1913–1917.
Baldrian, P., and Gabriel, J. 2003. Lignocellulose degradation by Pleurotus ostreatus in the presence of cadmium. *FEMS Microbiology Letters* 220:235–240.
Bengtsson, L., Johansson, B., Hackett, T.J., McHale, L., and McHale, A.P. 1995. Studies on the biosorption of uranium by *Talaromyces emersonii* CBS 814.70 biomass. *Applied Microbiology and Biotechnology* 42:807–811.
Bruins, M.R., Kapil, S., and Oehme, F.W. 2000. Microbial resistance to metals in the environment. *Ecotoxicology and Environmental Safety* 45:198–207.
Chatziefthimiou, A.D., Crespo-Medina, M., Wang, Y., Vetriani, C., and Barkay, T. 2007. The isolation and initial characterization of mercury resistant chemolithotrophic thermophilic bacteria from mercury rich geothermal springs. *Extremophiles* 11:469–479.
Chen, L., and Chen, Q. 2003. Industrial application of UF membrane in the pretreatment for RO system. *Journal of Membrane Science and Technology* 4:009.
Choi, A., Wang, S., and Lee, M. 2009. Biosorption of cadmium, copper, and lead ions from aqueous solutions by *Ralstonia* sp. and *Bacillus* sp. isolated from diesel and heavy metal contaminated soil. *Geosciences Journal* 13:331–341.
Collins, J.C. 1981. Zinc. In *The Effect of Heavy Metal Pollution on Plants*, ed. N.W. Lepp, 145–170. London: Applied Science Publishers.
Crawford, R.L., and Crawford, D.L. 1996. *Bioremediation: Principles and Applications*. Cambridge: Cambridge University Press.
Crusberg, T.C., Mark, S.S., and Dilorio, A. 2004. Biomineralization of heavy metals. In *Fungal Biotechnology in Agriculture, Food and Environmental Applications, ed.* D.K. Arora, 409–418. New York: Marcel Dekker.
Dupraz, C., Pamela Reid, R., Braissant, O., Decho, A.W., Sean Normanc, R., and Visscher, P.T. 2009. Processes of carbonate precipitation in modern microbial mats. *Earth-Science Reviews* 96:141–162.
Fu, F., and Wang, Q. 2011. Removal of heavy metal ions from wastewaters: A review. *Journal of Environmental Management* 92:407–418.
Gadd, G.M. 2017. Geomicrobiology of the built environment. *Nature Microbiology* 2. doi: 10 .1038/nmicrobiol.2016.275.

Gadd, G.M., Bahri-Esfahani, J., Li, Q., Rhee, Y.J., Wei, Z., Fomina, M., and Liang, X. 2014. Oxalate production by fungi: Significance in geomycology, biodeterioration and bioremediation. *Fungal Biology Reviews* 28:36–55.

Gorobets, S., Gorobets, O., Ukrainetza, A., Kasatkina, T., and Goyko, I. 2004. Intensification of the process of sorption of copper ions by yeast of *Saccharomyces cerevisiae* 1968 by means of a permanent magnetic field. *Journal of Magnetism and Magnetic Materials* 272–276:2413–2414.

Gugliandolo, C., Lentini, V., Spano, A., and Maugeri, T.L. 2012. New bacilli from shallow hydrothermal vents of Panarea Island (Italy) and their biotechnological potential. *Journal of Applied Microbiology* 112:1102–1112.

Gunatilake, S.K. 2015. Methods of removing heavy metals from industrial wastewater. *Journal of Multidisciplinary Engineering Science Studies* 1:12–18.

Hasyim, R., Imai, T., Reungsang, A., and Thong, S.O. 2011. Extreme-thermophilic biohydrogen production by an anaerobic heat treated digested sewage sludge culture. *International Journal of Hydrogen Energy* 36(14):8727–8734.

Hutchens, E. 2009. Microbial selectivity on mineral surfaces: Possible implications for weathering processes. *Fungal Biology Reviews* 23:115–121.

Kaur, G., and Satyanarayana, T. 2004. Production of extracellular pectinolytic, cellulolytic and xylanoytic enzymes by thermophilic mould *Sporotrichum* thermophile Apinis in solid state fermentation. *Indian Journal of Biotechnology* 3:552–557.

Kumar, K.K., Deshpande, B.S., and Ambedkar, S.S. 1993. Production of extracellular acidic lipase by *Rhizopus arrhizus* as a function of culture conditions. *Hindustan Antibiotics Bulletin* 35:33–42.

Kuthubutheen, A.J. 1981. Occurrence and growth of mesophilic and thermophilic fungi in palm oil mill effluent. *Transactions of the British Mycological Society* 77:420–423.

Lewis, D., and Kiff, R.J. 1988. The removal of heavy metals from aqueous effluents by immobilised fungal biomass. *Environmental Technology Letters* 9:991–998.

Li, Q., Csetenyi, L., Paton, G.I., and Gadd, G.M. 2015. $CaCO_3$ and $SrCO_3$ bioprecipitation by fungi isolated from calcareous soil. *Environmental Microbiology* 17:3082–3097.

Lian, B., Wang, B., Pan, M., Liu, C., and Teng, H.H. 2008. Microbial release of potassium from K-bearing minerals by thermophilic fungus *Aspergillus fumigatus*. *Geochimica et Cosmochimica Acta* 72:87–98.

Liang, X., Hillier, S., Pendlowski, H., Gray, N., Ceci, A., and Gadd, G.M. 2015. Uranium phosphate biomineralization by fungi. *Environmental Microbiology* 17:2064–2075.

López-Maldonado, E.A., Oropeza-Guzman, M.T., Jurado-Baizaval, J.L., and Ochoa-Terán, A. 2014. Coagulation-flocculation mechanisms in wastewater treatment plants through zeta potential measurements. *Journal of Hazardous Materials* 279:1–10. doi: 10.1016/j.jhazmat.2014.06.025.

Lowenstam, H.A. 1981. Minerals formed by organisms. *Science* 211:1126–1131.

Maheshwari, R., Bharadwaj, G., and Bhat, M.K. 2000. Thermophilic fungi: Their physiology and enzymes. *Microbiology and Molecular Biology Reviews* 64:461–488.

Nies, D.H. 1999. Microbial heavy-metal resistance. *Applied Microbiology and Biotechnology* 51:730–750.

Noel, M., and Combes, D. 2003. Effects of temperature and pressure on *Rhizomucor miehei* lipase stability. *Journal of Biotechnology* 102:23–32.

Oves, M., Khan, S., Qari, H., Felemban, N., and Almeelbi, T. 2016. Heavy metals: Biological importance and detoxification strategies. *Journal of Bioremediation and Biodegradation* 7:334. doi: 10.4172/2155-6199.1000334.

Pilon-Smits, E.A., Quinn, C.F., Tapken, W., Malagoli, M., and Schiavon, M. 2009. Physiological functions of beneficial elements. *Current Opinion in Plant Biology* 12:267–274.

Rome, L.D., and Gadd, G.M. 1991. Use of pelleted and immobilized yeast and fungal biomass for heavy metal and radionuclide recovery. *Journal of Industrial Microbiology* 7:97–104.

Rosling, A., Roose, T., Hermann, A.M., Davidson, F.A., Finlay, R.D., and Gadd, G.M. 2009. Approaches to modelling mineral weathering by fungi. *Fungal Biology Reviews* 23:138–144.

Sar, P., Kazy, S.K., Paul, D., and A. Sarkar. 2013. Metal bioremediation by thermophilic microorganisms. In *Thermophilic Microbes in Environmental and Industrial Biotechnology*, eds. T. Satyanarayana, J. Littlechild, and Y. Kawarabayasi, 171–201. Netherlands: Springer.

Sayer, J.A., and Gadd, G.M. 1997. Solubilization and transformation of insoluble metal compounds to insoluble metal oxalates by *Aspergillus niger*. *Mycological Research* 101:653–661.

Shim, H.Y., Lee, K.S., Lee, D.S., Jeon, D.S., Park, M.S., Shin, J.S., Lee, Y.K., Goo, J.W., Kim, S.B., and Chung, D.Y. 2014. Application of electrocoagulation and electrolysis on the precipitation of heavy metals and particulate solids in washwater from the soil washing. *Journal of Agricultural Chemistry and Environment* 3:130–138.

Singh, B. 2014. *Myceliophthora thermophila* syn. *Sporotrichum thermophile*: A thermophilic mould of biotechnological potential. *Critical Reviews in Biotechnology* 15:1–11.

Singh, B., and Satyanarayana, T. 2011. Microbial phytases in phosphorus acquisition and plant growth promotion. *Physiology and Molecular Biology of Plants* 17:93–103.

Singh, H. 2006. *Mycoremediation: Fungal Bioremediation*. Hoboken, NJ: Wiley.

Somkuti, G.A., Babel, F.J., and Somkuti, A.C. 1969. Lipase of *Mucor pusillus*. *Applied Microbiology* 17:606–610.

Spain, A.M., and Krumholz, L.R. 2011. Nitrate-reducing bacteria at the nitrate and radionuclide contaminated Oak Ridge integrated field research challenge site: A review. *Geomicrobiology Journal* 28:418–429.

Taha, M., Adetutu, E., Shahsavari, E., Smith, A., and Ball, A. 2014. Azo and anthraquinone dye mixture decolourization at elevated temperature and concentration by a newly isolated thermophilic fungus, *Thermomucor indicae-seudaticae*. *Journal of Environmental Chemical Engineering* 2:415–423.

Trivunac, K., and Stevanovic, S. 2006. Effects of operating parameters on efficiency of cadmium and zinc removal by the complexation–filtration process. *Desalination* 198:282–287.

Volesky, B. 1990. Removal and recovery of heavy metals by biosorption. In *Biosorption of Heavy Metals*, ed. B. Volesky, 7–43. Boca Raton, FL: CRC Press.

Wani, P.A., and Khan, M.S. 2010. *Bacillus* species enhance growth parameters of chickpea (*Cicer arietinum* L.) in chromium stressed soils. *Food and Chemical Toxicology* 48:3262–3267.

Watkin, E.L.J., Keeling, S.E., Perrot, F.A., Shiers, D.W., Palmer, M.L., and Watling, H.R. 2009. Metals tolerance in moderately thermophilic isolates from a spent copper sulfide heap, closely related to *Acidithiobacillus caldus*, *Acidimicrobium ferrooxidans* and *Sulfobacillus thermosulfidooxidans*. *Journal of Industrial Microbiology and Biotechnology* 36:461–465.

Wei Z., Kierans M., and Gadd G.M. 2012. A model sheet mineral system to study fungal bioweathering of mica. *Geomicrobiology Journal* 29:323–331.

Weiner, S., and Dove, P.M. 2003. An overview of biomineralization processes and the problem of the vital effect. *Reviews in Mineralogy and Geochemistry* 54:1–29.

Zaidi, A., Wani, P., Khan, A., and Saghir, M., eds. 2012. *Toxicity of Heavy Metals to Legumes and Bioremediation*. XII, 248. Springer-Verlag Wien.

11 Biocatalysts of Thermophilic Fungi

11.1 INTRODUCTION

The ubiquitous occurrence of thermophilic fungi make them important from both ecological and industrial points of view. Temperature is one of the environmental factors that plays an immensely important and often decisive role in the survival, growth, distribution, and diversity of organisms on the surface of the earth. Cooney and Emerson (1964) distinguished the heterogeneous group of thermophilic and thermotolerant fungi on the basis of their minimum and maximum temperatures for growth. Thermophilic fungi, because of their higher temperature tolerance, have wider applications in industry and in bioconversion technology, including cheap and safe disposal by composting of the enormous quantities of animal waste from intensive farming, the conversion of animal waste to fungal protein for recycling as food for cattle and poultry, the bioconversion of lignocellulosic crops and industrial waste, and the bioremediation of toxic waste products for the cleanup of contaminated land.

From an ecological perspective, the occurrence of thermophilic fungi in the aquatic sediment of lakes and rivers, as first reported by Tubaki et al. (1974), is mysterious in view of the low temperature (6°C–7°C) and low level of oxygen (average 10 ppm, <1.0 ppm at a depth of 31 m) available at the bottom of lakes. A number of thermophilic fungi survive stresses such as increased water pressure, the absence of oxygen, and desiccation (Mahajan et al., 1986). Undoubtedly, the thermophilic fungi owe their ubiquity and common occurrence in large measure to this special ability to occupy a temperature niche that most other fungi cannot inhabit. More attempts are needed, however, to provide evidence for their active involvement in the substrates from which they are being reported.

It is being widely realized that considering the population growth rate and the stock of available energy reserves in the form of coal, gas, and oil, it will be impossible to satisfy world energy demands in the twenty-first century. Since biomass is replenishable and inexhaustible, increasing attention is being focused on bioenergy technologies. Biomass in the form of cellulose, hemicellulose, and lignin represents an important energy and material source (Satyanarayana et al., 1988). One of the very widely attempted approaches for the utilization of biomass is the enzyme-catalyzed hydrolysis of cellulose and hemicellulose. The thermophilic fungi are among the numerous microorganisms that are capable of producing extracellular hydrolases. Among the other properties of enzymes, the most important is that they can act within a fixed range of temperature, beyond which enzymes are inactive or their proteins get denatured. Therefore, temperature is an important factor as far as the activity of enzymes is concerned. For this reason, thermophilic fungi assume greater importance for enzymatic studies.

One of the technical problems in industrial fermentation is the maintenance of temperature at an optimal level during the entire cultivation period. The use of thermophilic fungal strains offers an effective solution to this problem. Studies on the properties of an array of enzymes produced by thermophilic fungi show them to have superior thermal stability over their mesophilic counterparts (Maheshwari et al., 2000; Turner et al., 2007; Ahirwar et al., 2017). Thermophilic fungi are particularly valuable for use in bioconversion technologies, as they reduce the need for cooling during growth. This obviously reduces the costs associated with the installation, maintenance, and operation of the heat exchange equipment normally required for dissipating heat produced in large-scale mesophilic fermentations (Dix and Webster, 1995). An additional advantage of working at high temperatures is that the end product is virtually free of contamination by mesophiles. There are now indications that industrial processes involving thermophiles may offer potential for lower capital and operating costs in some fermentations.

Thermophilic fungi have been playing a role in nature's economy ever since they evolved on this earth. Their importance in the human economy has been realized from their ability to efficiently degrade organic matter, acting as biodeteriorants and natural scavengers; to produce extracellular as well as intracellular enzymes, amino acids, antibiotics, phenolic compounds, polysaccharides, and sterols of biotechnological importance; and to produce nutritionally enriched feeds and single-cell proteins (SCPs), as well as from their suitability as agents of bioconversion, for example, their role in the preparation of mushroom compost. Their involvement in genetic manipulations is a much more recent development. Nevertheless, such studies as have been done (Cooney and Emerson, 1964; Emerson, 1968; Tansey and Brock, 1978; Mehrotra, 1985; Sharma, 1989; Johri and Ahmad, 1992; Satyanarayana et al., 1992; Sharma and Johri, 1992; Singh et al., 2016) indicate that thermophilic fungi appear to be nature-borne biotechnologists. *Biocatalyst* is a general term referring to the whole microorganism or its bioactive products. One of the major industrial biocatalysts from thermophilic fungi is enzymes, which act as catalysts controlling

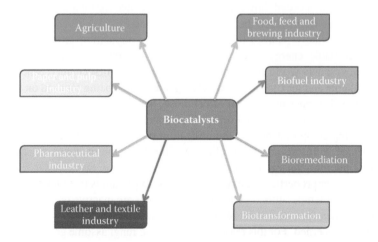

FIGURE 11.1   Potential industrial applications of thermophilic fungal biocatalysts.

biochemical processes. Figure 11.1 depicts a range of applications of biocatalysts from thermophilic fungi in various industries. The following section highlights some of the potential biotechnological applications of these fungi in various industries. Their detrimental activities are discussed at the end of this chapter.

11.2   EXTRACELLULAR THERMOSTABLE ENZYMES PRODUCED BY THERMOPHILIC FUNGI

Enzymes are biomolecules produced by living organisms to accelerate metabolic processes in the body and carry out a large number of biochemical interconversions. These are more akin to chemical catalysts in a chemical reaction, which help to accelerate the biochemical reactions extracellularly as well as intracellularly and are generally referred to as biocatalysts. Although uses of enzymes have been well known to humankind for centuries, the term *enzyme* (Greek ἔνζυμον, meaning "in leaven") was coined by Wilhelm Friedrich Kühne in 1887 for the first time (Kühne, 1877). Since ancient times, Egyptians have used enzymes for the preservation of food and beverages. Use of enzymes for making cheese dates back to 400 BC, when Homer's *Iliad* mentioned the use of a kid's stomach for making cheese. In his historic work on biogenesis (spontaneous generation of microbes) in 1783, Lazzaro Spallanzani suspected the importance of enzymes when he mentioned that there is a life-generating force inherent to certain kinds of inorganic matter that causes living microbes to create themselves given sufficient time (Vallery and Devonshire, 2003). Similarly, in 1812 Gottlieb Sigismund Kirchhoff, while working on the conversion of starch into glucose, highlighted the role of these biomolecules as catalysts (Asimov, 1982). The first enzyme, diastase, was discovered by French chemist Anselme Payen in 1833; its catalytic property to hydrolyze starch was later acknowledged by Swedish scientist Jöns Jacob Berzelius in 1835 (Payen and Persoz, 1833). Since then, a large number of enzymes have been discovered from living organisms, a major proportion of which come from microorganisms. The majority of industrial enzymes produced by thermophilic fungi are hydrolytic in action, being used for the degradation of various natural substances.

As mentioned in the previous section, one of the technical problems in industrial fermentation is the maintenance of temperature at optimal levels for the entire cultivation period. The use of thermophilic strains can be an effective solution to this problem (Grajek, 1987). Cost-effectiveness would be enhanced as a result of longer lifetimes of microorganisms and/or enzymes, particularly for immobilized systems. The greatest advantage of the use of thermostable enzymes in biotechnological processes operating at high temperature is the highly reduced chance of contamination with mesophilic microbes, in addition to decreasing the viscosity of the reaction medium and increasing the bioavailability and solubility of organic compounds (Gomes et al., 2016). Because of these and many more advantages, thermophilic fungi appear to be suitable candidates in biotechnological applications. Thermophilic fungi ubiquitously occurring in various natural environment habitats are known to produce a number of extracellular thermostable enzymes. Thermostable enzymes and the microorganisms producing them have been excellent topics for research during the last three decades; however, the interest in thermophiles and how their proteins function at high temperatures began in the 1960s by the pioneering work of Brock and his colleagues (Brock, 1986).

Enzymes from thermophilic fungi have been receiving considerable attention chiefly because of their potential to catalyze reactions at elevated temperatures in various industrial processes. These fungi produce an array of thermostable enzymes, including cellulases, amylases, proteases, phosphatases (including phytase), xylanases, lipases, and several other enzymes that are useful in industries like food and feed, textiles, detergents, leather, dairy, and pharmaceuticals. Depending on their intended use, the industrial enzyme market can be classified into three different sections: (1) technical enzymes, (2) food enzymes, and (3) animal feed enzymes. The largest section is of technical enzymes, where enzymes used for detergents and the pulp and paper industry constitute 52% of the total world market (Business Communications Company, 2004). The leading enzymes in this section are hydrolytic enzymes, such as proteases and amylases, which comprise 20% and 25% of the total market, respectively (Business Communications Company, 2004). Some of these enzymes, along with their properties and industrial applications, are discussed in this section.

11.2.1 Cellulases

Cellulose is the earth's most abundant renewable organic compound, with a wide range of applications. It may be found in relatively pure form, as in cotton, while in cell walls it is associated with hemicellulose, pectin, and lignin. Cellulose is a photosynthetic product and is found in nature in large quantities as agricultural, industrial, and urban wastes (Zhu et al., 1982). Cellulose is a linear polymer composed of D-glucose units linked to each other by β-1,4-glycosidic bonds, and it is present in the cell wall in the form of microfibrils. The number of glucose units per molecule ranges from as little as 15 to as high as 10,000–14,000. It is believed that cellulose molecules are deposited in an anti-parallel manner in cell walls. Cellulose is composed of two types of regions: crystalline and amorphous. Portions of several cellulose chains align themselves parallel to each other and form a highly crystalline region of cellulose called micelles. In contrast, portions of cellulose chains outside the micelles are not oriented in any specific manner and form the amorphous regions of cellulose (Agrios, 1988). Generally, crystalline cellulose is considered to be a superior carbon source for the induction of cellulose enzymes in thermophilic fungi compared with its amorphous form (McHale and Coughlan, 1981; Fracheboud and Canevascini, 1989). Thermophilic fungi stand out in the production of cellulase, especially when cultured under solid-state fermentation (SSF) using cellulosic residues as substrates (Pereira et al., 2015). The cellulase enzyme comprises a multicomponent enzyme system that catalyzes the hydrolysis of cellulose to smaller oligosaccharides (Figure 11.2).

It is now well established that the cellulose enzyme system in fungi is a multicomponent enzyme system consisting of three different enzymes:

1. Endoglucanase or carboxymethyl cellulase (endo-1,4-β-D-glucanase, EC 3.2.1.4), which nicks β linkages at random, usually in amorphous parts of

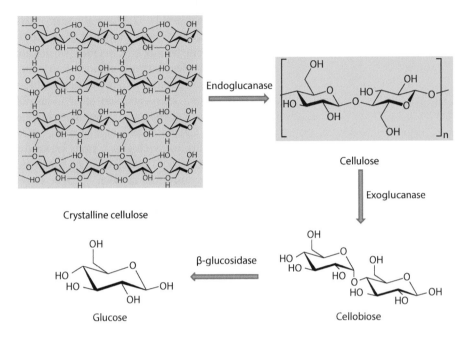

FIGURE 11.2 Schematic representation of sequential action of different cellulases on cellulose polymer.

the cellulose polymer, to decrease the length of the cellulose chain, thereby exposing reducing and nonreducing ends.

2. Exoglucanase or avicelase (exo-1,4-β-D-glucanase, EC 3.2.1.91), which acts on these ends to liberate cellooligosaccharides and cellobiose units from either the nonreducing or the reducing chain ends, usually from the crystalline part of the cellulose. This enzyme is comparatively rare, having been identified from celluloses of a few microorganisms. It is invariably associated with endoglucanases, and both the enzymes act synergistically.

3. β-Glucosidase (EC 3.2.1.21), which cleaves cellobiose and also small cellooligosaccharides, releasing glucose, thereby completing the enzymatic hydrolysis of cellulose (Sukumaran et al., 2009; Singhania et al., 2010). It is not considered a cellulose enzyme, as it has no direct action on cellulose; however, it enhances cellulase activity by acting on cellobiose, releasing glucose (Figure 11.2). β-Glucosidase also acts on cellodextrins and other β-linked glucose dimers (Grajek, 1987; Taj-Aldeen and Jaffar, 1992).

The world market for industrial enzymes was estimated at US$4411.6 million in 2013. This market is expected to reach US$7652.0 million by 2020. Globally, the maximum demand for enzymes is in the pharmaceutical industry, followed by food and feed and textiles. In all these industries, thermostable and thermotolerant cellulases play a pivotal role. Approximately 20% of the global enzyme market comprises glycoside hydrolases (GHs) that catalyze the release of monosaccharide from cellulose, hemicellulose, pectin, and starch (Bajpai, 2014). Traditionally, many GHs (cellulases, amylases, xylanases, and pectinases) have been produced from meso-philic microbes. In contrast, cellulase enzymes from thermophilic microorganisms are prized in view of their high titer and stability at elevated temperatures (Rao et al., 2013). In terms of end use, the cellulose market is segmented into health care, textiles, paper and pulp, food and beverages, wastewater, detergents, and other industrial applications. Cellulase is also being recognized as an alternative to antibiotics in the treatment of biofilms produced by *Pseudomonas*. Cellulases have also been employed for increasing the tensile strength of a sulfite pulp; recovering agar from seaweeds; saccharifying delignified cellulosic wastes (Virk et al., 1996); extracting green tea components; removing seed coats for isolating proteins from soybean, coconut, sweet potato, and corn starch; producing vinegar or pectin from the fruits or pulp of citrus; bleaching kraft pulp; and isolating protoplasts from fungi and higher plants (Ghose and Pathak, 1973; Doughlan, 1985; Satyanarayana et al., 1988). Recent potential applications of cellulases have been in the saccharification of organic wastes and city refuse, thereby contributing to minimizing the pollution problem.

It is now well established that the extracellular cellulase enzymes determine the extent of solubilization of cellulose present in wastes from agro-residues and other plant-derived wastes. The quest for the isolation of fungi capable of producing extracellular cellulases began in the late 1960s and early 1970s in an attempt to develop a process for enzymatic conversion of cellulose into glucose (Mandels, 1975; Mandels and Sternberg, 1976). The main focus was on the isolation of thermophilic fungi, as they produced enzymes with stability at elevated temperatures and better process operations.

A number of thermophilic fungi with potential cellulolytic abilities have been isolated from various substrates by several workers throughout the world. Fergus (1969) reported the cellulolytic activity of some thermophilic fungi. He observed that *Chaetomium thermophile* var. *coprophile*, *Chaetomium thermophile* var. *dissi-tum*, *Humicola grisea* var. *thermoidea*, *Humicola insolens*, *Myriococcum albomyces*, *Sportotrichum thermophile*, and *Torula thermophila* were able to degrade filter paper and utilize soluble carboxymethyl cellulose (CMC). Further, he reported that *Malbranchea pulchella* var. *sulfurea*, *Stlbella thermophila*, and *Talaromyces ther-mophilus* were unable to degrade filter paper; however, their cell-free extracts con-tained the Cx enzyme that could hydrolyze soluble CMC to reducing sugars. Tansey (1971) obtained similar results based on the clearing of acid-swollen cellulose. He reported rates of cellulolytic activity of thermophilic fungi at 45°C that were two to three times higher than those of the mesophilic *Chaetomium globossum* and *Trichoderma viride* at 25°C.

Flannigan and Sellars (1972) evaluated 30 thermophilous fungi for their ability to produce CMCase and found that 16 were able to degrade cellulose. Similarly, Rosenberg (1978) tested 21 species of thermophilic and thermotolerant fungi for their cellulolytic activity. It was observed by the author that *Chaetomium thermophile* var. *coprophile*, *Chaetomium thermophile* var. *dissitum*, *Humicola grisea*, *Humi-cola insolens*, *Myriococcum albomyces*, *Sportotrichum thermophile*, *Malbranchea pulchella*, *Allescheria terrestris*, *Allescheria fumigatus*, *Talaromyces thermophilus*, *Torula thermophila*, *Thielavia thermophile*, and *Chrysosporium pruinosum* showed positive cellulolytic activity. On the contrary, Jain et al. (1979) observed that *Humicola lanuginosa, Mucor meihei, Malbranchea pulchella* var. *sulfurea*, and *Talaromyces dupontii* did not degrade cellulose and filter paper when used as a sole source of carbon, but showed filter paper–degrading activity when grown on wheat straw. Based on these observations, they suggested that these fungi have some specific requirements for cellulose production that were fulfilled by growing on wheat straw. *Humicola lanuginosa* (syn. *Thermomyces lanuginosus*) was unable to synthesize cellulose, as observed by Chang and Hudson (1967), Fergus (1969), and Deacon (1985). However, some recent reports (Lee et al., 2014) indicate that *Thermomyces lanuginosus* is able to produce cellulase (Table 11.1).

Of the 15 thermophilc fungi tested by Srivastava et al. (1981), 6 were found to be celluloytic. They observed variations in the decomposition of cellulose by these fungi. Some of the fungal isolates were high decomposers, while others were weak decomposers. Thermophilic fungi such as *Mucor miehei, Mucor pusillus*, and *Rhizopus rhizopodiformis*, which were considered secondary sugar fungi, are confirmed to be moderately cellulolytic (Johri and Pandey, 1982). Tong and Cole (1982) observed *Thermoascus aurantiacus* as the most active cellulose producer among the several thermophilic fungi evaluated by them. Sandhu et al. (1985) screened 98 strains of thermophilous fungi for their ability to produce cellulases. Most of these fungi belonged to the genera *Acremonium, Aspergillus, Chaetomium, Penicillium, Thermoascus*, and *Thielavia*. They also revealed that cellulases produced by true thermophiles were more stable than those produced by thermotolerant and microthermophiles.

Grajek (1987), while working on comparative studies on the production of cellulases by six thermophilic fungi in liquid and SSF, observed that the best cellulose activities were depicted by *Thermoascus aurantiacus* and *Sporotrichum thermophile*. It was also observed that with the exception to *Allescheria terrestris*, all strains produced higher enzyme activities per unit of culture volume, with the advantage to SSF. Contrary to Grajek's observation, Morrison et al. (1987) observed high titers of β-glucosidase and low levels of endoglucanase by *Talaromyces emersonii* on lactose-containing liquid media. They also observed increased levels of cellulose in an ultraviolet (UV)-irradiated mutant of *Talaromyces emersonii* during growth on cellulose-, lactose-, and glucose-containing media. Stutzenberger (1990) reviewed thermostable fungal β-glucosidases from thermophilic fungi. The production of cellulases by thermophilic fungi grown on *Leptochloa fusca* straw was studied by Latif et al. (1995). During the screening of thermophilic fungi for wide-ranging lytic activity, a number of extracellular polysaccharases, including endoglucanase and β-glucosidase, were produced by a coal mine isolate of *Malbranchea cinnamomea* (Virk et al., 1996). Protoplast-regenerated variants (protoclones) of *Malbranchea cinnamomea* exhibited a marginal increase in the levels of polysaccharases under similar growth conditions. *Chaetomium thermophilum* var. *coprophilum* produced large quantities of extracellular as well as intracellular β-glucosidase when grown on cellulose or cellobiose (Venturi et al., 2002). The purification and characterization of cellulases from *Humicola grisea* and *Aspergillus fumigatus* have been studied by Takashima et al. (1996) and Ximenes et al. (1996).

An excellent review on thermophilic fungi by Maheshwari et al. (2000) highlighting various components of cellulase systems and their properties is highly useful

TABLE 11.1

Thermostable Cell-Free Enzymes Produced by Thermophilic Fungi for Various Biotechnological Applications

Enzyme	Organism	Optimal Temperature (°C)	Optimal pH	Application	Reference
Cellulase EC 3.2.1.4	*Thermoascus aurantiacus*	60	4.5	Detergent additives and biomass saccharification for ethanol production	Kawamori et al. (1987)
	Myceliophthora thermophila	65	4.8		Bhat and Maheshwari (1987)
	Sporotrichum thermophile	65	5.4		Kaur and Satyanarayana (2004)
	Thermomyces lanuginosus	65	6.0		Lin et al. (1999)
	Scytalidium thermophilium	–	–		Moriya et al. (2003)
	Melanocarpus albomyces	–	–		Hirvonen and Papageorgiou (2003)
	Talaromyces emersonni	75–80	5.5–5.8		Murray et al. (2004)
	Chaetomium thermophile var. *coprophile*	60	6.0		Ganju et al. (1990)
	var. *dissitum*	55	5.0		Erikson and Goksoyr (1977)
	Humicola insolens	50	5.6		Rao and Murthy (1988)
	Humicola grisea var. *thermoidea*	60	4.0–4.5		Filho (1996)
	Thermomucor indicae-seudaticae N31	75	–		Pereira et al. (2014)
Amylase EC 3.2.1.1	*Malbranchea cinnamomea*	65	6.5	Starch hydrolysis and dextrin production	Chadha et al. (1997), Han et al. (2013)
	Malbranchea sulfurea	–	–		Gupta and Gautam (1993)
	Thermomyces lanuginosus	70	4.4–6.6		Mishra and Maheshwari (1996), Chadha et al. (1997) Nguyena et al. (2002)
	Scytalidium thermophilum	65–70	5.5		Roy et al. (2000), Aquino et al. (2001)
	Thermomucor indicae-seudaticae	60	7.0		Kumar and Satyanarayana (2003)
	Humicola grisea	–	–		Campos and Felix (1995)

(Continued)

TABLE 11.1 (CONTINUED)
Thermostable Cell-Free Enzymes Produced by Thermophilic Fungi for Various Biotechnological Applications

Enzyme	Organism	Optimal Temperature (°C)	Optimal pH	Application	Reference
	Humicola grisea var. *thermoidea*	55	5.0		Tosi et al. (1993)
Lipase	*Thermomyces lanuginosus*	60	9.0	Dairy products, detergent additive, biodiesel production, leather industry, and pharmaceutical and fine chemical industries	Arima et al. (1968), Zheng et al. (2011)
EC 3.1.1.3	*Rhizomucor miehei*	–	–		Boel et al. (1988), Rao and Divakar (2002)
	Rhizopus arrhizus	–	–		Kumar et al. (1993)
	Absidia corymbifera	–	–		
	Humicola grisea	–	–		Satyanarayana (1978)
	M. pusillus	–	–		
	Humicola lanuginosa	–	–		Omar et al. (1987)
Protease	*Rhizomucor pusillus*	55	3.0–6.0	Baking, brewing, detergents, and leather industry	Arima et al. (1968)
EC 3.4.21.34	*Rhizomucor miehei*	–	4.1–5.6		Ottesen and Rickert (1970), Preetha and Boopathy (1997)
	Malbranchea pulchella var. *sulfurea*	45	8.5		Voordouw et al. (1974), Stevenson and Gaucher (1975)
	Thermomyces lanuginosus	70	5.0–9.0		Li et al. (1997), Jensen et al. (2002)
	Scytalidium thermophilum	37–45	6.5–8.5		Ifrij and Ogel (2002)
	Thermomucor indicae-seudaticae	65	5.5		Silva et al. (2014)
	Humicola lanuginosa	70	8.5		Stevenson and Gaucher (1975)
	Myceliophthora sp.	45	9.0		Zanphorlin et al. (2011)

(Continued)

TABLE 11.1 (CONTINUED)

Thermostable Cell-Free Enzymes Produced by Thermophilic Fungi for Various Biotechnological Applications

Enzyme	Organism	Optimal Temperature (°C)	Optimal pH	Application	Reference
Xylanase EC 3.2.1.8	*Penicillium dupontii*	60	2.5	Pulp and paper industry and biomass saccharification for ethanol production	Emi et al. (1976)
	Thermoascus aurantiacus	75–80	5.0		Ernest et al. (1987), Gomes et al. (1994), Kalogeris et al. (1998)
	Rhizomucor miehei	60–65	5.0–6.5		Song et al. (2013), Robledo et al. (2015)
	Thermomyces lanuginosus	50	6.0		Singh et al. (2000), Lee et al. (2014)
	Scytalidium thermophilum	80	6.0		Du et al. (2013)
	Myceliophthora thermophila	60–70	5.0		Katapodis et al. (2003), Moretti et al. (2012)
	Sporotrichum thermophile	70	5.0		Katapodis et al. (2003)
	Malbranchea cinnamomea	80			Fan et al. (2014)
	Chaetomium thermophile	70	4.8–6.0		Maheshwari et al. (2000)
	Melanocarpus albomyces	65	6.6		Prabhu and Maheshwari (1999), Narang et al. (2001)
	Talaromyces emersonii	60	2.5		Tuohy et al. (1993)
	Talaromyces byssochlamydoides	70	5.0		Maheshwari et al. (2000)
	Humicola lanuginosa	65	6.0		Anand et al. (1990)
	Chaetomium cellulolyticum	–	6.0–7.0		Baraznenoka et al. (1999)
Laccase EC 1.10.3.2	*Myceliophthora thermophila*	60	6.5	Pulp and paper industry and biomass	Berka et al. (1997)
	Chaetomium thermophilum	50	5–10		Chefetz et al. (1998)

(*Continued*)

TABLE 11.1 (CONTINUED)

Thermostable Cell-Free Enzymes Produced by Thermophilic Fungi for Various Biotechnological Applications

Enzyme	Organism	Optimal Temperature (°C)	Optimal pH	Application	Reference
Phytase EC 3.1.3.8	Sporotrichum thermophile	60	5.0	Saccharification for ethanol production	Singh and Satyanarayana (2009)
	Aspergillus fumigatus	55	5.5	Additive in monogastric feed	Wang et al. (2007)
	Thermomyces lanuginosus	65	7.0		Berka et al. (1998)
	Mycetiophthora thermophila	45	5.5		Mitchell et al. (1997)
	Rhizomucor pusillus	50	8.0		Chadha et al. (2004)
	Thermoascus aurantiacus	45	5.5		Nampoothiri et al. (2004)
Glucoamylase EC 3.2.1.3	Scytalidium thermophilum	70	5.5	Starch processing, food industry, and production of glucose syrup	Aquino et al. (2001), Ferreira-Nozawa et al. (2008), Du et al. (2013)
	Humicola grisea var. thermoidea	55	5.0		Tosi et al. (1993)
	Thermomucor indicae-seudaticae	60	7.0		Kumar and Satyanarayana (2007)
	Humicola lanuginosa	20–25	–		Taylor et al. (1978)
	Scytalidium thermophilum	60	6.5		Cereia et al. (2000)
	Thermomyces lanuginosus	70	5.0		Thorsen et al. (2006)
	Talaromyces emersonii	70	4.0–4.5		Nielsen et al. (2002)
	Chaetomium thermophilum	65	–		Chen et al. (2007)
Cellobiose dehydrogenase EC 1.1.99.18	Sporotrichum thermophile	60–65	4.0	Degradation of cellulose and lignin in pulp	Canevascini et al. (1991), Coudray et al. (1982)
	Humicola insolens	–	7.5–8.0		Igarashi et al. (1999), Schou et al. (1998)

(Continued)

TABLE 11.1 (CONTINUED)
Thermostable Cell-Free Enzymes Produced by Thermophilic Fungi for Various Biotechnological Applications

Enzyme	Organism	Optimal Temperature (°C)	Optimal pH	Application	Reference
α-D-glucuronidase EC 3.2.1.139	*Thermoascus aurantiacus*	65	4.5	Pulp and paper industry	Khandke et al. (1989b)
Polygalacturonase EC 3.2.1.15	*Aspergillus fumigatus*	50	–	Food industry, textile	Phutela et al. (2005)
	Thermoascus aurantiacus	60–65	4.5–5.5	processing, degumming	Inamdar (1987), Martins et al. (2013)
	Sporotrichum thermophile	–	–	of plant rough fibers,	Kaur et al. (2004)
	Rhizomucor pusillus	–	–	and treatment of pectic	Siddiqui et al. (2012)
	Acremonium cellulolyticus	50	–	wastewaters	Gao et al. (2014)
D-Glucosyltransferase EC 2.4.1.24	*Talaromyces dupontii*	70	4.5	Used for the production of α-transglucosidase	Bousquet et al. (1998)
Phosphatase EC 3.1.3.1	*Acremonium alabamensis*	–	–	Label for enzyme	Satyanarayana et al. (1985)
	Rhizopus rhizopodiformis	–	–	immunoassay, and in	Satyanarayana et al. (1985)
	Scytalidium thermophilum	70–75	9.5–10	dairy industry as a marker of pasteurization in cow milk	Guimaraes et al. (2001)
Pectinase EC 3.2.1.15	*Aspergillus fumigatus*	50	–	Paper and pulp industry,	Phutela et al. (2005)
	Thermomucor indicae-seudaticae N31	45	–	food industry, textile	Martin et al. (2010)
	Thermomyces lanuginosus	–	–	processing, and poultry	Puchart et al. (1999)
	Myceliophthora thermophila	55	7.0	feed	Kaur and Satyanarayana (2004)
	Rhizomucor pusillus	55	5.0		Siddiqui et al. (2012)
	Thermoascus aurantiacus	65	5.5		Martins et al. (2013)

for academicians and industrialists. More recently, biocatalysts from thermophilic fungi, including cellulose enzymes, have been reviewed by Singh (2014), Singh and Satyanarayana (2014), Singh et al. (2016), and Schuerg et al. (2017). The heterologous expression of genes of thermophilic fungi in more suitable industrial hosts has been studied. Karnaouri et al. (2014) functionally expressed an endoglucanse gene from the thermophilic fungus *Myceliophthora thermophila* in a methylotrophic yeast, *Pichia pastoris*. The purified enzyme depicted high enzymatic activity on substrates containing β-1,4-glycosidic bonds, such as CMC, barley β-glucan, and cellooligosaccharides.

The ever-increasing world energy demand has forced scientists to develop cost-effective processes for the production of renewable biofuels, such as biogas, bioethanol, fuel cells, and biodiesel. Bioethanol is one of the most common renewable fuels and is regarded as the "energy for the future." In countries like the United States and Brazil, ethanol is being produced from starch and sucrose (cane sugar), respectively. Cellulases are increasingly employed for the conversion of lignocellulosic materials into fermentable sugars, such as pentoses and hexoses. It has been estimated that 25%–50% of the total production cost of cellulosic ethanol is attributed to cellulase enzyme, which is used for the saccharification of lignocellulosic materials. The main reasons are the low yield of enzymatic hydrolysis and the low stability of commercial enzymes (Celluclast®, Accellerase 1000®, and Spezyme®) marketed today (Gomes et al., 2016). These enzymes are stable up to 50°C, beyond which little or no activity is retained (Pribowo et al., 2012). However, they tolerate neither the climatic conditions of tropical countries nor the conditions in industrial application (Dias et al., 2012; Moretti et al., 2014). Thus, the need to develop commercial thermostable cellulases usually harnessed from thermophilic microorganisms is imminent in view of the rising demand of ethanol as a substitute energy biofuel.

11.2.2 AMYLASES

Thermostable starch-hydrolyzing enzymes from microorganisms play a great role in the industrial production of glucose from starch. Starch represents one of nature's renewable polymeric heterogeneous molecules from plants and is composed of a linear amylose and a branched amylopectin. As a reserve compound in plants, it is mostly located, in the form of starch granules, in the cytosol of cells and in seeds. Owing to its complex structure, amylose is insoluble in water and needs to be liquefied at high temperature in order to make it a usable substrate for biocatalytic hydrolysis. Both amylose and amylopectin account for 15%–25% and 75%–85% of the polysaccharide, respectively. Amylose is composed of 1000–6000 α-1,4-glucopyranose residues, whereas amylopectin comprises 6×10^6 molecules of glucose, with the actual number of monomers depending on the plant source, and is water soluble.

Amylases are classified into three subclasses, α-, β-, and γ-amylase, of which α-amylase is the most important biocatalyst. α-Amylase (EC 3.2.1.1, α-1,4-glucan-4-glucanohydrolase) is an endo-acting enzyme, randomly cleaving α-1,4 linkages in the interior of starch molecules. A-Amylases have been assigned to GH families 13, 57, and 119. Owing to their stability at elevated temperatures, amylases derived from thermophilic organisms, particularly thermophilic fungi, are advantageous in biotechnological processes. Various applications of amylases in different industries, such as in food, bread making, paper, textiles, sweeteners, glucose and fructose syrups, fruit juices, detergents, fuel ethanol from starches, alcoholic beverages, digestive aids, and stain removers in dry cleaning, make them precious enzymes, particularly in biotechnology industries (Gurung et al., 2013). They are particularly useful in starch processing industries, degrading starch into smaller, more valuable products, such as glucose, maltose, fructose, or dextrin. There is also a growing demand for tailor-made amylases to be produced in bulk quantities for various biotechnological applications.

Amylases are produced by different species of microorganisms; however, for commercial application amylases derived from bacteria and fungi are most suitable. For major groups of enzymes, thermostability is a desirable property. Current research focuses on thermostable amylolytic enzymes from thermophilic fungi. Fungal α-amylases are usually preferred, as they are produced from fungi that are generally recognized as safe (GRAS). Several thermophilic fungi, such as *Thermomyces lanuginosus*, *Malbranchea sulfurea*, *Mucor meihei*, *Mucor pusillus*, *Sporotrichum thermophile*, and *Torula thermophila*, are known to produce high levels of extracellular amylolytic enzymes under both surface and submerged conditions (Adams and Deploey, 1976; Subrahmanyam et al., 1977; Rao et al., 1979, 1981; Jensen et al., 1987, 1988; Kanlayakrit et al., 1987; Haasum et al., 1991; Jensen and Olsen, 1992).

Fergus (1969) reported amylase activity in 6 of the 15 thermophilic fungi that he tested by flooding with Lugol's iodine solution. *Humicola insolens*, *Humicola lanuginosa*, *Humicola stellata*, *Malbranchea pulchella* var. *sulfurea*, *Mucor pusillus*, and *Talaromyce thermophilus* produced considerable amounts of amylase. Adams and Deploey (1976) related amylase production by *Mucor meihei* and *Mucor pusillus* to the type of carbohydrate present in the medium. The highest level of amylase activity was observed when these fungi were incubated in media containing starch as the only source of carbohydrate. However, amylase production could be induced when the media contained other carbohydrates, such as glucose, fructose, and lactose, but the quantity of amylase produced was less than when starch was used as a substrate. Similarly, the effect of temperature and pH on the stability of amylase has been reported by Deploey et al. (1982). Jensen et al. (1987) obtained an eightfold increase in amylase activity by *Thermomyces lanuginosus* when low-molecular-weight dextrans were used in contrast to starch as the carbon source. Detailed studies pertaining to physiological properties of purified α-amylase have been published for *Mucor pusillus* (Somkuti and Steinberg, 1980), *Talaromyces emersonii* (Bunni et al., 1989), and *Thermomyces lanuginosus* (Jensen et al., 1988; Jensen and Olsen, 1992). The thermostable amylase from the genus *Bacillus* and the thermolabile fungal amylases of *Aspergillus niger* and *Aspergillus oryzae* are the most widely used (Glazer and Nikaido, 1995; Carlsen et al., 1996). α-Amylases of the thermophilic fungus *Thermomyces lanuginosus* strain IISc 91 have been characterized by Mishra and Maheshwari (1996). As the partial amino acid sequence of α-amylases from this strain depicted a single N-terminal amino acid, the authors suggested that the native enzyme is a homodimer with a molecular mass of ~42 kDa. This was the first report by these authors of a dimeric form of α-amylases in fungi. The enzyme produced exceptionally high levels of maltose from raw potato starch (Mishra and Maheshwari, 1996). Thus, the thermophilic fungus *Thermomyces lanuginosus* is an excellent producer of amylase. Jensen et al. (2002) and Kunamneni et al. (2005) have also purified the α-amylase from this fungus with a high degree of thermostability. A novel α-amylase from the thermophilic fungus *Malbranchea cinnamomea*, displaying optimal activity at pH 6.5 and 65°C with a molecular mass of 60.3 kDa, was studied by Han et al. (2013). The enzyme exhibited stability over a pH range of 5.0–10.0 and also showed wide substrate specificity.

11.2.3 GLUCOAMYLASE

Glucoamylase (EC 3.2.1.3) belongs to the class of amylolytic enzymes hydrolyzing α-1,4-glycosidic linkages and, less often, α-1,6-glycosidic linkages from the non-reducing end of starch, producing β-D-glucose. It is commonly used for the manufacture of high-fructose corn syrup (HFCS), in the processing of starch, and in food industry for the clarification of fruit juices. Several thermophilic fungi are known to produce glucoamylase in culture, including *Thermomyces lanuginosus*, *Humicola grisea* var. *thermoidea*, *Humicola lanuginose*, *Thermomucor indicae-seudaticae*, *Scytalidium thermophilum*, and *Myceliophthora thermophila* (Taylor et al., 1978; Rao et al., 1979; 1981; Tosi et al., 1993; Aquino et al., 2001; Kumar and Satyanarayana, 2007). Haasum et al. (1991) studied the production of glucoamylase in *Thermomyces lanuginosus* in a synthetic medium and observed that although maltose had been reported to be a better inducer of glucoamylase in this fungus, growth in a starch medium allowed both glucoamylase and α-amylase to be produced simultaneously. Despite the variation in the molecular mass of the enzyme as reported by different authors using different techniques—57 kDa by gel filtration and sodium dodecyl sulfate–polyacrylamide gel electrophoresis (SDS-PAGE) (Romanelli et al., 1975), 70–77 kDa by SDS-PAGE (Jensen et al., 1988), 45 kDa by SDS-PAGE (Moloney et al., 1985), and 72 kDa by SDS-PAGE (Duo-Chuan et al., 1998)—it was found to be thermostable at 60°C for 7 h (Mishra and Maheshwari, 1996). Chen et al. (2007) cloned a gene from *Chaetomium thermophilum* encoding thermostable glucoamylase and successfully expressed it in *Pichia pastoris*.

11.2.4 XYLANASES

Hemicellulose is the major component in plant materials and accounts for about 35% of the total dry weight. Hemicellulose, unlike cellulose, is a class of highly branched heteropolysaccharides. Hydrolysis of hemicelluloses using microbial enzymes is an important step toward utilizing them. Hemicelluloses are classified based on the type of monomers united to form heteropolymers and are named xylans, mannans, galactans, or arabinans (Deutschmann and Dekker, 2012). The xylanolytic enzyme complex includes an endo-1,4-β-D-xylanase (EC 3.2.1.8), which randomly cleaves the β-1,4-glycosidic bonds within the xylan molecule, producing xylooligosaccharides, which are converted into xylose by 1,4-β-D-xylosidase (EC 3.2.1.37). Further, β-xylosidases catalyze the breakdown of xylobiose and attack the nonreducing ends, thereby releasing xylose (Bajpai, 2009). This complex was completely characterized in *Talaromyces byssochlamydoides* Yh-50 (Yoshioka et al., 1981). Various research papers and reviews suggest that the majority of thermophilic fungi produce endoxylanases of the xylan-degrading enzyme complex (Maheshwari et al., 2000).

The potential application of xylanase stems from its ability to degrade xylan, a major constituent of woody plants. Xylanases are used commercially for the production of xylose from plant biomass, to liquefy coffee mucilage for the manufacture of liquid coffee, and to change the organoleptic and rheological properties of must and wines (Satyanarayana et al., 1992). Since this enzyme also catalyzes isomerization of glucose to fructose, it is commercially used in the production of HFCS as a "glucose isomerase" (Jensen and Rugh, 1987). Xylanases can be used in the pretreatment of cellulosic substrates in order to remove hemicelluloses so that a greater surface area of cellulose may be available for the action of cellulases. Both endo- and exoxylanases have been extensively employed in various biotechnological processes, for example, biobleaching of paper and pulp, production of animal feed, clarification of beer and fruit juices (Katapodis et al., 2006), and bread making (Kulkarni et al., 1999). They have also been widely used for the production of packaging films based on xylan (Hansen and Plackett, 2008) and the production of biodegradable surfactants (Damez et al., 2007). Recently, xylanases have also been used for the production of bioactive xylooligosaccharides, which are considered prebiotics (Rustiguel et al., 2010). The application of xylanases in biomass saccharification to produce second-generation ethanol is a recent innovation of the biorefineries (Kim and Kim, 2014) that will accelerate the demand of these enzymes 100-fold during the next decade. Recently, much emphasis has been put on human health–promoting bioactive compounds of natural origin. Xylanases have been exploited for the production of prebiotic compounds, such as bioactive xylooligosaccharides (Rustiguel et al., 2010). A high yield of xylooligosaccharides has been obtained from corncobs by the hydrolytic action of thermostable xylanase from *Paecilomyces thermophila* J18 at 70°C (Teng et al., 2010). Xylanases are also produced on a commercial scale as additives in feed for poultry (Niehaus et al., 1999), as well as additives to wheat flour for improving the quality of baked products (Wong and Saddler, 1992).

Several studies have reported the production of thermostable xylanase from thermophilic fungi. In particular, thermophilic fungi such as *Absidia ramosa, Acremonium albamensis, Aspergillus fumigatus, Thermomyces lanuginosus, Melanocarpus albomyces, Rhizomucor pusillus, Rhizomucor miehei, Thermomucor indicae-seudaticae, Sporotrichum thermophile, Thermoascus aurantiacus*, and *Thielavia terrestris* have been reported to produce xylanases with activity from 50°C to 80°C (Margaritis et al., 1983; Satyanarayana and Johri, 1983; Maheshwari and Kamalam, 1985; Matsuo and Yasui, 1985, 1988; Dubey and Johri, 1987; Grajek, 1987; Tan et al., 1987; Merchant et al., 1988; Satyanarayana et al., 1988; Gilbert et al., 1992, 1993; Banerjee et al., 1994, 1995; Gomes et al., 1994; Hoq et al., 1994; Saha, 2002; Damaso et al., 2003; Lee et al., 2009; Maalej et al., 2009; Zhang et al., 2011; Sadaf and Khare, 2014; Robledo et al., 2015).

Flannigan and Sellars (1972) reported that thermophilic fungi, that is, *Absidia ramosa, Aspergillus fumigatus, Humicola lanuginose, Mucor pusillus*, and *Thermoascus aurantiacus*, strongly degraded arabinoxylan. Similarly, Satyanarayana and

Johri (1983) screened six isolates of thermophilic fungi from paddy straw compost for their xylanolytic activity and observed that *Sporotrichum thermophile* and *Acremonium albamensis* exhibited strong xylanolytic activity in comparison with other isolates. The enzyme was produced in xylose- or xylose-containing hemicelluloses, indicating its inducible nature. Maheshwari and Kamalam (1985) investigated xylanase production by *Melanocarpus albomyces* and reported a higher enzyme titer in shake flasks than in a static fermenter. Extracellular xylanase of five isolates of thermophilic *Sporotrichum* sp. and *Myceliophthora thermophilum* was studied by Dubey and Johri (1987) in media containing wheat straw, wheat bran, paddy straw, paddy husk, sugarcane bagasse, and local grass. They observed that *Sporotrichum* sp. strain 1 and wheat straw as substrate were ideal for higher xylanase production.

A comparison of two xylanases (xylanase II from *Thielavia terrestris* 255B and the 32-KDa xylanase from *Thermoascus crustaceus* 235E) was made by Gilbert et al. (1993) to determine if they had different and complementary modes of action when they hydolyzed various types of xylans. An assessment of the combined hydrolytic action of the two xylanases did not reveal any synergistic action. However, the two xylanases were able to remove about 12% of the xylan remaining in an aspen kraft pulp. Similarly, Banerjee et al. (1995) evaluated 13 thermophilic fungal strains for the production of xylanase. However, only five species, *Malbranchea pulchella* var. *sulfurea*, *Sporotrichum thermophile*, *Thielavia terrestris*, *Humicola insolens*, and *Acremonium albamensis*, produced high levels of xylanolytic enzymes. Pereira et al., (2015) isolated 32 filamentous fungi with the ability to grow at 45°C and screened 26 isolates for the production of xylanase. They found that the thermophilic *Myceliophthora thermophila* JCP 1-4 was the best producer of xylanase and can be potentially exploited in the saccharification of sugarcane bagasse. Recently, Ahirwar et al. (2017) isolated 19 thermophilic and thermotolerant fungi from various self-heated habitats based on their ability to grow at 45°C and identify potential xylanase and mannanase producers. They observed that the highest xylanase production was obtained by the cultivation of *Malbranchea cinnamomea* NFCCI 3724, followed by *Melanocarpus albomyces*, *Aspergillus terreus*, and *Myceliophthora thermophila* NFCCI 3725.

In the recent past, genome-wide studies of xylanases from different thermophilic fungi have yielded information that can be exploited for the cloning and heterologous expression of these enzymes. N-terminal disulfide bridges between positions 1 and 24 of the *Thermomyces lanuginosus* DSM 10635 GH11 xylanase have been engineered by Wang et al. (2012) and subsequently produced in *Escherichia coli*. The authors observed that the engineered enzyme version exhibited increased optimal temperature, higher activity, and thermostability in alkaline conditions.

Song et al. (2013) cloned two novel xyloglucanase genes from the thermophilic fungus *Rhizomucor miehei* having open reading frames of 729 bp encoding 242 amino acids. Both of these genes, without signal peptides, were cloned and successfully expressed in *Escherichia coli*, having similar molecular masses. Enzyme A depicted optimal activity at pH 6.5 and 65°C, whereas enzyme B showed optimal activity at pH 5.0 and 60°C. Similarly, three xylanase genes (*xynA*, *xynB*, and *xynC*) were identified in *Humicola insolens* Y1 by Du et al. (2013). When these genes were expressed in *Pichia pastoris*, they showed optimal activity at pH 6.0–7.0 and

70°C–80°C. They further reported that xynA exhibited better alkaline adaptation and thermostability and showed higher catalytic efficiency and wider substrate specificity than the other two genes.

11.2.5 Lipases

Lipases (EC 3.1.1.3) are a subclass of the esterases that catalyze the hydrolysis of triacylglycerols and the synthesis of esters from glycerols and long-chain fatty acids. In living organisms, lipases perform essential roles in the digestion, transport, and processing of dietary lipids (Gurung et al., 2013). In recent years, interest has increased in the use of lipases, which have applications in the fields of nutrition, food technology, clinical medicine, detergents, and analytical chemistry. Lipase in commercially available microbial protease preparations has advantages in flavor development in some types of cheeses (Somkuti and Babel, 1968). Microbial lipases from thermophilic fungi are also used in digestive aids, dry cleaning, and sewage disposal (Satyanarayana and Johri, 1981). Lipases from thermophilic fungi have found diverse industrial applications, ranging from their use in the textile industry for desizing cotton fabrics and denim during their commercial preparation to their role in the detergent industry, as they are mainly used in laundry and household dishwasher detergents. In laundry detergents, thermophilic fungal lipases are alkaline and thermostable at pH 10–11, and the temperatures range from 30°C to 60°C. In fact, *Humicola lanuginosa* lipase (Lipolase, Novo Industri A/S) is used in detergent formulations in conjunction with other microbial enzymes, such as protease, amylase, and cellulase. Lipases are also used in the dairy industry for the ripening of cheese and modification of milk fats. Lipase-producing thermophilic fungi have also found application in the treatment of effluents rich in fatty materials originating from the oil industries. Thermophilic fungi such as *Rhizopus arrhizus* (Kumar et al., 1993), *Thermomyces lanuginosus* strain Y-38, *Rhizomucor miehei*, and *Thermomyces lanuginosus* (Noel and Combes, 2003) are reported to produce lipases in submerged fermentation that can be used for treating the effluent from the oil industries. As stated above, lipase can hydrolyze and degrade triglycerides; immobilized lipase is a promising alternative biocatalyst in the transesterification process for biodiesel production, which may lead to energy and chemical savings, as well as avoiding soap formation (Noureddini et al., 2005).

Several thermophilic fungi, such as *Mucor pusillus*, *Humicola lanuginosa*, *Humicola grisea* var. *thermoidea*, and *Talaromyces thermophilus*, are reported to produce lipase (Somkuti and Babel, 1968; Somkuti et al., 1969; Arima et al., 1972). Oso (1974a) isolated seven thermophilic fungi from stacks of oil palm kernels in Nigeria: *Chaetomium thermophile* var. *coprophile* and var. *dissitum*, *Humicola lanuginosa*, *Mucor pusillus*, *Talaromyces emersonii*, *Thermoascus aurantiacus*, and *Thermoascus crustaceus*. Later, he (Oso, 1974b) observed that six isolates could utilize a number of fatty acids as the sole source of carbon. Adams and Deploey (1978) studied extracellular enzyme production by 10 thermophilic fungi. They observed that all the isolates produced a detectable amount of lipase. Similarly, Ogundero (1980) studied the production of lipases by thermophilic fungi, namely, *Mucor pusillus* and *Talaromyces thermophilus*, associated with Nigerian groundnuts.

The author observed the highest activity of lipase at 45°C, with an optimum pH ranging from 5.5 to 6.0. Satyanarayana and Johri (1981) evaluated 14 thermophilic fungi isolated from paddy straw compost for their ability to produce lipases. They observed that only eight species, *Absidia corymbifera, Acremonium albamensis, Aspergillus fumigatus, Chaetomium thermophile, Humicola lanuginosa, Mucor pusillus, Sporotrichum thermophile*, and *Thielavia minor*, were able to degrade butter and coconut oil that contained saturated fatty acids with conjugated double bonds. Their observation supported the role of lipase-producing microorganisms in the decomposition process, besides the fact that thermophiles are, in general, more active than their mesophilic counterparts. Lipases of thermophilic fungi have been elegantly reviewed by Johri et al. (1985), Satyanarayana and Johri (1992), and Satyanarayana et al. (1992). A lipase gene from *Thermomyces lanuginosus* HSAUP0380006 was cloned by Zheng et al. (2011) using reverse transcriptase polymerase chain reaction (RT-PCR) and rapid amplification of cDNA ends (RACE) amplification, which predicted a 292-residue protein with a 17-amino-acid signal peptide. Subsequently, the authors achieved a high-level expression for recombinant lipase in *Pichia pastoris* GS115 under the control of a strong AOX1 promoter. The enzyme was purified to homogeneity and had a molecular mass of 33 kDa. The enzyme was found to be stable at 60°C with an optimum pH of 9.0 and stability from pH 8.0 to 12.0. The cell-free enzyme hydrolyzed and synthesized esters efficiently.

11.2.6 PROTEASES

Proteases (EC 3.4.23.6) are enzymes that degrade proteins and are classified on the basis of (1) the critical amino acid required for the catalytic activity (e.g., serine protease), (2) the optimum pH for their activity (e.g., acidic, neutral, or alkaline), (3) their site of cleavage (e.g., aminopeptidase), and (4) their requirement of a free thiol group (e.g., thiol proteinase). Proteases have been extensively used in food processing industries, such as the dairy industry to enhance cheese ripening, and the baking industry to modify gluten quality in biscuits, cookies, and crackers; therefore, various proteases are used to treat "bucky" dough for improving elasticity (van Oort, 2010). Proteases have been frequently employed in meat tenderization, the production of fish protein hydrolysates, viscosity reduction, skin removal, and roe processing. They are also used in detergent industries for protein stain removal, in the leather industry for the removal of unwanted proteins during the soaking and liming process, and in cosmetics and personal care to impart a gentle peeling effect in skin care products. Microbial proteases have found application as a substitute for papain, pepsin, and rennin in the food industry (Johri and Ahmad, 1992; Satyanarayana et al., 1992).

Several thermophilic fungi, such as *Chaetomium thermophile, Mucor pusillus, Mucor miehei, Penicillium dupontii, Sporotrichum thermophile*, and *Malbranchea pulchella* var. *sulfurea*, produce extracellular proteases in liquid cultures (Somkuti and Babel, 1968; Karavaeva et al., 1975; Ong and Gaucher, 1976; Khan et al., 1979, 1983; Mehrotra, 1985). The extracelllular proteases of thermophilic fungi have greater thermostability than their mesophilic counterparts (Ong and Gaucher, 1973). An extensive screening of about 800 microorganisms led Arima et al. (1968) to isolate high-protease-producing *M. pusillus,* which produced an enzyme with a high ratio of milk clotting to proteolytic activity. Later, Ottesen and Rickert (1970) isolated and characterized an acid protease produced by *M. miehei.* The milk-clotting activity of the enzyme was found to be due to its selective attack on the κ-casein fraction, which stabilizes the casein micelle in milk. It is observed that both *M. pusillus* and *M. miehei* rennins hydrolyzed peptide bonds in synthetic peptides with an aromatic amino acid as the carboxyl donor (Arima et al., 1970; Sternberg, 1972). An extracellular protease was produced by *Scytalidium thermophilum* when it was grown on microcrystalline cellulose and was found to be optimally active at pH 6.5–8.0 and 37°C–45°C (Ifrij and Ogel, 2002). Similarly, Zanphorlin et al. (2011) isolated and characterized an alkaline protease from the thermophilic *Myceliophthora* sp. The enzyme was identified as serine protease and had a molecular mass of 28.2 kDa, as determined by matrix-assisted laser desorption/ionization time-of-flight mass spectrometry (MALDI-TOF MS), with optimal activity at pH 9.0 and 45°C. The gamut of proteases varies between different species, as well as with nutrient availability and pH conditions in the culture medium. Protease activity particularly increases when the supply of available nitrogen is limited.

Owing to advances in fungal genetic engineering and genomics, the production of proteases in heterologous hosts offered the possibility of their large-scale production. Gray et al. (1986) reported recombinant production of *Mucor miehei* protease in *Aspergillus nidulans*, which was found to be similar in specific activity to that produced by *Mucor miehei*. However, the recombinant enzymes pose a major safety risk and are often associated with respiratory allergies; in the case of proteases, the minor risk involves skin and eye irritation.

11.2.7 PECTINASES

Pectic substances such as pectin are composed of polygalacturonic acid linked by α-1,4 linkages, which play an important role in maintaining the integrity of the primary cell wall in plants and are the main components of the middle lamella. Pectinase broadly refers to enzymes such as pectozyme, pectolyase, and polygalacturonase, constituting about 25% of the global food enzyme market. Further, pectinases comprise enzymes that catalyze the degradation of pectic substances by depolymerization and deesterification reactions. Pectinases include protopectinases A and B; endopolygalacturonases (EC 3.2.1.15) that hydrolyze glycosidic α-1,4 bonds along the polygalacturonan network in a random fashion; exopolygalacturonases (EC 3.2.1.67) that hydrolyze the nonreducing end of the polymer, generating galacturonic acid; endopolygalacturonase lyase (EC 4.2.2.2) and exopolygalacturonase lyase (EC 4.2.2.9), which cleave α-1,4 bonds at the end of pectin, liberating unsaturated products from the reducing end of the polymer; and pectin methyl esterase (EC 3.1.1.11), which catalyzes the hydrolysis of the methyl esters of galacturonic acid (Pedrolli et al., 2009).

The pectic polysaccharides have long been used as bioactive food ingredients, as a supplement in children's food, and as a thickening agent in the jelly industry. Pectinases are employed in the maceration and solubilization of fruit pulps and their clarification (Naidu and Panda, 1998). Other applications of pectinases include bioscouring of cotton fibers, textile processing, use in the paper and pulp industry, degumming of plant bast fibers, treatment of wastewaters containing pectic substances, coffee and tea fermentation, and poultry feed.

Thermophilic fungi known to produce pectinases include *Penicillium dupontii, Humicola stellata, Humicola lanuginosa, Humicola insolens, Mucor pusillus* (Craveri et al., 1967), *Talaromyces thermophilus* (Tong and Cole, 1975), Rhizomucor pusillus (Siddiqui et al., 2012), and *Thermoascus aurantiacus* (Martins et al., 2013). Secretion of pectinolytic enzymes by thermophilic fungi, that is, Sporotrichum thermophile, *Mucor miehei, Papulaspora thermophila, Talaromyces leycettanus,* and *Thermoascus aurantiacus,* was reported by Adams and Deploey (1978). *Sporotrichum thermophile* (syn. *Myceliophthora thermophila*) has recently emerged as a strong candidate for the production of a large number of thermostable enzymes of biotechnological potential (Singh et al., 2016; Singh and Satyanarayana, 2016) and biomolecules such as thiol protease inhibitors (Yaginuma et al., 1989), antimicrobial xylooligosaccharides (Christakopoulos et al., 2003), and fructooligosaccharises (Katapodis et al., 2004).

11.2.8 PHYTASES

Phytases (*myo*-inositol hexakisophosphate phosphohydrolase, EC 3.1.3.8) are the phosphatases that catalyze the hydrolysis of phytic acid (*myo*-inositol hexakisophosphate) to inorganic phosphate and *myo*-inositol phosphate derivatives such as mono-, di-, tri-, tetra-, and pentaphosphates. Phosphorus is a major essential macronutrient required for biological growth and development and is an important constituent of the cell and its components. Phytic acid, an anti-nutritional factor present in cereals, legumes, nuts, and oil seeds, is the principal storage form of phosphorus. As phytic acid is indigestible in monogastric animals due to the lack of phytate-degrading enzymes in their gastrointestinal tract, it needs to be hydrolyzed in order to increase the level of phosphorus in animal feed. Due to these problems, considerable interest in phytate-hydrolyzing enzymes from microorganisms has been generated in the food- and feed-producing industries. Supplementing the food with phytase will improve the availability of phosphorus and hence its mobilization (Singh et al., 2011). Treatment of the feed with phytase is known to increase the available phosphate, which helps phosphate uptake and reduces the risks associated with phosphate runoff from feedlots. Phytases from thermophilic fungi are commercially desirable due to their thermostability and high reaction rates. In addition to their potential application in the supplementation of food and feed, phytase can also be used in pharmaceuticals and soil amendments and for plant growth promotion.

Several thermophilic fungi are known to produce phytases in SSF and submerged fermentation (SmF) including *Chaetomium thermophilum* (ATCC5 8420), *Rhizomuco miehei* (ATCC 22064), *Thermomucor indicae-seudaticae* (ATCC 28404), and *Myceliophthora thermophila* (ATCC 48102), as reported by Mitchell et al. (1997). Chadha et al. (2004) isolated and screened a number of thermophilic fungi (*Rhizomucor pusillus, Scytalidium thermophilum, Melanocarpus albomyces, Chaetomium thermophile,* and *Thermomyces lanuginosus*) for the production of phytase. Nampoothiri et al. (2004) reported the production of phytase from *Thermoascus aurantiacus* in a medium containing glucose, starch, and wheat bran. Similarly, *Sporotrichum thermophile* produced HAP-phytase in solid-state (Singh and Satyanarayana 2006a) and submerged (Singh and Satyanarayana 2006b) fermentations. Recently, Bala et al. (2014) described the production of phytase from *Humicola nigriscens*, a thermophilic mold. Phytase genes from thermophilic fungi, such as *Thermomyces lanuginosus* (Berka et al., 1998), *Myceliophthora thermophila*, and *Talaromyces thermophilus* (Wyss et al., 1999), have been cloned and overexpressed in heterologous hosts.

11.2.9 PHOSPHATASES

Phosphatases are a group of hydrolytic enzymes that catalyze phosphomonoesters, making inorganic phosphorus available to organisms. On the basis of their pH optima, phosphatases have been classified as alkaline phosphatase (EC 3.1.3.1) and acid phosphatase (EC 3.1.3.2). Phosphatases are widely distributed in nature and generally exhibit broad substrate specificity. Besides their role in the food and feed industry, phosphatases are also employed in the dairy industry for ascertaining the pasteurization of milk (Singh and Satyanarayana, 2013).

A large number of thermophilic fungi, including Rhizopus micromyces, Rhizomucor pusillus, Talaromyces thermophilus, Papulaspora thermophila, Thermomyces lanuginosus, Acremonium thermophile, Thermoascus aurantiacus, and Chrysosporium thermophilum, were screened by Bilai et al. (1985) for the production of acid phosphatases; however, only two strains produced acid phosphatase. Similarly, Satyanarayana et al. (1985) also screened a number of thermophilic fungi for the production of phosphatases. However, only two species, namely, Acremonium albamensis and Rhizopus rhizopodiformis, produced acid phosphatase, while other fungi produced both acid and alkaline phosphatases. Guimaraes et al. (2001) investigated the production of an extracellular (conidial) and an intracellular (mycelial) alkaline phosphatase from Scytalidium thermophilum. The enzyme was later purified using DEAE-cellulose and Concanavalin A–Sepharose chromatography.

Both phytases and phosphatases act in a coordinated manner. While phytases cleave phytic acid selectively, phosphatase breaks down the inositol phosphate intermediates independently, thereby accelerating the total dephosphorylation process.

11.2.10 LACCASES

Laccases (EC 1.10.3.2) are copper-containing enzymes that belong to the class of oxidoreductase and catalyze the oxidation of phenolic compounds and aromatic amines, which is accompanied by the reduction of oxygen to water (Giardina et al., 2010). They are also known as polyphenol oxydases, which detoxify litter phenolics. Enzyme formulations containing laccase have been commercially produced using *Aspergillus oryzae* and *Trichoderma reesei*. Even though the role of laccases in lignin degradation is debatable, they have tremendous applications in biotechnological processes. For instance, industrial laccases are employed in the bleaching of textiles (Vinod, 2001) and wood pulp (Widsten and Kandelbauer, 2008). They are also used in the clearing (degradation of lignin) of fruit juice, beer, and wine (Minussi et al., 2002); hair dyes; the degradation of plastics; the decontamination of soils (Kunamneni et al., 2008); and biofuel production (Kudanga and Le Roes-Hill, 2014).

Few species of thermophilic fungi have been reported to produce laccase enzyme. Chefetz et al. (1998) reported laccase enzyme in culture filtrates of *Chaetomium thermophilum*. They purified the enzyme using ultrafiltration, anion-exchange chromatography, and affinity chromatography. The enzyme was found to be a glycoprotein having a molecular mass of 77 kDa, optimal activity at pH 5–10, and stability at 50°C ($t_{1/2}$ = 12 h). The authors further observed that the enzyme polymerized a low-molecular-weight water-soluble organic matter fraction isolated from compost into a high-molecular-weight product. Similarly, Lloret et al. (2012) reported that the immobilized laccase of *Myceliophthora thermophila* on Eupergit C and Eupergit C 250 L degraded various endocrine-disrupting chemicals. Berka et al. (1997) characterized a gene encoding an extracellular laccase from *Myceliophthora thermophile*, which was cloned and expressed in *Aspergillus oryzae*. Subsequently, the recombinant enzyme (r-MtL) was purified from culture broth with a two- to fourfold-higher yield (11–19 mg/L) than native (MtL) laccase. The r-MtL depicted optimal activity at pH 6.5, and it also retained full activity when incubated at 60°C for 20 min. Babot et al. (2011) increased the brightness of eucalyptus pulp after treatment with laccase of *Sporotrichum thermophile*; however, the greatest improvements were attained with methyl syringate as laccase mediator, with a simultaneous decrease in kappa number.

11.2.11   α-D-Glucuronidase

α-D-glucuronidase (EC 3.2.1.139) belongs to the family of xylanolytic enzymes that hydrolyzes the 1,2-linked α-glycosidic between xylose and glucuronic acid and finds its application mostly in the pulp and paper industry. Khandke et al. (1989a) investigated the ability of crude culture filtrate protein obtained from thermophilic fungus *Thermoascus aurantiacus* to degrade larch wood xylan. It was observed that the enzyme completely degraded the xylan to produce xylose and 4-O-methyl-α-D-glucuronic acid. Later, the authors (Khandke et al., 1989b) purified and characterized α-D-glucuronidase from the culture filtrate protein of the fungus. The enzyme was found to be a single polypeptide of 118 kDa showing optimal activity at pH 4.5 and 65°C. The fungus, however, produced three extracellular enzymes, xylanase, α-D-glucuronidase, and β-glucosidae, which acted in tandem to completely degrade larch wood xylan. The authors also observed that α-D-glucuronidase was not as thermostable as other enzymes (xylanase and cellulase) obtained from the same culture filtrates and lost 50% of its original activity within 6 h of incubation at 60°C.

11.2.12   Cellobiose Dehydrogenase

Cellobiose dehydrogenases (CDHs) (EC 1.1.99.18) are produced by many plant biomass-degrading and phytopathogenic fungi, but the biological role of these enzymes remains uncertain (Zamocky et al., 2006). It is believed that the enzyme is able to enhance cellulose oxidation by lytic polysaccharide monooxygenases (LPMOs) (Langston et al., 2011). CDHs were isolated from thermophilic *Sporotrichum thermophile* (Canevascini, 1988; Canevascini et al., 1991) and *Humicola insolens* (Schou et al., 1998). The enzyme is reported to be a hemoflavoprotein, having a molecular mass of 92–95 kDa and stability at 60°C–65°C. However, the pH optima of CDHs from *Humicola insolens* was 7.5–8.0, in contrast to pH ~4.0 for *Sporotrichum thermophile*. van Noort et al. (2013) investigated the genome of *Chaetomium thermophilum* and observed that there are fewer genes that encode complex carbohydrate-degrading enzymes, in particular, in thermophilic mold genomes than their mesophilic counterpart. The genome of *Chaetomium thermophilum* encodes three CDHs, while those of *Chaetomium globosum* and *Neurospora crassa* encode only two, which depicts a higher cellulolytic ability of the former.

11.2.13   D-GLUCOSYLTRANSFERASE

D-Glucosyltransferase or transglucosidase (EC 2.4.1.24) belongs to the class of transferases. Bousquet et al. (1998) investigated thermophilic mold, *Talaromyces dupontii*, grown on maltodextrins at 37°C for the production of α-transglucosidase. The enzyme was purified by using ammonium sulfate precipitation, hydrophobic interaction chromatography, ion-exchange chromatography, and chromatofocusing, and had a molecular mass of 170 kDa. The enzyme was found to be extremely stable at 60°C ($t_{1/2}$ = 73 h), showing optimum activity at 4.5 pH and 70°C temperature. D-Glucosyltransferase can be effectively used for the synthesis of α-alkylglucoside by transferring glucosyl moieties from maltooligosaccharide donors to butanol in biphasic medium, as at high temperature the solubility of alcohol in the aqueous phase can be enhanced.

11.2.14   DNASE

The production of extracellular nuclease by thermophilic fungi is scanty throughout the literature. Adams and Deploey (1978) attempted, for the first time, to examine the production of extracellular nucleases from thermophilic fungi. They tested 10 thermophilic fungi and found that all produced an extracellular DNase and 6 produced an extracellular RNase. Since then, very little effort to characterize extracellular nucleases produced by thermophilic fungi has been made. Recently, Landry et al. (2014) studied DNase from a novel unidentified thermophilic fungus TM-417 using both traditional and innovative purification techniques. The fungus was found to be closely related to *Chaetomium* sp. and produced DNase optimally at 55°C. The enzyme was purified 145-fold using a novel affinity membrane purification system retaining 25% of the initial enzyme activity. Further, electrophoresis of the purified enzyme resulted in a single protein band, indicating DNase homogeneity. The enzyme was also found to be inducible in the presence of DNA and/or deoxyribose. Production of DNase in the culture medium was influenced by growth conditions, with static growth of the organism resulting in significantly higher DNase production, in contrast to agitated growth.

11.3   INTRACELLULAR OR CELL-ASSOCIATED THERMOSTABLE ENZYMES PRODUCED BY THERMOPHILIC FUNGI

Besides the extracellular enzymes, thermophilic fungi are known to produce enzymes that occur in the cytosol, in periplasmic space, on the surface of cells, or within the multilayered structure of the cell wall (Table 11.2). However, the precise location of the enzyme is rarely known, and therefore not much data are available on the

TABLE 11.2
Cell-Bound Enzymes Produced by Various Thermophilic Fungi

Enzyme	Organism	Thermal Stability $(t_{1/2})$	Optimal pH	Application	Reference
Trehalase EC 3.2.1.28	*Rhizopus microsporus* var. *rhizopodiformis*	45	–	Known to possess unique property of stabilizing membranes and enzymes	Aquino et al. (2001)
	Scytalidium thermophilum	65	–		Kadowaki et al. (1996)
	Humicola grisea var. *thermoidea*	60	5.5		Cardello et al. (1994)
	Malbranchea cinnamomea	55	5.5		Pereira et al. (2011)
	Thermomyces lanuginosus	50	5–5.5		Bharadwaj and Maheshwari (1999)
Invertase EC 3.2.1.26	*Thermomyces lanuginosus*	Unstable	–	Helps in overall human disease prevention, physical rejuvenation, and anti-aging process; production of noncrystallizable sugar syrup from sucrose; production of alcoholic beverages, lactic acid, glycerol, etc.; also used in drug and pharmaceutical industries	Basha and Palanivelu (1998)
	Thermomyces lanuginosus	55	–		Basha and Palanivelu (2000)
	Aspergillus caespitosus	55	–		Alegre et al. (2009)
	Humicola lanuginosa	–	–		Satyanarayana and Johri (1981)
β-D-glucosidase EC 3.2.1.21	*Chaetomium thermophile* var. *coprophile*	Unstable	NR	Used in synthesis of glycosides; widely used as surfactants, food colorants and flavoring agents, sweeteners, antioxidants, and anti-inflammatory, antimicrobial, and cardiac-related drugs	Lusis and Becker (1973)
	Thermoascus aurantiacus	75	–		Leite et al. (2007)
	Thermomucor indicae-seudaticae N31	75	–		Pereira et al. (2014)
	Sporotrichum thermophile	15 min at 48°C	–		Meyer and Canevascini (1981)

(Continued)

TABLE 11.2 (CONTINUED)
Cell-Bound Enzymes Produced by Various Thermophilic Fungi

Enzyme	Organism	Thermal Stability ($t_{1/2}$)	Optimal pH	Application	Reference
β-D-glucosidase EC 3.2.1.21	*Humicola grisea* var. *thermoidea*	12 min at 55°C	–		Peralta et al. (1990)
	Thermomyces lanuginosus	Moderately stable at 56°C	6.7–7.2		Fischer et al. (1995)
ATP sulfurylase EC 2.7.7.4	*Penicilium dupontii*	5.5 min at 70°C	–	Used in pyrosequencing	Renosto et al. (1985)
Protein disulfide isomerase EC 5.3.4.1	*Humicola insolens*	–	–	Used for fundamental studies and for industrial applications	Sugiyama et al. (1993)
Lipoamide dehydrogenase EC 1.6.4.3	*Malbranchea pulchella* var. *sulfurea*	170 min at 65°C	–	Plays a vital role in energy metabolism in eukaryotes	McKay and Stevenson (1979)

Note: NR = Not Reported.

production of intracellular enzymes from thermophilic fungi. Further, the difficulty in disrupting mycelium and extracting cellular protein from it hindered studies on these enzymes. Nevertheless, this section highlights the properties and applications of enzymes produced intracellularly by thermophilic fungi.

11.3.1 TREHALASE

Trehalase (EC 3.2.1.28) acts on the nonreducing disaccharide sugar trehalose (commonly known as fungal sugar or mycose), which is known to possess the unique property of stabilizing membranes and enzymes (Crowe et al., 1992) against drying and thermal denaturation, and therefore is the main driving force of thermophily in fungi. Trehalose is a disaccharide composed of two glucose units linked by an α-1,1-glycosidic bond (Figure 11.3). It is expected that trehalose may be present in large amounts in thermophilic fungi. Bharadwaj and Maheshwari (1999) compared the thermal and kinetic properties of trehalase from thermophilic (*Thermomyces lanuginosus*) and mesophilic (*Neurospora crassa*) fungi. The authors observed that trehalases from both fungi had pH optima between 5.0 and 5.5. Further, *T. lanuginosus* trehalase was a monomeric protein having a molecular mass of 145 kDa, in contrast to the 92 kDa of *N. crassa*, and the enzyme was a homotetramer. Both trehalases were found to be glycoproteins, having carbohydrate contents of 20% (*T. lanuginosus*) and 43% (*N. crassa*) and an optimum temperature of 50°C,

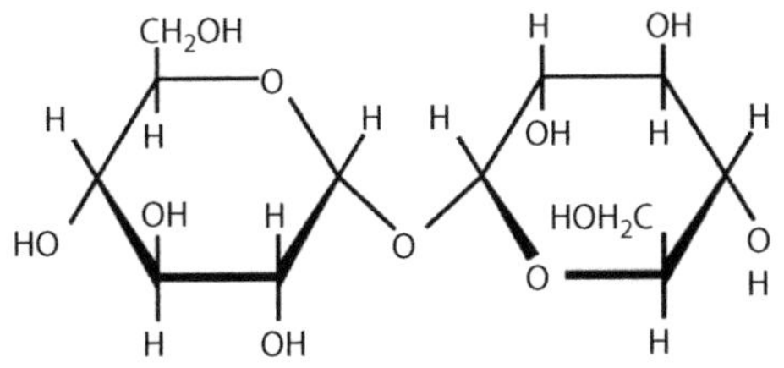

FIGURE 11.3 Chemical structure of trehalose (the disaccharide is composed of two glucose molecules joined by α1,1-glycosidic bond).

with similar thermostabilities at this temperature ($t_{1/2} > 6$ h) (Prasad and Maheshwari, 1979). However, the catalytic efficiency (k_{cat}/K_m) of *N. crassa* trehalase was one order of magnitude higher than that of *T. lanuginosus* trehalase. The authors' results showed that trehalase from *N. crassa* is comparatively more stable and a better catalyst than that from thermophilic *T. lanuginosus* trehalase.

It is presumed that enzymes from different parts of a microbe are the same. However, in *Humicola grisea* var. *thermoidea*, trehalases from conidia and mycelium were quite different, as observed by Zimmermann et al. (1990). The authors observed that trehalase obtained from conidia was a hexamer comprising three different polypeptides and was localized on the surface. In contrast, the mycelial trehalase was a homotrimer and present in the cytosol. Further, the conidial and mycelial trehalases also differed in their carbohydrate contents (56% and 12%, respectively) and K_m (2.5 and 0.86 mM, respectively) but exhibited similar pH (5.5) and temperature (60°C) optima.

From the above discussion, it is evident that trehalases are bound to the mycelia or conidial wall or are present intracellularly in the cytosol. However, it is interesting to note that *Scytalidium thermophilum* also produced trehalase extracellularly in the medium when grown on starch (Kadowaki et al., 1996). However, both forms (intracellular and extracellular) were pentamers with similar subunit molecular masses (65 kDa) and temperature optima (60°C), and both contained large amounts of carbohydrates. The simultaneous presence of two different trehalases has been recognized in several fungal species. It is believed that neutral trehalases mobilize cytosolic trehalose under the control of chemical and nutrient signals, while acid trehalases act as "carbon scavengers" and utilize exogenous trehalose as a carbon source under the control of carbon catabolic regulatory circuits (Jorge et al., 1997). An extracellular trehalase was also biochemically characterized from *Malbranchea pulchella* var. *sulfurea* by Pereira et al. (2011). The observation that *Scytalidium thermophilum* trehalase was associated with the cell wall under one growth condition and secreted under another may be explained by the dynamic nature of the cell wall, whose composition changes in response to environmental and growth conditions and affects the retention or secretion of wall proteins (Chaffin et al., 1998).

11.3.2 INVERTASE

Invertase (EC 3.2.1.26), also known as β-fructofuranosidase, is a glycoprotein with an optimum pH of 4.5 and temperature stability at 50°C that cleaves the terminal nonreducing β-fructofuranoside residues. Its ability to break down sucrose into glucose and fructose makes it a vital component for the digestion of complex sugars into blood sugar (glucose). Invertase not only plays a key role in the digestive process but also helps in overall human disease prevention, physical rejuvenation, and the anti-aging process. Invertase is widely distributed in nature and mostly characterized in plants and microorganisms (Kulshrestha et al., 2013).

One of the major applications of invertase enzyme is in the production of noncrystallizable sugar syrup from sucrose. Invert syrup possesses hygroscopic properties, which makes it useful in the manufacturing of soft-centered candies and fondants as humectants (Kotwal and Shankar, 2009). Invertase is also required for the production of alcoholic beverages, lactic acid, glycerol, and so forth, produced by the fermentation of sucrose-containing substrates. In addition, invertase is also used in drug and pharmaceutical industries, such as in the manufacturing of artificial honey and the plasticizing agents used in cosmetic preparations. Being an antimicrobial agent and an antioxidant, invertase helps in the prevention of bacterial infestations and gut fermentation due to oxidation. In a study of 18,000 patients in Europe, honey was proved effective for treating respiratory tract infections, such as bronchitis, asthma, and allergies.

A few thermophilic fungi are reported to produce the invertase enzyme. Maheshwari et al. (1983) studied the distinctive behavior of invertase in *Thermomyces lanuginosus* and also in *Penicillium dupontii*. They reported that invertase was an inducible enzyme; that is, it is produced in the presence of sucrose in the growth medium, and its activity is highly unstable in cell extracts. In contrast, invertase from mesophilic fungi retains activity for long periods under storage conditions. However, Chaudhuri et al. (1999) observed that the induced enzyme activity begins to decline before any substantial quantity of sucrose is utilized or an appreciable amount of biomass is formed. They also depicted that invertase in *T. lanuginosus* is localized in the hyphal tips of the mycelium. The authors showed that in the early stages of growth, the number of tips per unit mass of mycelium is maximal and corresponds with the maximal invertase activity. At later stages of growth, however, the number of tips per unit mass of mycelium decreases, as the increase in mass is mainly contributed to by cell elongation and cell wall thickening. As a result, invertase activity showed an obvious decline with growth. On the contrary, invertase in mesophilic *Neurospora crassa* is evenly distributed all along the hyphae, mostly bound to the hyphal wall. Thus, invertase activity steadily increases with an increase in growth. As suggested, the localization of invertase in the hyphal tips of *T. lanuginosus* needs to be further validated by immunological techniques. Moreover, cloning of the invertase gene from this fungus and its heterologous expression in a suitable host are needed in order to understand the behavior of the enzyme in this fungus.

11.3.3 β-Glycosidase

β-Glycosidases (EC 3.2.1.21) play an important role in the synthesis of glycosides and catalyze the hydrolysis of glycosidic bonds to liberate monosaccharides and oligosaccharides of lower molecular weight, compared with the native simple and complex carbohydrate substrates. β-Glycosides are widely used as surfactants

(Busch et al., 1994), food colorants and flavoring agents (Sakata et al., 1998), sweeteners (Shibata et al., 1991), antioxidants, and anti-inflammatory (Gomes et al., 2002), antimicrobial (Zhou, 2000), and cardiac-related drugs (Ooi et al., 1985).

Among the thermostable glycosidases used for the synthesis of glycosides, β-glucosidase from the hyperthermophilic *Pyrococcus furiosus* is the most significant (Kengen et al., 1993); it is comparatively easy to grow, and the enzyme is found to be stable at 100°C for 85 h. Later, this enzyme was cloned and overexpressed in *Escherichia coli* by Voorhorst et al. (1995). Lusis and Becker (1973) compared the properties of partially purified preparations of cell-bound and extracellular β-glucosidases from *Chaetomium thermophile* var. *coprophile*. However, they observed that the extracellular enzyme was more stable than the cell-bound enzyme. Similarly, Fischer et al. (1995) purified an inducible β-galactosidase from *Thermomyces lanuginosus* using fractional salt precipitation, hydrophobic interaction chromatography, and anion-exchange chromatography. They observed that the enzyme was most active at pH 6.7–7.2 and was only moderately stable at 56°C, losing all its activity within 1 h. The enzyme was found to be dimeric, having a subunit molecular mass of 75–80 kDa. Bhat et al. (1993) reported the production of intracellular β-glucosidase by *Sporotrichum thermophile* when it was grown in a medium containing cellulose. Multiple forms of β-glucosidase produced by this fungus are known, and they differ in their molecular mass, pH and temperature optima, thermostability, affinity for β-glycosides, and transglycosylation activity (Canevascini and Meyer, 1979; Meyer and Canevascini, 1981; Gaikwad and Maheshwari, 1994).

11.3.4 ATP SULFURYLASE

ATP sulfurylase (EC 2.7.7.4), also known as ATP:sulfate adenylyltransferase, catalyzes the first and second reactions in the assimilation of inorganic sulfate: ATP + $SO_4^{2-} \rightarrow PP_i + APS$ (adenosine-5′-phosphosulfate) and ATP + APS → ADP + PAPS (3′-phosphoadenosine-5′-phosphosulfate). The reaction is reversible: ATP is formed from APS and pyrophosphate (PPi). This enzyme participates in three metabolic pathways: purine metabolism, seleno–amino acid metabolism, and sulfur metabolism. It is also used in pyrosequencing.

Little information is available on the production of ATP sulfurylase by thermophilic fungi. Renosto et al. (1985) studied the catalytic and chemical properties of sulfate-activating enzymes, that is, ATP sulfurylase and APS kinase, from a mesophilic *Penicillium chrysogenum* and thermophilic *Penicillium dupontii*. They reported that ATP sulfurylase from *P. duponti* showed optimum temperature at 70°C, in contrast to 55°C for that from *P. chrysogenum*. It was also observed that the ATP sulfurylase from the thermophilic source was about 90 times more heat stable than that of the mesophile enzyme, and was also more stable in response to urea and acidic pH. Both enzymes were shown to be hexamers, comprising nearly equal-sized subunits (69 kDa), and the amino acid composition was also quite identical. The authors also observed that at 50°C, the specific activity of *P. duponti* ATP sulfurylase was approximately double that of the *P. chrysogenum* enzyme, demonstrating that the thermophile enzyme was a better catalyst in contrast to the mesophile enzyme.

11.3.5 Protein Disulfide Isomerase

Isomerases catalyze the transfer of groups from one position to another in the same molecule. In other words, these enzymes change the structure of a substrate by rearranging its atoms. Protein disulfide isomerase (EC 5.3.4.1) is an intracellular enzyme that in the endoplasmic reticulum facilitates disulfide interchange by shuffling the disulfide bonds to rapidly establish the most thermodynamically stable pairing (Maheshwari et al., 2000). This enzyme helps in the refolding and renaturation of scrambled protein *in vitro*, for fundamental studies and industrial applications. Sugiyama et al. (1993) isolated a thermostable protein disulfide isomerase from mycelia extract of *Humicola insolens*. The enzyme was purified using anion-exchange chromatography, concanavalin A affinity chromatography, and reverse-phase high-pressure liquid chromatography. The enzyme was found to be a homodimer having a molecular mass of 60 kDa and was similar to the yeast and bovine protein disulfide isomerase. Later, Kajino et al. (1994) cloned the cDNA of *Humicola insolens* protein disulfide isomerase and expressed it in *Bacillus brevis*. The recombinant enzyme was secreted into the culture medium and was purified to homogeneity in two steps by anion-exchange chromatography and hydrophobic interaction chromatography with a high yield (corresponding to 2.1 g of bovine liver enzyme per liter). They observed that the recombinant protein disulfide isomerase was smaller by 1 kDa, as the recombinant protein produced by the bacterial host is not glycosylated. Considering the similar properties of native and recombinant protein disulfide isomerase, it has been suggested that glycosylation of the enzyme is not necessary for its activity and stability (Kajino et al., 1998).

11.3.6 Lipoamide Dehydrogenase

Lipoamide dehydrogenase (EC 1.6.4.3) is also known as dihydrolipoamide dehydrogenase or dihydrolipoyl dehydrogenase (DLD). It is a component of the multi-enzyme pyruvate dehydrogenase complex and plays a vital role in energy metabolism in eukaryotes. It catalyzes the reduction of NAD^+ with dihydrolipoamide. The enzyme associates into tightly bound homodimers required for its enzymatic activity. McKay and Stevenson (1979) purified lipoamide dehydrogenase from *Malbranchea pulchella*, a thermophilic fungus, using salt precipitation, affinity chromatography, and ion-exchange chromatography. They observed that the lipoamide dehydrogenase from *Malbranchea pulchella* var. *sulfurea* was quite similar to the enzyme purified from several other sources. While studying the thermal stability of the enzyme, the authors observed that it showed considerable resistance to thermal denaturation ($t_{1/2}$, 170 min at 65°C). However, the activity was completely lost in 30 days at 4°C even in concentrated solutions (>2 mg/mL).

11.4 BIOACTIVE COMPOUNDS FROM THERMOPHILIC FUNGI

A number of thermophilic fungi have been reported to produce bioactive compounds, such as antibiotics that have antibacterial and antifungal properties. The antibacterial antibiotics include penicillin G, malbranchins A and B, 6-aminopenicillanic acid, sillucin, miehein, and vioxanthin, which are active against both Gram-positive and Gram-negative bacteria. The antifungal antibiotics include myriocin from *Myriococcum albomyces* and thermozymocidin from *Thermoascus aurantiacus* (Mehrotra, 1985; Satyanarayana et al., 1992).

Thermophilic fungal species of *Mucor*, *Talaromyces*, and others have been reported to produce organic acids (lactic and citric acids) and extracellular phenolic compounds. Some thermophilic fungi are known to secrete various amino acids, like alanine, glutamic acid, and lysine (Subrahmanyam, 1985).

Svahn et al. (2012) studied 61 strains of filamentous fungi isolated from highly antibiotic-contaminated river sediments for their antimicrobial activity. Among these strains, *Aspergillus fumigatus* showed antimicrobial activity against methicillin-resistant *Staphylococcus aureus*, extended-spectrum β-lactamase-producing *Escherichia coli*, vancomycin-resistant *Enterococcus faecalis*, and *Candida albicans*. Gliotoxin was found to be the prominent secondary metabolite in *Aspergillus fumigatus*. Earlier, Guo et al. (2011) isolated six indole alkaloids (talathermophilins) from the thermophile *Talaromyces thermophilus* strain YM3-4. They observed that compounds 1 and 2 were new analogues to precursor notoamide E, whereas compound 3 was a novel analogue of preechinulin and compound 4 was found as a naturally occurring cyclo(glycyl-tryptophyl) for the first time. The metabolite profile of this thermophilic fungus exhibited a biosynthetic pathway for talathermophilins. Macrocyclic PKS-NRPS hybrid metabolites represent a unique family of natural products, mainly from bacteria, with profound biological activities. However, their distribution in fungi has rarely been reported, and little is known about their nematicidal activity. Guo et al. (2012) reported thermolides with potential nematocidal PKS-NRPS hybrid metabolites from a thermophilic fungus, *T. thermophilus*. These PKS-NRPS hybrid metabolites possess a 13-membered lactam-bearing macrolactone, thermolides A–F (1–6). They observed that thermolides 1 and 2 exhibited potent inhibitory activity against three notorious nematodes with LC50 (lethal concentration, 50%) values of 0.5–1 g/mL, which was found to be as active as the commercial nematicide avermectins. Similarly, Chu et al. (2010) isolated and identified prenylated indole alkaloids, namely, talathermophilins A and B, from *T. thermophilus* strain YM1-3 using nuclear magnetic resonance (NMR) and MS analyses. The ratio of A to B in the culture broths was found to be unexpectedly rather constant; it remained unchanged despite the addition of exogenous A or B, indicating that talathermophilins might have a special function in this fungus. The bioactive secondary metabolites in *Talaromyces* species have recently been reviewed by Zhai et al. (2016).

Similarly, the thermophilic fungus *Myceliophthora thermophila* has been reported to produce a large number of bioactive molecules displaying an array of industrial applications. Estatins A ($C_{18}H_{25}N_5O_5$) and B ($C_{18}H_{25}N_5O_6$), possessing thiol protease inhibitory activity, have been isolated from the culture filtrate of *Myceliophthora thermophila* M4323 by Yaginuma et al. (1989). These estatins have been characterized as having an agmatine, trans-epoxysuccinic acid, and L-phenylalanine or L-tyrosine moieties in their structure, and they are water soluble. The authors observed that the estatins were specific inhibitors against thiol proteases, for example, papain, ficin, and bromelain. Further, the authors reported that the estatins suppressed immunoglobulin E (IgE) antibody production in mice, but not IgG, a property that makes them useful in many pharmaceutical applications. The thermophilic *Humicola* sp. have been used in the biosynthesis of metal nanoparticles, such as silver nanoparticles, which were found to be active against the MDA-MB-231 human breast carcinoma cell line (Syed et al., 2013), and gadolinium oxide nanoparticles, which were bioconjugated with chemically modified anti-cancer drug taxol for the treatment of cancer.

11.5 SINGLE-CELL PROTEIN PRODUCTION

SCP pertains to sources of mixed protein extracted from pure or mixed cultures of microorganisms grown on inexpensive agricultural residues or wastes. SCPs can be used as a substitute for protein-rich foods for humans and animals alike. The microbial biomass so produced contains essential nutrients other than protein. Lignocellulosic materials are the earth's most abundant renewable organic carbon source and are composed of 65%–72% utilizable polysaccharides (hemicelluloses and cellulose) and 10%–30% lignin. Filamentous fungi, especially thermophilic ones, are good decomposers of organic matter and are amenable to growing in SSF and yielding high biomass. The importance of cellulolytic thermophiles over their mesophilic counterparts is realized due to their higher rate of cellulose breakdown, good sources of proteins, higher specific growth rates, and activity over a wide range of temperatures, 20°C–50°C (Seal and Eggins, 1976).

The fungal protein, or mycoprotein, is attracting the attention of food and feed scientists and protein engineers. *Chaetomium cellulolyticum* and *Sporotrichum pulverulentum* are the most widely used organisms for upgrading animal feed and producing SCPs from lignocellulosic wastes (Thomke et al., 1980). Thermophilic *Aspergillus fumigatus*, *Sporotrichum thermophile*, and *Thermoascus aurantiacus* produced increased (twofold) protein contents on sugar beet pulp and molasses solution (Sundman et al., 1981; Grajek, 1988). Ghai et al. (1979) used sag waste from the canning industry as a substrate for the production of SCPs using the thermotolerant fungi *Chaetomium cellulolyticum* and *Actinomucor* sp. El-Refai et al. (1991) used *Myceliophthora thermophila* for the production of SCPs using crude orange waste as the sole source of carbon. Oil-free orange waste supported the maximum growth of the fungus and produced SCPs after 20 days, whereas the yields were low in media containing delignified or depectinized orange waste. The resultant fungal SCP contained a high proportion of crude protein (41%) with all the essential amino acids.

11.6 TOOLS FOR GENETIC RECOMBINATION

Considering the enormous industrial applications of thermostable enzymes from thermophilic fungi, protein engineering approaches have been employed to enhance their thermostability, high or low pH tolerance, and biocatalytic properties. Recent technological advances in genetic manipulation of thermophilic fungi by induced mutants, using both chemical and physical mutagens, have accelerated the production of enzymes like cellulases by these fungi. In order to enhance cellulase production, Morrison et al. (1987) isolated a mutant of *Talaromyces emersonii* CBS 814.70

named UV7 by UV irradiation. This mutant was capable of increased cellulase production during growth on cellulose-, lactose-, and glucose-containing media. Moloney et al. (1983) obtained a morphological mutant UCG 42 by treating the strain with nitrosoguanidine. This mutant had an improved filter paper hydrolyzing activity. This is an instance of the applicability of thermophilic fungi in an experimental system for genetic manipulation to achieve the desired objectives. Fähnrich P and Irrgang K (1981) obtained a hybrid 7S/7 from the progeny of a cross between two morphological and physiological variants of *Chaetomium cellulolyticum* (i.e., variants 4S and 7S). The hybrid strain 7S/7 was capable of producing more CMCase than either of the parent strains.

The possibility of genetic manipulation for achieving genetic recombination following protoplast fusion was reported by Gupta and Gautam (1995). Protoplasts of a catabolite repression-resistant strain of *Malbranchea sulfurea* and a mutant of it overproducing amylase were isolated and fused via electrofusion. The yield of the hybrids obtained was 5×10^{-5}. One stable hybrid, DGCS1, was insensitive to glucose repression and produced approximately twice the α-amylase activity as either of its parents. The amount of DNA in DGCS 1 was also twice that of either parent strain. Heinzelman et al. (2009) employed a structure-guided recombination approach (SGRA) for engineering chimeric thermostable cellobiohydrolase II (CBH II) from *Sporotrichum thermophilum* and *Chaetomium thermophilum*, and also from the mesophilic model cellulolytic fungus *Trichoderma reesei*. The chimeric enzymes depicted a broader optimum pH range and higher thermostability than the parent enzymes. Komor et al. (2012) devised computational methods in order to envisage mutations associated with the chimeric enzymes with enhanced thermostability in *Talaromyces emersonii* CBH I. They observed that chimeric enzymes generated from 43 predicted mutations showed a 10°C higher optimum temperature for activity and a higher activity on microcrystalline cellulose than the native strains. Similarly, Wang et al. (2012) engineered an N-terminal disulfide bridge between positions 1 and 24 of the *Thermomyces lanuginosus* DSM 10635 GH11 xylanase and expressed it in *Escherichia coli*. They designed diverse mutations so as to give more flexibility to the N-terminus, thus allowing the formation of the disulfide bridge. The engineered enzyme displayed increased (by 8°C) optimal temperature and higher activity and thermostability in alkaline conditions.

11.7 DETRIMENTAL ACTIVITIES

Ever since the dawn of agriculture, man has depended on the storage of agricultural produce for use in time of need. The storage fungi do not invade agricultural produce before harvest to any appreciable degree (Wallace, 1973). But during storage, thermophilic fungi can cause deterioration of agricultural produce, such as cereal grains, groundnuts, palm kernels, hay, wood chips, baggasse, and peat (Sharma, 1989). During storage, such products undergo a process of heating that may, under some conditions, advance to the plateau where spontaneous combustion occurs. A number of thermophilic fungi from stored cereal grains have been reported. The thermophilic fungi commonly associated with stored grains are *Absidia* spp., *Aspergillus* spp., *Mucor pusillus*, *Thermomyces lanuginosus*, and *Thermoascus aurantiacus*

(Mehrotra, 1985; Sharma, 1989). These fungi can cause reduction in the rate of germination, discoloration, damage to the seed, and spoilage of the stored grains due to microbial activity. The thermophilic fungi of stored grains are receiving the attention of workers because of their toxigenic and pathogenic potential (Mehrotra, 1985). Moreover, in view of the recent surge in global warming, thermophilic and thermotolerant fungi are anticipated to dominate the aflatoxin- and mycotoxin-producing fungi (Paterson and Lima, 2017). Thermophilic and thermotolerant fungi produce a range of secondary metabolites, or potential mycotoxins and patulins that may become a new threat. In addition, *Aspergillus fumigatus* will appear more frequently as a serious human pathogen because it is (1) thermotolerant and (2) present on crops; hence, this is an even greater problem (Paterson and Lima, 2017).

The implication of thermophilic fungi in the spoilage of groundnuts and palm kernels is attributed to their strong lipolytic activity. A wide range of thermophilous fungi, namely, *Aspergillus fischieri, Aspergillus fumigatus, Chaetomium globosum, Chrysosporium thermophilum, Humicola lanuginosa, Mucor pusillus, Thermoasus aurantiacus, Penicillium duponti, Paecilomyces varioti, Scopulariopsis fusca, Absidia blackesleana, Absidia ramosa*, and *Thermomyces ibadanensis*, are reported from moldy groundnuts and palm kernels (Kuku and Adeniji, 1976; Ogundcro, 1980). All thermophilous isolates can utilize a variety of lipids as a carbon source.

Commercial wood chips stored outside in piles spontaneously generate heat, which causes deterioration of chips and can lead to spontaneous ignition (Schmidt, 1969). The thermophilic and thermotolerant fungi isolated from wood chip piles include *Chaetomium thermophile* var. *coprophile, Chaetomium thermophile* var. *dissitum, Humicola grisea* var. *thermoidea, Humicola lanuginosa, Sporotrichum thermophile, Talaromyces emersonii, Talaromyces thermophilus*, and *Thermoascus aurantiacus* (Tansey, 1971). Economic losses from the spoilage of bagasse (a fibrous residue of sugarcane after the extraction of juice) and peat due to self-heating and combustion can be significant. The presence of residual sugar in bagasse makes it prone to microbial spoilage (Paturau, 1969). During storage, thermophilic microorganisms degrade the polysaccharides, causing spoilage of the cellulosic waste. Inhalation of spores released by the microorganisms can cause bagassosis, a disease of the respiratory tract. The chief fungal species present in stored bagasse include *Absidia corymbifera, Chrysosporium keratinophilum, Paecilomyces varioti*, and *Phialophora lignicola*. The fungi reported were found to be thermotolerant rather than thermophilic (Sharma, 1989).

There are numerous reports of thermophilic fungi as pathogens of humans and other warm-blooded animals (Hughes and Crosier, 1973; Tansey and Brock, 1978). The thermophilic *Mucor pusillus* in particular is a pathogen that causes a variety of mycoses (Meyer and Armstrong, 1973; Meyer et al., 1973). Thermotolerant fungi such as *Absidia ramosa* (Nottebrock et al., 1974) and *Aspergillus fumigatus* (Rippon, 1974) are more frequently reported as pathogens than are thermophiles. The thermotolerant fungus *Dactylaria gallopava* has been found to be a cause of epidemics in young turkeys and chickens (Blalock et al., 1973; Tansey, 1973; Ranck et al., 1974; Waldrip et al., 1974). However, the medical implications of thermophilous fungi appear very unimportant compared with their agricultural activities.

Peritonitis is one of the most serious complications of peritoneal dialysis, and fungal peritonitis is rarely observed. Oz et al. (2010) observed colonization of a peritoneal catheter with a thermophilic fungus, *Thermoascus crustaceous*, in a 50-year-old male patient undergoing chronic peritoneal dialysis. The authors anticipated broadening of the spectrum of pathogenic fungi in humans, including thermophilic fungi.

The foregoing account of thermophilic fungi suggests a need for more work to be done to elucidate their role in enzyme immobilization technology, biocatalytic properties, biodegradation of wastes, and biodeterioration of stored agricultural products. Further, specific methods for detecting multiple metabolites in thermophilic fungi may be required as an initial step to study their importance. Interdisciplinary approaches in seeking solutions to these problems are warranted and must be looked at for their social, economic, legal, and political consequences.

REFERENCES

Adams, P.R., and Deploey, J.J. 1976. Amylase production by *Mucor pusillus* and *Mucor miehei*. *Mycologia* 68:934–938.

Adams, P.R., and Deploey, J.J. 1978. Enzymes produced by thermophilic fungi. *Mycologia* 70:906–910.

Agrios, G.N. 1988. *Plant Pathology*. New York: Academic Press.

Ahirwar, S., Soni, H., Prajapati, B.P., and Kango, N. 2017. Isolation and screening of thermophilic and thermotolerant fungi for production of hemicellulases from heated environments. *Mycology* 1–10. doi: 10.1080/21501203.2017.1337657.

Alegre, A.C.P., Lourdes, M., Moraes, T., Polizeli, M., Terenzi, F., Jorge, J.A., and Guimaraes, L.H. 2009. Production of thermostable invertases by *Aspergillus caespitosus* under submerged or solid state fermentation using agroindustrial residues as carbon source. *Brazilian Journal of Microbiology* 40:612–622.

Anand, L., Krishnamurthy, S., and Vithayathil, P.J. 1990. Purification and properties of xylanase from the thermophilic fungus, *Humicola lanuginose* (Griffon and Maublanc) Bunce. *Archives of Biochemistry and Biophysics* 276:546–553.

Aquino, A.C., Jorge, J.A., Terenzi, H.F., and Polizeli, M.L. 2001. Thermostable glucose-tolerant glucoamylase produced by the thermophilic fungus *Scytalidium thermophilum*. *Folia Microbiol (Praha)* 46:11–16.

Arima, K., Iwasaki, S., and Tamura, G. 1968. Milk clotting enzymes from microorganisms. V. Purification and crystallization of *Mucor* rennin from *Mucor pusillus* Lindt. *Applied Microbiology* 16:1727–1733.

Arima, K., Liu, W.H., and Beppu, T. 1972. Studies on the lipase of thermophilic fungus *Humicola lanuginosa*. *Agricultural and Biological Chemistry* 36:893–895.

Arima, K., Yu, J., and Iwasaki, S. 1970. Milk-clotting enzyme from *Mucor pusillus* var. *Lindt*. *Methods in Enzymology* 19:446–459.

Asimov, I. 1982. *Asimov's Biographical Encyclopedia of Science and Technology*.

Babot, E.D., Rico, A., Rencoret, J., Kalum, L., Lund, H., Romero, J., del Río, J.C., Martínez, A.T., and Gutiérrez, A. 2011. Towards industrially-feasible delignification and pitch removal by treating paper pulp with *Myceliophthora thermophila* laccase and a phenolic mediator. *Bioresource Technology* 102:6717–6722.

Bajpai, P. 2009. Xylanases. In *Encyclopedia of Microbiology*, ed. M. Schaechter and J. Lederberg, 600–612. San Diego: Academic Press.

Bajpai P. 2014. Purification of xylanases. In *Xylanolytic Enzymes*, ed. P. Bajpai, 53–61. Amsterdam: Academic Press.

Bala, A., Sapna, Jain, J., Kumari, A., and Singh, B. 2014. Production of an extracellular phytase from a thermophilic mould *Humicola nigrescens* in solid state fermentation and its application in dephytinization. *Biocatalysis and Agricultural Biotechnology* 3: 259–264.

Banerjee, S., Archana, A., and Satyanarayana, T. 1994. Xylose metabolism in a thermophilic mould *Malbranchea pulchella* var. *sulfurea* TMD-8. *Current Microbiology* 29:349–352.

Banerjee, S., Archana, A., and Satyanarayana, T. 1995. Xylanolytic activity and xylose utilization by thermophilic molds. *Folia Microbiologica* 40:279–282.

Baraznenoka, V.A., Becker, E.G., Ankudimova, N.V., and Okunevbet, N.N. 1999. Characterization of neutral xylanases from *Chaetomium cellulolyticum* and their biobleaching effect on eucalyptus pulp. *Enzyme and Microbial Technology* 25:651–659.

Basha, S.Y., and Palanivelu, P. 1998. Short communication: Enhancement in activity of an invertase from the thermophilic fungus *Thermomyces lanusinosus* by exogenous proteins. *World Journal of Microbiology and Biotechnology* 14:603–605.

Basha, S.Y., and Palanivelu, P. 2000. A novel method for immobilization of invertase from the thermophilic fungus, *Thermomyces lanuginosus*. *World Journal of Microbiology and Biotechnology* 16:151–154.

Berka, R.M., Rey, M.W., Brown, K.M., Byun, T., and Klotz, A.V. 1998. Molecular characterization and expression of a phytase gene from the thermophilic fungus *Thermomyces lanuginosus*. *Applied and Environmental Microbiology* 64:4423–4427.

Berka, R.M., Schneider, P., Golightly, E.J., Brown, S.H., Madden, M., Brown, K.M., Halkier, T., Mondorf, K., and Xu, F. 1997. Characterization of the gene encoding an extracellular laccase of *Myceliophthora thermophila* and analysis of the recombinant enzyme expressed in *Aspergillus oryzae*. *Applied and Environmental Microbiology* 63: 3151–3157.

Bharadwaj, G., and Maheshwari, R. 1999. A comparison of thermal and kinetic parameters of trehalases from a thermophilic and a mesophilic fungus. *FEMS Microbiology Letters* 181:187–193.

Bhat, K.M., Gaikwad, J.S., and Maheshwari, R. 1993. Purification and characterization of an extracellular b-glucosidase from the thermophilic fungus *Sporotrichum thermophile* and its influence on cellulase activity. *Journal of General Microbiology* 139:2825–2832.

Bhat, K.M., and Maheshwari, R. 1987. *Sprotrichum thermophile*, growth, cellulose degradation, and cellulase activity. *Applied and Environmental Microbiology* 53:2175–2182.

Bilai, T.I., Chernyagina, T.B., Dorokhov, V.V., Poedinok, N.L., Zakharchenko, V.A., Ellanskaya, I.A., and Lozhkina, G.A. 1985. Phosphatase activity of different species of thermophilic and mesophilic fungi. *Mikrobiologia* 47:53–56.

Blalock, H.G., Georg, L.K., and Derieux, W.T. 1973. Encephalitis in turkey poults due to *Dactylaria (Diplorhinotrichum) gallopava*: A case report and its experimental reproduction. *Avian Diseases* 17:197–204.

Boel, E., Huge-Jensen, B., Christensen, M., Thim, L., and Fiil, N.P. 1988. Rhizomucor miehei triglyceride lipase is synthesized as a precursor. *Lipids* 23:701–701.

Bousquet, M.P., Willemot, R.-M., Monsan, P., and Boures, E. 1998. Production, purification, and characterization of thermostable a-transglucosidase from *Talaromyces duponti*—Application to a-alkylglucoside synthesis. *Enzyme and Microbial Technology* 23:83–90.

Brock, T.D. 1986. Introduction, an overview of the thermophiles. In *Thermophiles: General, Molecular and Applied Microbiology*, ed. T.D. Brock, 1–16. New York: John Wiley & Sons.

Bunni, L., McHale, L., and McHale., A.P. 1989. Production, isolation and partial characterization of an amylase system produced by *Talaromyces emersonii* CBS 814.70. *Enzyme and Microbial Technology* 11:370–375.

Busch, P., Hensen, H., Khare, J., and Tesmann, H. 1994. Alkylpolyglycosides—A new cosmetic concept for milderness. *Agro Food Industry Hi-Tech* 5:20–28.

Business Communications Company. 2004. *Enzymes for industrial applications.* RC-147U. http://www.bccresearch.com.

Campos, L., and Felix, C.R. 1995. Purification and characterization of glucoamylase from *Humicola grisea. Applied and Environmental Microbiology* 61:2436–2438.

Canevascini, G. 1988. Cellobiose dehydrogenase from *Sporotrichum thermophile. Methods in Enzymology* 160:443–448.

Canevascini, G., Borer, P., and Dreyer, J.-L. 1991. Cellobiose dehydrogenases of *Sporotrichum (Chrysosporium) thermophile. European Journal of Biochemistry* 198:43–52.

Canevascini, G., and Meyer, H.P. 1979. β-Glucosidase in the cellulolytic fungus *Sporotrichum thermophile* Apinis. *Experimental Mycology* 3:203–214.

Cardello, L., Terenzi, H.F., and Jorge, J.A. 1994. A cystolic trehalase from the thermophilic fungus *Humicola grisea* var. *thermoidea. Microbiology* 140:1671–1677.

Carlsen, M., Nielsen, J., and Villadsen, J. 1996. Growth and amylase production by *Aspergillus oryzae* during continuous cultivations. *Journal of Biotechnology* 45:81–93.

Cereia, M., Terenzi, H.F., Jorge, J.A., Greene, L.J., Rosa, J.C., Lourdes, M.D., and Polizeli, T.M. 2000. Glucoamylase activity from the thermophilic fungus *Scytalidium thermophilum.* Biochemical and regulatory properties. *Journal of Basic Microbiology* 40:83–92.

Chadha, B.S., Gulati, H., Minhas, M., Saini, H.S., and Singh, N. 2004. Phytase production by the thermophilic fungus *Rhizomucor pusillus. World Journal of Microbiology and Biotechnology* 20:105–109.

Chadha, B.S., Singh, S., Vohra, G., and Saini, H.S. 1997. Shake culture studies for the production of amylases by *Thermomyces lanuginosus. Acta Microbiologica et Immunologica Hungarica* 44:181–185.

Chaffin, W.L., López-Ribot, J.L., Casanova, M., Gozalbo, D., and Martínez, J.P. 1998. Cell wall and secreted proteins of *Candida albicans.* Identification, function and expression. *Microbiology and Molecular Biology Reviews* 62:130–180.

Chang, Y., and Hudson, H.J. 1967. The fungi of wheat straw compost. II. Biochemical and physiological studies. *Transactions of the British Mycological Society* 50:649–666.

Chaudhuri, A., Bharadwaj, G., and Maheshwari, R. 1999. An unusual pattern of invertase activity development in the thermophilic fungus *Thermomyces lanuginosus. FEMS Microbiology Letters* 177:39–45.

Chefetz, B., Chen, Y., and Hadar, Y. 1998. Purification and characterization of laccase from *Chaetomium thermophilium* and its role in humification. *Applied and Environmental Microbiology* 64:3175–3179.

Chen, J., Zhang, Y.Q., Zhao, C.Q., Li, A.N., Zhou, Q.X., and Li, D.C. 2007. Cloning of a gene encoding thermostable glucoamylase from *Chaetomium thermophilum* and its expression in *Pichia pastoris. Journal of Applied Microbiology* 103:2277–2284.

Christakopoulos, P., Katapodis, P., Kalogeris, E., Kekos, D., Macris, B.J., Stamatis, H., and Skaltsa, H. 2003. Antimicrobial activity of acidic xylo-oligosaccharides produced by family 10 and 11 endoxylanases. *International Journal of Biological Macromolecules* 31:171–175.

Chu, Y.S., Niu, X.M., Wang, Y.L., Guo, J.P., Pan, W.Z., Huang, X.W., and Zhang, K.Q. 2010. Isolation of putative biosynthetic intermediates of prenylated indole alkaloids from a thermophilic fungus *Talaromyces thermophilus. Organic Letters* 12:4356–4359.

Cooney, D.G., and Emerson, R. 1964. *Thermophilic Fungi: An Account of Their Biology, Activities, and Classification.* San Francisco: W.H. Freeman and Co.

Coudray, M.-R., Canevascini, G., and Meier, H. 1982. Characterization of a cellobiose dehydrogenase in the cellulolytic fungus *Sporotrichum (Chrysosporium) thermophile. Biochemical Journal* 203:277–284.

Craveri, R., Craveri, A., and Guicciardi, A. 1967. Research on the properties and activities of enzymes of eumycete thermophilic isolates of soil. *Annals of Microbiology and Enzymology* 17:1–30.

Crowe, J., Hoekstra, F., and Crowe, L.M. 1992. Anhydrobiosis. *Annual Review of Plant Physiology* 54:579–599.

Damaso, M.C.T., Almeida, M.S., Kurtenbach, E., Martins, O.B., Pereira, N., Andrade, C.M.M.C., and Albano, R.M. 2003. Optimized expression of a thermostable xylanase from *Thermomyces lanuginosus* in *Pichia pastoris*. *Applied and Environmental Microbiology* 69:6064–6072.

Damez, C., Bouquillon, S., Harakat, D., Hénin, F., Muzart, J., Pezron, I., and Komunjer, L. 2007. Alkenyl and alkenoyl amphiphilic derivatives of d-xylose and their surfactant properties. *Carbohydrate Research.* 342:152–162

Deacon, J.W. 1985. Decomposition of filter paper cellulose by thermophilic fungi acting singly, in combination, and in sequence. *Transactions of the British Mycological Society* 85:663–669.

Deploey, J.J., Nasta, M., and Adams, P.R. 1982. Quantitative determinations of the temperature and pH stability of extracellular amylase obtained from *Mucor pusillus*. *Mycologia* 74:847–850.

Deutschmann, R., and Dekker, R.F.H. 2012. From plant biomass to bio-based chemicals: Latest developments in xylan research. *Biotechnology Advances* 30:1627–1640.

Dias, M.O.S., Junqueira, T.L., Cavalett, O., Cunha, M.P., Jesus, C.D.F., Rossell C.E.V., Maciel Filho, R., and Bonomi, A. 2012. Integrated versus stand-alone second generation ethanol production from sugarcane bagasse and trash. *Bioresource Technology* 103: 152–161.

Dix, N.J., and J. Webster. 1995. *Fungal Ecology*. London: Chapman & Hall.

Doughlan, M.P. 1985. The properties of fungal and bacterial cellulases with comment on their production and application. *Biotechnology & Genetic Engineering Reviews* 3:39–110.

Du, Y., Shi, P., Huang, H., Zhang, X., Luo, H., Wang, Y., and Yao, B. 2013. Characterization of three novel thermophilic xylanases from *Humicola insolens* Y1 with application potentials in the brewing industry. *Bioresource Technology* 130:161–167.

Dubey, A.K., and Johri, B.N. 1987. Xylanolytic activity of thermophilic *Sporotrichum* sp. and *Myceliophthora thermophilum*. *Proceedings of the Indian Academy of Sciences (Plant Science)* 97:247–255.

Duo-Chuan, L., Yi-Jung, Y., You-Liang, P., and Chong-Yao, S. 1998. Purification and characterization of extracellulae glucoamylase from the thermophilic *Thermomyces lanuginosus*. *Mycological Research* 102:568–572.

El-Refai, A.H., Ghanem, K.M., and El-Sabaery, A.H. 1991. Single cell protein production from orange waste by *Sporotrichum thermophile* cultivated under optimal conditions. *Microbios* 16:63–67.

Emerson, R. 1968. Thermophiles. In *The Fungi*, ed. G.C. Ainsworth and A.S. Sussman, 105–128. New York: Academic Press.

Emi, S., Myers, D.V., and Iacobucci, G.A. 1976. Purification and properties of the thermostable acid protease of *Penicillium duponti*. *Biochemistry* 15:842–848.

Eriksen, J., and Goksoyr, J. 1997. Cellulases from Chaetomium thermophile var. dissitum. *European Journal of Biochemistry* 77:445–450.

Fähnrich, P., and Irrgang, K. 1981. Cellulase and protein production by *Chaetomium cellulolyticum* strains grown on cellulosic substrates. *Biotechnology Letters* 3:201–206.

Fan, G., Yang, S., Yan, Q., Guo, Y., Li, Y., and Jiang, Z. 2014. Characterization of a highly thermostable glycoside hydrolase family 10 xylanase from *Malbranchea cinnamomea*. *International Journal of Biological Macromolecules* 70:482–489.

Fergus, C.L. 1969. The production of amylase by some thermophilic fungi. *Mycologia* 61:1171–1175.

Ferreira-Nozawa, M.S., Rezende, J.L., Guimarães, L.H.S., Terenzi, H.F., Jorge, J.A., and Polizeli, M.L.T.M. 2008. Mycelial glucoamylases produced by the thermophilic fungus

Scytalidium thermophilum strains 15.1 and 15.8: Purification and biochemical characterization. *Brazilian Journal of Microbiology* 39:344–352.

Filho, E.X.F. 1996. Purification and characterization of a β-glucosidase from solid-state cultures of *Humicola grisea* var. *thermoidea*. *Canadian Journal of Microbiology* 42:1–5.

Fischer, L., Scheckermann, C., and Wagner, F. 1995. Purification and characterization of a thermotolerant b-galactosidase from *Thermomyces lanuginosus*. *Applied and Environmental Microbiology* 61:1497–1501.

Flannigan, B., and Sellars, P.N. 1972. Activities of thermophilous fungi from barley kernels against arabinoxylan and carboxymethyl cellulose. *Transactions of the British Mycological Society* 58:338–341.

Fracheboud, D., and Canevascini, G. 1989. Isolation, purification, and properties of the exocellulase from *Sporotrichum* (*Chrysosporium*) thermophile. *Enzyme and Microbial Technology* 11:220–229.

Gaikwad, J.S., and Maheshwari, R. 1994. Localization and release of β-glucosidase in the thermophilic and cellulolytic fungus, *Sporotrichum thermophile*. *Experimental Mycology* 18:300–310.

Ganju, R.K., Murthy, S.K., and Vithayathil, P.J. 1990. Purification and functional characteristics of an endocellulase from *Chaetomium thermophile* var. *coprophile*. *Carbohydrate Research* 197:245–255.

Gao, M.T., Yano, S., and Minowa, T. 2014. Characteristics of enzymes from *Acremonium cellulolyticus* strains and their utilization in the saccharification of potato pulp. *Biochemical Engineering Journal* 83:1–7.

Ghai, S.K., Kahlon, S.S., and Chahal, D.S. 1979. Single cell protein from canning industry waste: Sag waste as substrate for thermotolerant fungi. *Indian Journal of Experimental Biology* 17:789–791.

Ghose, T.K., and Pathak, A.N. 1973. Cellulases-2: Applications. *Process Biochemistry* 8:20–21.

Giardina, P., Faraco, V., Pezzella, C., Piscitelli, A., Vanhulle, S., and Sannia, G. 2010. Laccases: A never-ending story. *Cellular and Molecular Life Sciences* 67:369–385.

Gilbert, M., Breuil, C., Yaguchi, M., and Saddler, J.N. 1992. Purification and characterization of a xylanase from the thermophilic ascomycete *Thielavia terrestris* 255B. *Applied Biochemistry and Biotechnology* 34:247–259.

Gilbert, M., Yaguchi, M., Watson, D.C., Wong, K.K.Y., Breuil, C., and Saddler, J.N. 1993. A comparison of two xylanases from the thermophilic fungi *Thielavia terrestris* and *Thermoascus crustaceus*. *Applied Microbiology and Biotechnology* 40:508–514.

Glazer, A.N., and Nikaido, H. 1995. *Microbial Biotechnology: Fundamentals of Applied Microbiology*. New York: W.H. Freeman and Company.

Gomes, D.C.F., Alegrio, L.V., Leon, L.L., and de Lima, M.E.F. 2002. Total synthesis and anti-leishmanial activity of some curcumin analogues. *Arzneimittel Forschung* 52:695–698.

Gomes, D.J., Gomes, J., and Steiner, W. 1994. Production of highly thermostable xylanase by a wild strain of thermophilic fungus *Thermoascus aurantiacus* and partial characterization of the enzyme. *Journal of Biotechnology* 37:11–22.

Gomes, E., de Souza, A.R., Orjuela, G.L., Da Silva, R., de Oliveira, T.B., and Rodrigues, A. 2016. Applications and benefits of thermophilic microorganisms and their enzymes for industrial biotechnology. In *Gene Expression Systems in Fungi: Advancements and Applications, Fungal Biology*, ed. M. Schmoll and C. Dattenböck, 459–492.

Grajek, W. 1987. Comparative studies on the production of cellulases by thermophilic fungi in submerged and solid-state fermentation. *Applied Microbiology and Biotechnology* 26:126–129.

Grajek, W. 1988. Production of protein by thermophilic fungi from sugar-beet pulp in solid-state fermentation. *Biotechnology and Bioengineering* 32:255–260.

Gray, G.L., Hayenga, K., Cullen, D., Wilson, L.J., and Norton, S. 1986. Primary structure of *Mucor miehei* aspartyl protease: Evidence for a zymogen intermediate. *Gene* 48:41–53.

Guimaraes, L.H.S., Terenzi, H.F., Jorge, J.A., and Polizeli, M.L.T.M. 2001. Thermostable conidial and mycelial alkaline phosphatases from the thermophilic fungus *Scytalidium thermophilum*. *Journal of Industrial Microbiology and Biotechnology* 27:265–270.

Guo, J.P., Tan, J.L., Wang, Y.L., Wu, H.Y., Zhang, C.P., Niu, X.M., Pan, W.Z., Huang, X.W., and Zhang, K.Q. 2011. Isolation of talathermophilins from the thermophilic fungus *Talaromyces thermophilus* YM3-4. *Journal of Natural Products* 74:2278–2281.

Guo, J.P., Zhu, C.Y., Zhang, C.P., Chu, Y.S., Wang, U.L., Zhang, J.X., Wu, D.K., Zhang, K.Q., and Niu, X.M. 2012. Thermolides, potent nematocidal PKS-NRPS hybrid metabolites from thermophilic fungus *Talaromyces thermophilus*. *Journal of the American Chemical Society* 134:20306–20309.

Gupta, A.K., and Gautam, S.P. 1995. Improved production of extracellular α-amylase, by the thermophilic fungus *Malbranchea sulfurea*, following protoplast fusion. *World Journal of Microbiology and Biotechnology* 11:193–195.

Gurung, N., Ray, S., Bose, S., and Rai, V. 2013. A broader view: Microbial enzymes and their relevance in industries, medicine, and beyond. *BioMed Research International*. doi: 10.1155/2013/329121.

Haasum, I., Eriksen, S.H., Jensen, B., and Olsen, J. 1991. Growth and glucoamylase production by the thermophilic fungus *Thermomyces lanuginosus* in a synthetic medium. *Applied Microbiology and Biotechnology* 34:656–660.

Han, P., Zhou, P., Hu, S., Yang, S., Yan, Q., and Jiang, Z. 2013. A novel multifunctional α-amylase from the thermophilic fungus *Malbranchea cinnamomea*: Biochemical characterization and three-dimensional structure. *Applied Biochemistry and Biotechnology* 170:420–435.

Hansen, N.M.L., and Plackett, D. 2008. Sustainable films and coatings from hemicelluloses: A review. *Biomacromolecules* 9:1493–1505.

Heinzelman, P., Snow, C.D., Wu, I., Nguyen, C., Villalobos, A., Govindarajan, S., Minshull, J., and Arnold, F.H. 2009. A family of thermostable fungal cellulases created by structure-guided recombination. *Proceedings of the National Academy of Sciences of the United States of America* 106:5610–5615.

Hirvonen, M., and Papageorgiou, A.C. 2003. Crystal structure of a family 45 endoglucanase from *Melanocarpus albomyces*: Mechanistic implication based on the free cellubiose bound forms. *Journal of Molecular Biology* 10:329–403.

Hoq, M.M., Hempel, C., and Deckwer, W.D. 1994. Cellulase-free xylanase by *Thermomyces lanuginosus* RT9: Effect of agitation, aeration, and medium components on production. *Journal of Biotechnology* 37:49–58.

Hughes, W.T., and Crosier, J.W. 1973. Thermophilic fungi in the mycoflora of man and environmental air. *Mycopathologia et Mycologia Applicata* 49:147–152.

Ifrij, H., and Ogel, Z.B. 2002. Production of neutral and alkaline extracellular proteases by the thermophilic fungus, *Scytalidium thermophilum*, grown on microcrystalline cellulose. *Biotechnology Letters* 24:1107–1110.

Igarashi, K., Verhagen, M.F.J.M., Samejima, M., Schülein, M., Eriksson, K.-E.L., and Nishino, T. 1999. Cellobiose dehydrogenase from the fungi *Phanerochaete chrysosporium* and *Humicola insolens*. *Journal of Biological Chemistry* 274:3338–3344.

Inamdar, A. 1987. *Polygalacturonase from Thermoascus aurantiacus: Isolation and functional characteristics*. PhD thesis, Indian Institute of Science, Bangalore.

Jain, M.K., Kapoor, K.K., and Mishra, M.M. 1979. Cellulase activity, degradation of cellulose and lignin, and humus formation by thermophilic fungi. *Transactions of the British Mycological Society* 73:85–89.

Jensen, B., Nebelong, P., Olsen, J., and Reeslev, M. 2002. Enzyme production in continuous cultivation by the thermophilic fungus, *Thermomyces lanuginosus*. *Biotechnology Letters* 24:41–45.

Jensen, B., and Olsen, J. 1992. Physicochemical properties of a purified alpha-amylase from the thermophilic fungus *Thermomyces lanuginosus*. *Enzyme and Microbial Technology* 14:112–116.

Jensen, B., Olsen, J., and Allermann, K. 1987. Effect of media composition on the production of extracellular amylase from the thermophilic fungus *Thermomyces lanuginosus*. *Biotechnology Letters* 9:313–316.

Jensen, B., Olsen, J., and Allermann, K. 1988. Purification of extracellular amylolytic enzymes from the thermophilic fungus *Thermomyces lanuginosus*. *Canadian Journal of Microbiology* 34:218–223.

Jensen, V.J., and Rugh, S. 1987. Industrial-scale production and application of immobilized glucose isomerase. *Methods in Enzymology* 136:356–370.

Johri, B.N., and Ahmad, S. 1992. Thermophilic fungi at cross roads of biotechnological development. In *Fungal Biotechnology*, ed. H.C. Dubey, 45–73. New Delhi: Today and Tomorrow's Printers.

Johri, B.N., Jain, S., and Chauhan, S. 1985. Enzymes from thermophilic fungi: Proteases and lipases. *Proceedings of the Indian Academy of Sciences (Plant Science)* 94:175–196.

Johri, B.N., and Pandey, A.R. 1982. Cellulose breakdown by *Rhizopus rhizopodiformis* (Cohn) Zopf., a thermophilic mould. *Indian Journal of Microbiology* 22:76–78.

Jorge, J.A., Polizeli, M.L., Thevelein, J.M., and Terenzi, H.F. 1997. Trehalases and trehalose hydrolysis in fungi. *FEMS Microbiology Letters* 154:165–171.

Kadowaki, M.K., Polizeli, M.L.T.M., Terenzi, H.F., and Jorge, J.A. 1996. Characterization of trehalase activities from the thermophilic fungus *Scytalidium thermophilum*. *Biochimica et Biophysica Acta* 1291:199–205.

Kajino, T., Miyazaki, C., Asami, O., Hirai, M., Yamada, Y., and Udaka, S. 1998. Thermophilic fungal protein disulfide isomerase. *Methods in Enzymology* 290:50–59.

Kajino, T., Sarai, K., Imaeda, T., Idekoba, C., Asami, O., Yamada, Y., Hirai, M., and Udaka, S. 1994. Molecular cloning of a fungal cDNA encoding protein disulfide isomerase. *Bioscience, Biotechnology, and Biochemistry* 58:1424–1429.

Kalogeris, E., Christakopoulos, P., Kekos, D., Macris, B.J. 1998. Studies on the solid state production of thermostable endoxylanases from *Thermoascus aurantiacus*, characterization of two isozymes. *Journal of Biotechnology* 60:155–163.

Kanlayakrit, W., Ishimatsu, K., Nakao, M., and Hayashida, S. 1987. Characteristics of raw-starch-digesting glucoamylase from thermophilic *Rhizomucor pusillus*. *Journal of Fermentation Technology* 65:379–385.

Karavaeva, N.N., Zakirov, M.Z., and Mukhiddinova, N.G. 1975. Partial purification and some properties of protease from *Torula thermophilia*. *Biokhimiya* 47:1625–1630.

Karnaouri, A.C., Topakas, E., and Christakopoulos, P. 2014. Cloning, expression, and characterization of a thermostable GH7 endoglucanase from *Myceliophthora thermophila* capable of high consistency enzymatic liquefaction. *Applied Microbiology and Biotechnology* 98:231–242.

Katapodis, P., Christakopoulou, V., and Christakopoulos, P. 2006. Optimization of xylanase production by *Sporotrichum thermophile* using corn cobs and response surface methodology. *Engineering in Life Sciences* 6:410–415.

Katapodis, P., Kalogeris, E., Kekos, D., Macris, B.J., and Christakopoulos, P. 2004. Biosynthesis of fructooligosaccharides by *Sporotrichum thermophile* during submerged batch cultivation in high sucrose media. *Applied Microbiology and Biotechnology* 63:378–382.

Katapodis, P., Vrsanska, M., Kekos, D., Nerinckx, W., Biely, P., Claeyssens, M., Macris, B.J., and Christakopoulos, P. 2003. Biochemical and catalytic properties of an endoxylanase purified from the culture filtrate of *Sporotrichum thermophile*. *Carbohydrate Research* 338:1881–1890.

Kaur, G., Kumar, S., and Satyanarayana, T. 2004. Production, characterization and application of a thermostable polygalacturonase of a thermophilic mould *Sporotrichum thermophile* Apinis. *Bioresource Technology* 94:239–243.

Kaur, G., and Satyanarayana, T. 2004. Production of extracellular pectinolytic, cellulolytic and xylanolytic enzymes by a thermophilic mould *Sporotrichum thermophile* Apinis in solid state fermentation. *Indian Journal of Biotechnology* 3:552–557.

Kawamori, M., Takayama, K., and Takasawa, S. 1987. Production of cellulase by a thermophillic fungus *Thermoascus aurantiacus* A-131. *Agricultural and Biological Chemistry* 51:647–654.

Kengen, S.W.M., Luesink, E.J., Stams, A.J.M., and Zehnder, A.J.B. 1993. Purification and characterization of an extremely thermostable β-glucosidase from the hyperthermophilic archaeon *Pyrococcucus furious*. *European Journal of Biochemistry* 213:305–312.

Khan, M.R., Blain, J.A., and Patterson, J.D.E. 1979. Partial purification of *Mucor pusillus* intracellular proteases. *Applied and Environmental Microbiology* 45:94–96.

Khan, M.R., Blain, J.A., and Patterson, J.D.E. 1983. Extracellular protease of *Mucor pusillus* intracellular proteases. *Applied and Environmental Microbiology* 37:719–724.

Khandke, K.M., Vithayathil, P.J., and Murthy, S.K. 1989a. Degradation of larchwood xylan by enzymes of a thermophilic fungus, *Thermoascus aurantiacus*. *Archives of Biochemistry and Biophysics* 274:501–510.

Khandke, K.M., Vithayathil, P.J., and Murthy, S.K. 1989b. Purification and characterization of an α-D-glucuronidase from a thermophilic fungus, *Thermoascus aurantiacus*. *Archives of Biochemistry and Biophysics* 274:511–517.

Kim, T.H., and Kim, T.H. 2014. Overview of technical barriers and implementation of cellulosic ethanol in the U.S. *Energy* 66:13–19.

Komor, R.S., Romero, P.A., Xie, C.B., and Arnold, F.H. 2012. Highly thermostable fungal cellobiohydrolase I (Cel7A) engineered using predictive methods. *Protein Engineering, Design and Selection* 25:827–833.

Kotwal, S.M., and Shankar, V. 2009. Immobilized invertase. *Biotechnology Advances* 27: 311–322.

Kudanga, T., and Le Roes-Hill, M. 2014. Laccase applications in biofuels production: Current status and future prospects. *Applied Microbiology and Biotechnology* 98:6525–6542.

Kühne, W. 1877. Über das Verhalten verschiedener organisirter und sog. Ungeformter Fermente. *Verhandlungen des Heidelberg Naturhistorisch-Medicinischen Vereins, Neue Folge* 1:190–193.

Kuku, F.O., and Adeniji, M.O. 1976. The effect of moulds on the quality of Nigerian palmkernels. *International Biodeterioration & Biodegradation* 12:37–41.

Kulkarni, N., Shendye, A., and Rao, M. 1999. Molecular and biotechnological aspects of xylanases. *FEMS Microbiology Reviews* 23:411–456.

Kulshrestha, S., Tyagi, P., Sindhi, V., and Yadavilli, K.S. 2013. Invertase and its applications: A brief review. *Journal of Pharmacy Research* 7:792–797.

Kumar, K.K., Deshpande, B.S., and Ambedkar S.S. 1993. Production of extracellular acidic lipase by *Rhizopus arrhizus* as a function of culture conditions. *Hindustan Antibiotics Bulletin* 35:33–42.

Kumar, S., and Satyanarayana, T. 2003. Purification and kinetics of a raw starch hydrolyzing, thermostable and neutral glucoamylase of a thermophilic mold *Thermomucor indicae-seudaticae*. *Biotechnology Progress* 19:936–944.

Kumar, S., and Satyanarayana, T. 2007. Economical glucoamylase production by alginate-immobilized *Thermomucor indicae-seudaticae* in cane molasses medium. *Letters in Applied Microbiology* 45:392–397.

Kunamneni, A., Permaul, K., and Singh, S. 2005. Amylase production in solid state fermentation by the thermophilic fungus *Thermomyces lanuginosus*. *Journal of Bioscience and Bioengineering* 100:168–171.

Kunamneni, A., Plou, F.J., Ballesteros, A., and Alcalde, M. 2008. Laccases and their applications: A patent review. *Recent Patents on Biotechnology* 2:10–24.

Landry, K.S., Vu, A., and Levin, R.E. 2014. Purification of an inducible DNase from a thermophilic fungus. *International Journal of Molecular Sciences* 15:1300–1314.

Langston, J.A., Shaghasi, T., Abbate, E., Xu, F., Vlasenko, E., and Sweeney, M.D. 2011. Oxidoreductive cellulose depolymerization by the enzymes cellobiose dehydrogenase and glycoside hydrolase 61. *Applied and Environmental Microbiology* 77: 7007–7015.

Latif, F., Rajoka, M.I., and Malik, K.A. 1995. Production of cellulases by thermophilic fungi grown on *Leptochloa fusca* straw. *World Journal of Microbiology and Biotechnology* 11:347–348.

Lee, H., Lee, Y.M., Jang, Y., Lee, S., Lee, H., Ahn, B.J., Kim, G.H., and Kim, J.J. 2014. Isolation and analysis of the enzymatic properties of thermophilic fungi from compost. *Mycobiology* 42:181–184.

Lee, J.W., Park, J.Y., Kwon, M., and Choi, I.G. 2009. Purification and characterization of a thermostable xylanase from the brown-rot fungus *Laetiporus sulphureus*. *Journal of Bioscience and Bioengineering* 107:33–37.

Leite, R.S.R., Gomes, E., and da Silva, R. 2007. Characterization and comparison of thermostability of purified β-glucosidases from a mesophilic *Aureobasidium pullulans* and a thermophilic *Thermoascus aurantiacus*. *Process Biochemistry* 42:1101–1106.

Li, B., Nagalla, S.R., and Renganathan, V. 1997. Cellobiose dehydrogenase from *Phanerochaete chrysosporium* is encoded by two allelic variants. *Applied and Environmental Microbiology* 63:796–799.

Lin, J., Ndlovu L.M., Singh, S., and Pillay, B. 1999. Purification and biochemical characteristics of β-D-xylanase from a thermophilic fungus, *Thermomyces lanuginosus*-SSBP. *Biotechnology and Applied Biochemistry* 30:73–79.

Lloret, L., Hollmann, F., Eibes, G., Feijoo, G., Moreira, M.T., and Lema, J.M. 2012. Immobilisation of laccase on Eupergit supports and its application for the removal of endocrine disrupting chemicals in a packed-bed reactor. *Biodegradation* 23:373–386.

Lusis, A.J., and Becker, R.R. 1973. The β-glucosidase system of the thermophilic fungus *Chaetomium thermophile* var. *coprophile* n. var. *Biochimica et Biophysica Acta* 329:5–16.

Maalej, I., Belhaj, I., Masmoudi, N., and Belghith, H. 2009. Highly thermostable xylanase of the thermophilic fungus *Talaromyces thermophilus*: Purification and characterization. *Applied Biochemistry and Biotechnology* 158:200–212.

Mahajan, M.K., Johri, B.N., and Gupta, R.K. 1986. Influence of desiccation stress in a xerophilic thermophile *Humicola* sp. *Current Science* 55:928–930.

Maheshwari, R., Balasubramanyam, P.V., and Palanivelu, P. 1983. Distinctive behaviour of invertase in a thermophilic fungus, *Thermomyces lanuginosus*. *Archives of Microbiology* 134:255–260.

Maheshwari, R., Bharadwaj, G., and Bhat, M.K. 2000. Thermophilic fungi: Their physiology and enzymes. *Microbiology and Molecular Biology Reviews* 64:461–488.

Maheshwari, R., and Kamalam, P.T. 1985. Isolation and culture of a thermophilic fungus, *Melanocarpus albomyces*, and factors influencing the production and activity of xylanase. *Journal of General Microbiology* 131:3017–3027.

Mandels, M. 1975. Microbial sources of cellulase. *Biotechnology and Bioengineering Symposium* 5:81–105.

Mandels, M., and Sternberg, D. 1976. Recent advances in cellulase technology. *Journal of Fermentation Technology* 54:267–286.

Margaritis, A., Merchant, R., and Yaguchi, M. 1983. Xylanase, CM-cellulase and avicelase production by the thermophilic fungus *Sporotrichum thermophile*. *Biotechnology Letters* 5:265–270.

Martin, N., Guez, M.A., Sette, L.D., Da Silva, R., and Gomes, E. 2010. Pectinase production by a Brazilian thermophilic fungus *Thermomucor indicae-seudaticae* N31 in solid-state and submerged fermentation. *Mikrobiologiia* 79:321–328.

Martins, E.S., Leite, R.S., and da Silva, R. 2013. Purification and properties of poly-galacturonase produced by thermophilic fungus *Thermoascus aurantiacus* CBMAI-756 on solid-state fermentation. *Enzyme Research* 2013:438645. doi: 10.1155/2013/438645.

Matsuo, M., and Yasui, T. 1985. Properties of xylanase of *Malbranchea pulchella* var. *sulfurea* no. 48. *Agricultural and Biological Chemistry* 49:839–841.

Matsuo, M., and Yasui, T. 1988. Xylanase of *Malbranchea pulchella* var. *sulfurea*. *Methods in Enzymology* 160:671–674.

McHale, A., and Coughlan, M.P. 1981. The cellulolytic system of *Talaromyces emersonii*. Identification of the various components produced during growth on cellulosic media. *Biochimica et Biophysica Acta* 662:145–151.

McKay, D.J., and Stevenson, K.J. 1979. Lipoamide dehydrogenase from *Malbranchea pulchella*: Isolation and characterization. *Biochemistry* 18:4702–4707.

Mehrotra, B.S. 1985. Thermophilic fungi—Biological enigma and tools for the biotechnologist and biologist. *Indian Phytopathology* 38:211–229.

Merchant, R., Merchant, F., and Margaritis, A. 1988. Production of xylanase by the thermo-philic fungus *Thielavia terrestris*. *Biotechnology Letters* 10:513–516.

Meyer, H.P., and Canevascini, G. 1981. Separation and some properties of two intracellular b-glucosidases of *Sporotrichum* (*Chrysosporium*) *thermophile*. *Applied and Environmental Microbiology* 41:924–931.

Meyer, R.D., and Armstrong, D. 1973. Mucormycosis-changing status. *CRC Critical Reviews in Clinical Laboratory Sciences* 4:421–451.

Meyer, R.D., Kaplan, M.H., Ong, M., and Armstrong, D. 1973. Cutaneous lesions in dis-seminated mucormycosis. *Journal of the American Medical Association* 225:737–738.

Minussi, R., Pastore, G.M., and Duran, N. 2002. Potential applications of laccase in the food industry. *Trends in Food Science & Technology* 13:205–216.

Mishra, R., and Maheshwari, R. 1996. Amylases of the thermophilic fungus *Thermomyces lanuginosus*, their purification, properties, action on starch and response to heat. *Journal of Biosciences* 21:653–672.

Mitchell, D.B., Vogel, K., Weimann, B.J., Pasamontes, L., and van Loon, A.P.G.M. 1997. The phytase subfamily of histidine acid phosphatase; isolation of genes for two novel phytases from the *Aspergillus terreus* and *Myceliophthora thermophila*. *Microbiology* 143:245–252.

Moloney, A.P., Hackett, T.J., Considine, P.J., and Coughlan, M.P. 1983. Isolation of mutants of *Talaromyces emersonii* CBS 814.70 with enhanced cellulase activity. *Enzyme and Microbial Technology* 5:260–264.

Moloney, A.P., McCrae, S.I., Wood, T.M., and Coughlan, M.P. 1985. Isolation and charac-terization of 1,4-β-D-glucan glucanohydrolases of *Talaromyces emersonii*. *Biochemical Journal* 225:365–374.

Moretti, M.M.D.S., Bocchini-Martins, D.A., Nunes, C.D.C.C., Villena, M.A., Perrone, O.M., Silva, R., Boscolo, M., and Gomes, E. 2014. Pretreatment of sugarcane bagasse with microwaves irradiation and its effects on the structure and on enzymatic hydrolysis. *Applied Energy* 122:189–195.

Moretti, M.M.S., Martins, D.A.B., Silva, R., Rodrigues, A., Sette, L.D., and Gomes, E. 2012. Selection of thermophilic and thermotolerant fungi for the production of cellulases and xylanases under solid-state fermentation. *Brazilian Journal of Microbiology* 10: 1062–1071.

Moriya, T., Watanabe, M., and Sumida, N. 2003. Cloning and overexpression of avi2 gene encoding a major cellulase produced by *Humicola isolens* FERM, BP-5977. *Bioscience, Biotechnology, and Biochemistry* 67:1434–1437.

Morrison, J., McCarthy, U., and McHale, A.P. 1987. Cellulase production by *Talaromyces emersonii* CBS 814.70 and a mutant UV7 during growth on cellulose, lactose and glucose containing media. *Enzyme and Microbial Technology* 9:422–425.

Murray, P., Aro, N., and Collins, C. 2004. Expression in *Trichoderma reesei* and characterisation of thermostable family 3 beta-glucosidase from the moderately thermophilic fungus *Talaromyces emersonii*. *Protein Expression and Purification* 38: 248–257.

Naidu, G.S.N., and Panda, T. 1998. Production of pectolytic enzymes—A review. *Bioprocess Engineering* 19:355–361.

Nampoothiri, K.M., Tomes, G.J., Roopesh, K., Szakacs, G., Nagy, V., Soccol, C.R., and Pandey, A. 2004. Thermostable phytase production by *Thermoascus aurantiacus* in submerged fermentation. *Applied Biochemistry and Biotechnology* 118:205–214.

Narang, S., Sahai, V., and Bisaria, V.S. 2001. Optimization of xylanase production by *Melanocarpus albomyces* IIS68 in solid state fermentation using response surface methodology. *Journal of Bioscience and Bioengineering* 91:425–427.

Nguyena, Q.D., Judit, M., Claeyssens, R.M., Stals, I., and Hoschke, A. 2002. Purification and characterisation of amylolytic enzymes from thermophilic fungus *Thermomyces lanuginosus* strain ATCC 34626. *Enzyme and Microbial Technology* 31:345–352.

Niehaus, F., Bertoldo, C., Kähler, M., and Antranikian, G. 1999. Extremophiles as a source of novel enzymes for industrial application. *Applied Microbiology and Biotechnology* 51:711–729.

Nielsen, B.R., Lehmbeck, J., and Frandsen, T.P. 2002. Cloning, heterologous expression, and enzymatic characterization of a thermostable glucoamylase from *Talaromyces emersonii*. *Protein Expression and Purification* 26:1–8.

Noel, M., and Combes, D. 2003. Effects of temperature and pressure on *Rhizomucor miehei* lipase stability. *Journal of Biotechnology* 102:23–32.

Nottebrock, H., Scholer, H.J., and Wali, M. 1974. Taxonomy and identification of mucormycosis-causing fungi I. Synonymity of *Absidia ramosa* with *A. corymbifera*. *Sabouraudia* 12:64–74.

Noureddini, H., Gao, X., and Philkana, R.S. 2005. Immobilized *Pseudomonas cepacia* lipase for biodiesel fuel production from soybean oil. *Bioresource Technology* 96: 769–777.

Ogundero, V.W. 1980. Lipase activities of thermophilic fungi from mouldy groundnuts in Nigeria. *Mycologia* 72:118–126.

Omar, I.C., Nishio, N., and Nagai, S. 1987. Production of a thermostable lipase by *Humicola lanuginosa* grown on sorbitol-corn steep liquor medium. *Agricultural and Biological Chemistry* 51:2145–2151.

Ong, P.S., and Gaucher, G.M. 1973. Protease production by thermophilic fungi. *Canadian Journal of Microbiology* 19:129–133.

Ong, P.S., and Gaucher, G.M. 1976. Production, purification and characterization of thermomycolase, the extracellular serine protease of the thermophilic fungus *Malbranchea pulchella* var. *sulfurea*. *Journal of Microbiology* 22:165–176.

Ooi, Y., Hashimoto, T., Mitsuo, N., and Satoh, T. 1985. Enzymatic formation of β-galactosidase from *Aspergillus oryzae* and its application to the synthesis of chemically unstable cardiac glycosides. *Chemical and Pharmaceutical Bulletin* 33:1808–1814.

Oso, B.A. 1974a. Thermophilic fungi from stacks of oil palm kernels in Nigeria. *Zeitschrift für allgemeine Mikrobiologie* 14:593–601.

Oso, B.A. 1974b. Utilization of lipids as sole carbon sources by thermophilic fungi. *Zeitschrift für Allgemeine Mikrobiologie* 14:713–717.

Ottesen, M., and Rickert, W. 1970. The isolation and partial characterization of an acid protease produced by *Mucor miehei*. *Comptes-Rendus des Travaux du Laboratoire Carlsberg* 37:301–325.

Oz, Y., Kiraz, N., Ozkurt, S., and Soydan, M. 2010. Colonization of peritoneal catheter with a thermophilic fungus, *Thermoascus crustaceus*: A case report. *Medical Mycology* 48: 1105–1107.

Paterson, R.R.M., and Lima, N. 2017. Thermophilic fungi to dominate aflatoxigenic/ mycotoxigenic fungi on food under global warming. *International Journal of Environmental Research and Public Health* 14:199. doi: 10.3390/ijerph14020199.

Paturau, J.M. 1969. *Byproducts of the Cane Sugar Industry, an Introduction to Their Utilization*. Amsterdam: Elsevier.

Payen, A., and Persoz, J.F. 1833. Memoir on diastase, the principal products of its reactions, and their applications to the industrial arts. *Annales de Chimie et de Physique*. 53:73–92.

Pedrolli, D.B., Monteiro, A.C., Gomes, E., and Carmona, E.C. 2009. Pectin and pectinases: Production, characterization and industrial application of microbial pectinolytic enzymes. *Open Biotechnology Journal* 3:9–18.

Peralta, R.M., Terenzi, H.F., and Jorge, J.A. 1990. β-D-Glycosidase activities of *Humicola grisea*: Biochemical and kinetic characterization of a multifunctional enzyme. *Biochimica et Biophysica Acta* 1033:243–249.

Pereira, J.C., Leite, R.S.R., Alves-Prado, H.F., Bocchini-Martins, D.A., Gomes, E., and DaSilva, R. 2014. Production and characterization of β-glucosidase obtained by the solid-state cultivation of the thermophilic fungus *Thermomucor indicae-seudaticae* N31. *Applied Biochemistry and Biotechnology* 174:1–8.

Pereira, J.C., Marques, N.P., Rodrigues, A., Oliveira, T.B., Boscolo, M., Silva, R., Gomes, E., and Martins, D.A.B. 2015. Thermophilic fungi as new sources for production of cellulases and xylanases with potential use in sugarcane bagasse saccharification. *Journal of Applied Microbiology* 118:928–939.

Phutela, U., Dhuna, V., Sandhu, S., and Chadha, B.S. 2005. Pectinase and polygalacturonase production by a thermophilic *Aspergillus fumigatus* isolated from decomposting orange peels. *Brazilian Journal of Microbiology* 36:63–69.

Prabhu, K.A., and Maheshwari, R. 1999. Biochemical properties of xylanases from a thermophilic fungus, *Melanocarpus albomyces*, and their action on plant cell walls. *Journal of Biosciences* 24:461–470.

Prasad, A.R.S., and Maheshwari, R. 1979. Temperature-response of trehalase from mesophilic (*Neurospora crassa*) and a thermophilic fungus (*Thermomyces lanuginosus*). *Archives of Microbiology* 122:275–280.

Preetha, S., and Boopathy, R. 1997. Purification and characterization of a milk clotting protease from *Rhizomucor miehei*. *World Journal of Microbiology and Biotechnology* 13: 573–578.

Pribowo, A., Arantes, V., and Saddler, J.N. 2012. The adsorption and enzyme activity profiles of specific *Trichoderma reesei* cellulase/xylanase components when hydrolyzing steam pretreated corn stover. *Enzyme and Microbial Technology* 50:195–203.

Puchart, V., Katapodis, P., Biely, P., Kremnicky, L., Christakopoulos, P., Vršanska, M., Kekos, D., Macris, B.J., and Bhat, M.K. 1999. Production of xylanases, mannanases, and pectinases by the thermophilic fungus *Thermomyces lanuginosus*. *Enzyme and Microbial Technology* 24:355–361.

Ranck, R.M., Jr., Georg, L.K., and Wallace, D.H. 1974. Dactylariosis: A newly recognized fungus disease of chickens. *Avian Diseases* 18:4–20.

Rao, L.V., Chandel, A.K., Chandrasekhar, G., Rodhe, A.V., and Sridevi, J. 2013. Cellulases of thermophilic microbes. In *Thermophilic Microbes in Environmental and Industrial Biotechnology*, ed. T. Satyanarayana, J. Littlechild, and Y. Kawarabayasi, 671–688. Berlin: Springer.

Rao, P., and Divakar, S. 2002. Response surface methodological approach for *Rhizomucor miehei* lipase-mediated esterification of a-terpineol with propionic acid and acetic anhydride. *World Journal of Microbiology and Biotechnology* 18:345–349.

Rao, U.S., and Murthy, S.K. 1988. Purification and characterization of a b-glucosidase and endocellulase from *Humicola insolens*. *Indian Journal of Biochemistry and Biophysics* 25:687–694.

Rao, V.B., Maheshwari, R., Sastri, N.V.S., and Rao, P.V.S. 1979. A thermostable glucoamylase from the thermophilic fungus *Thermomyces lanuginosus*. *Current Science* 48:113–115.

Rao, V.B., Sastri, N.V.S., and Rao, P.S. 1981. Purification and characterization of a thermostable glucoamylase from the thermophilic fungus *Thermomyces lanuginosus*. *Biochemical Journal* 193:379–387.

Renosto, F., Schultz, T., Re, E., Mazer, J., Chandler, C.J., Barron, A., and Segel, I.H. 1985. Comparative stability and catalytic and chemical properties of the sulfate-activating enzymes from *Penicillium chrysogenum* (mesophile) and *Penicillium duponti* (thermophile). *Journal of Bacteriology* 164:674–683.

Rippon, J.W. 1974. *Medical Mycology*. Philadelphia: W.B. Saunders.

Robledo, A., Aguilar, C.N., Belmares-Cerda, R.E., Flores-Gallegos, A.C., Contreras-Esquivel, J.C., Montañez, J.C., and Mussatto, S.I. 2015. Production of thermostable xylanase by thermophilic fungal strains isolated from maize silage. *CyTA—Journal of Food* 14:302–308. doi: 10.1080/19476337.2015.110529.

Romanelli, R.A., Houston, C.W., and Barnett, S.M. 1975. Studies on thermophilic cellulolytic fungi. *Journal of Applied Microbiology* 30:276–281.

Rosenberg, S.L. 1978. Cellulose and lignocellulose degradation by thermophilic and thermotolerant fungi. *Mycologia* 70:1–13.

Roy, I., Sastry, M.S.R., Johri, B.N., and Gupta, M.N. 2000. Purification of alphaamylase isoenzymes from *Scytalidium thermophilum* on a fluidized bed of alginate beads followed by concanavalin A-agarose column chromatography. *Protein Expression and Purification* 20:162–168.

Rustiguel, C.B., Terenzi, H.F., Jorge, J.A., and Guimarães, L.H.S. 2010. A novel silver-activated extracellular β-D-fructofuranosidase from *Aspergillus phoenicis*. *Journal of Molecular Catalysis B: Enzymatic* 67:10–15.

Sadaf, A., and Khare, S.K. 2014. Production of *Sporotrichum thermophile* xylanase by solid state fermentation utilizing deoiled *Jatropha curcas* seed cake and its application in xylooligosachharide synthesis. *Bioresource Technology* 153:126–130.

Saha, B.C. 2002. Production, purification and properties of xylanase from a newly isolated *Fusarium proliferatum*. *Process Biochemistry* 37:1279–1284.

Sakata, I., Maruyama, I., Kobayashi, A., and Yamamoto, I. 1998. Production of phenethyl alcohol glycoside. *Japanese Kokai Tokkyo Koho*, Japan Patent JP 10052297 (CA 128229438).

Sandhu, D.K., Bagga, P.S., and Singh, S. 1985. Cellulolytic activity of thermophilous fungi isolated from soils. *Kavaka* 13:21–31.

Satyanarayana, T. 1978. *Thermophilic microorganisms and their role in composting process*. PhD thesis, Sagar University, Sagar, India.

Satyanarayana, T., Chavant, L., and Montant, C. 1985. Applicability of APIZYM for screening enzyme activity of thermophilic moulds. *Transactions of the British Mycological Society* 85:727–730.

Satyanarayana, T., Jain, S., and Johri, B.N. 1988. Cellulases and xylanases of thermophilic moulds. In *Perspectives in Mycology and Plant Pathology*, ed. V.P. Agnihotri, K. Sarbhoy, and D. Kumar, 25–60. New Delhi: Malhotra Publishing House.

Satyanarayana, T., and Johri, B.N. 1981. Lipolytic activity of thermophilic fungi of paddy straw compost. *Current Science* 50:680–682.

Satyanarayana, T., and Johri, B.N. 1983. Extracellular protease production of thermophilic fungi of decomposing paddy straw. *Tropical Plant Science Research* 1:137–140.

Satyanarayana, T., and Johri, B.N. 1992. Lipids of thermophilic moulds. *Indian Journal of Microbiology* 32:1–14.

Satyanarayana, T., Johri, B.N., and Klein, J. 1992. Biotechnological potential of thermophilic fungi. In *Handbook of Applied Mycology*, ed. D.K. Arora, R.P. Elander, and K.G. Mukerji, 729–761. New York: Marcel Dekker.

Schmidt, F.L. 1969. Observations of spontaneous heating toward combustion of commercial chip piles. *Tappi* 52:1700–1701.

Schou, C., Christensen, M.H., and Schülein, M. 1998. Characterization of a cellobiose dehydrogenase from *Humicola insolens*. *Biochemical Journal* 330:565–571.

Schuerg, T., Gabriel, R., Baecker, N., Baker, S.E., and Singer, S.W. 2017. *Thermoascus aurantiacus* is an intriguing host for the industrial production of cellulases. *Current Biotechnology* 6:89–97.

Seal, K.J., and Eggins, H.O.W. 1976. The upgrading of agricultural wastes by thermophilic fungi. In *Food from Wastes*, ed. G.G. Birch, K.J. Parker, and J.T. Wargan, 58–78. London: Applied Science Publishers.

Sharma, H.S.S. 1989. Economic importance of thermophilous fungi. *Applied Microbiology and Biotechnology* 31:1–10.

Sharma, H.S.S., and Johri, B.N. 1992. The role of thermophilic fungi in agriculture. In *Handbook of Applied Mycology*, ed. D.K. Arora, R.P. Elander, and K.G. Mukerji, 707–728. New York: Marcel Dekker.

Shibata, H., Sonoke, S., Ochiai, H., Nishihashi, H., and Yamada, M. 1991. Glucosylation of steviol and steviol glucosides in extracts from *Stevia rebaudiana* Bertoni. *Plant Physiology* 95:152–156.

Siddiqui, M.A., Pande, V., and Arif, M. 2012. Production, purification, and characterization of polygalacturonase from *Rhizomucor pusillus* isolated from decomposting orange peels. *Enzyme Research* 2012:138634. doi: 10.1155/2012/138634.

Silva, B.L., Geraldes, F.M., Murari, C.S., Gomes, E., and Da-Silva, R. 2014. Production and characterization of a milk-clotting protease produced in submerged fermentation by the thermophilic fungus *Thermomucor indicae-seudaticae* N31. *Applied Biochemistry and Biotechnology* 172:1999–2011.

Singh, B. 2014. *Myceliophthora thermophila* syn. *Sporotrichum thermophile*: A thermophilic mould of biotechnological potential. *Critical Reviews in Biotechnology* 15:1–11.

Singh, B., Kunze, G., and Satyanarayana, T. 2011. Developments in biochemical aspects and biotechnological applications of microbial phytases. *Biotechnology and Molecular Biology Reviews* 6:69–87.

Singh, B., Poças-Fonseca, M.J., Johri B.N., and Satyanarayana, T. 2016. Thermophilic molds: Biology and applications. *Critical Reviews in Microbiology* 42:985–1006.

Singh, B., and Satyanarayana, T. 2006a. Phytase production by thermophilic mold *Sporotrichum thermophile* in solid-state fermentation and its application in dephytinization of sesame oil cake. *Applied Biochemistry and Biotechnology* 133:239–250.

Singh, B., and Satyanarayana, T. 2006b. A marked enhancement in phytase production by a thermophilic mould *Sporotrichum thermophile* using statistical designs in a cost-effective cane molasses medium. *Journal of Applied Microbiology* 101:344–352.

Singh, B., and Satyanarayana, T. 2009. Characterization of a HAP-phytase from a thermophilic mould *Sporotrichum thermophile*. *Bioresource Technology* 100:2046–2051.

Singh, B., and Satyanarayana, T. 2013. Phytases and phosphatases of thermophilic microbes: Production, characteristics and multifarious biotechnological applications. In *Thermophilic Microbes in Environmental and Industrial Biotechnology*, ed. T. Satyanarayana, J. Littlechild, and Y. Kawarabayasi, 671–688. Berlin: Springer.

Singh, B., and Satyanarayana, T. 2014. Thermophilic fungi: Their ecology and biocatalysts. *Kavka* 42:37–51.

Singh, B., and Satyanarayana, T. 2016. Thermophilic mould *Sporotrichum thermophile*: Biology and potential biotechnological applications. *Kavka* 47:99–106.

Singh, S., Pillay, B., Prior, B.A. 2000. Thermal stability of beta-xylanases produced by different Thermomyces lanuginosus strains. *Enzyme and Microbial Technology* 26:502–508.

Singhania, R.R., Sukumaran, R.K., Patel, A.K., Larroche, C., and Pandey, A. 2010. Advancement and comparative profiles in the production technologies using solid-state and submerged fermentation for microbial cellulases. *Enzyme and Microbial Technology* 46:541–549.

Somkuti, G.A., and Babel, F.J. 1968. Purification and properties of *Mucor pusillus* acid protease. *Journal of Bacteriology* 95:1407–1414.

Somkuti, G.A., Babel, F.J., and Somkuti, A.C. 1969. Lipase of *Mucor pusillus*. *Applied Microbiology* 17:606–610.

Somkuti, G.A., and Steinberg, D.H. 1980. Thermoacidophilic extracellular amylase of *Mucor pusillus*. *Review in Industrial Microbiology* 21:327–333.

Song, S., Tang, Y., Yang, S., Yan, Q., Zhou, P., and Ziang, Z. 2013. Characterization of two novel family 12 xyloglucanases from the thermophilic *Rhizomucor miehei*. *Applied Microbiology and Biotechnology* 97:10013–10024.

Srivastava, R.B., Narain, R., and Mehrotra, B.S. 1981. Comparative cellulolytic ability of mesophilic and thermophilic fungi isolated in India from manure and plant refuse. *Indian Journal of Mycology and Plant Pathology* 11:66–72.

Sternberg, M. 1972. Bond specificity, active site and milk clotting mechanism of the *Mucor miehei* protease. *Biochimica et Biophysica Acta* 285:383–392.

Stevenson, K.J., and Gaucher, G.M. 1975. The substrate specificity of thermomycolase, an extracellular serine proteinase from the thermophilic fungus *Malbranchea pulchella* var. *sulfurea*. *Biochemical Journal* 151:527–542.

Stutzenberger, F. 1990. Thermostable fungal beta-glucosidases. *Letters in Applied Microbiology* 11:173–178.

Subrahmanyam, A. 1985. *Studies on morphology and biochemical activities of some thermophilic fungi*. DSc thesis, Kumaun University, Nainital, India.

Subrahmanyam, A., Mangallam, S., and Gopal Krisahn, K.S. 1977. Amyloglucosidase production by *Torula thermophila*. *Indian Journal of Experimental Biology* 15: 495–496.

Sugiyama, H., Idekoba, C., Kajino, T., Hoshino, F., Asami, O., Yamada, Y., and Udaka, S. 1993. Purification of protein disulfide isomerase from a thermophilic fungus. *Bioscience, Biotechnology, and Biochemistry* 57:1704–1707.

Sukumaran, R.K., Singhania, R.R., Mathew, G.M., and Pandey, A. 2009. Cellulase production using biomass feed stock and its application in lignocellulose saccharification for bioethanol production. *Renew Energy* 34:421–424.

Sundman, G., Kirk, T.K., and Chang, H.M. 1981. Fungal decolorization of kraft bleach plant effluent. *Tappi* 64:145–148.

Svahn, K.S., Go Ransson, U., El-Seedi, H., Bohlin, L., Joakim Larsson, D.G., Olsen, B., and Chryssanthou, E. 2012. Antimicrobial activity of filamentous fungi isolated from highly antibiotic contaminated river sediment. *Infection Ecology & Epidemiology* 2. doi: 10.3402/iee.v2i0.11591.

Syed, A., Saraswati, S., Kundu, G.C., and Ahmad, A. 2013. Biological synthesis of silver nanoparticles using the fungus *Humicola* sp. and evaluation of their cytoxicity using normal and cancer cell lines. *Spectrochimica Acta Part A: Molecular and Biomolecular Spectroscopy* 114:144–147.

Taj-Aldeen, S.J., and Jaffar, W.N. 1992. Cellulase activity of a thermotolerant *Aspergillus niveus* isolated from desert soil. *Mycological Research* 96:14–18.

Takashima, S., Nakamura, A., Masaki, H., and Uozumi, T. 1996. Purification and characterization of cellulases from *Humicola grisea*. *Bioscience, Biotechnology, and Biochemistry* 60:77–82.

Tan, L.U.L., Mayers, P., and Saddler, J.N. 1987. Purification and characterization of a thermostable xylanase from a thermophilic fungus *Thermoascus aurantiacus*. *Canadian Journal of Microbiology* 33:689–692.

Tansey, M.R. 1971. Isolation of thermophilic fungi from self-heated industrial wood chip piles. *Mycologia* 63:537–547.

Tansey, M.R. 1973. Isolation of thermophilic fungi from alligator nesting material. *Mycologia* 65:594–601.

Tansey, M.R., and Brock, T.D. 1978. Microbial life at high temperature ecological aspects. In *Microbial Life in Extreme Environments*, ed. D. Kushner, 159–216. London: Academic Press.

Taylor, P.M., Napier, E.J., and Fleming, I.D. 1978. Amyloglucosidase production and purification from thermophilic fungus *Humicola lanuginosa*. *Carbohydrate Research* 61:301–308.

Teng, C., Yan, Q., Jiang, Z., Fan, G., and Shi, B. 2010. Production of xylooligosaccharides from the steam explosion liquor of corncobs coupled with enzymatic hydrolysis using a thermostable xylanase. *Bioresource Technology* 101:7679–7682.

Thomke, S., Rundgren, M., and Eriksson, K.E. 1980. Nutritional evaluation of the white rot fungus *Sporotrichum pulverulentum* as feed stuff to rats, pigs and sheep. *Biotechnology and Bioengineering* 22:2285–2303.

Thorsen, T.S., Johnsen A.H., Josefsen, K., and Jensen, B. 2006. Identification and characterization of glucoamylase from the fungus *Thermomyces lanuginosus*. *Biochimica et Biophysica Acta* 1764:671–676.

Tong, C.C., and Cole, A.L. 1982. Cellulase production by the thermophilic fungus, *Thermoascus aurantiacus*. *Pertanika* 5:255–262.

Tong, C.C., and Cole, A.L.J. 1975. Physiological studies on the thermophilic fungus *Talaromyces thermophilus* Stolk. *Mauri Ora* 3:37–43.

Tosi, L.R.O., Terenzi, H.F., and Jorge, J.A. 1993. Purification and characterization of an extracellular glucoamylase from the thermophilic fungus *Humicola grisea* var. *thermoidea*. *Canadian Journal of Microbiology* 39:846–852.

Tubaki, K., Ito, T., and Matsuda, Y. 1974. Aquatic sediments as a habitat of thermophilic fungi. *Annals of Microbiology* 24:199–207.

Tuohy, M.G., Puls, J., Claeyssens, M., Vršanska, M., and Coughlan, M.P. 1993. The xylan-degrading enzyme system of *Talaromyces emersonii*: Novel enzymes with activity against aryl b-D-xylosides and unsubstituted xylans. *Biochemical Journal* 290:515–523.

Turner, P., Mamo, G., and Karlsson, E.N. 2007. Potential and utilization of thermophiles and thermostable enzymes in biorefining. *Microbial Cell Factories* 6:9. doi: 10.1186/1475-2859-6-9.

Vallery, R., and Devonshire, R.L. 2003. *Life of Pasteur*.

van Noort, V., Bradatsch, B., Arumugam, M., Amlacher, S., Bange, G., Creevey, C., Falk, S., Mende, D.R., Sinning, I., Hurt, E., and Bork, P. 2013. Consistent mutational paths predict eukaryotic thermostability. *BMC Evolutionary Biology*, 13:7. https://doi.org/10.1186/1471-2148-13-7.

van Oort, M. 2010. Enzymes in bread making. In *Enzymes in Food Technology*, ed. R.J. Whitehurst and M. van Oort, 103–143. Oxford: Blackwell.

Venturi, L.L., Polizeli, L.M., Terenzi, H.F., Furriel R., and Jorge, J.A. 2002. Extracellular beta-D-glucosidase from *Chaetomium thermophilum* var. *coprophilum*: Production, purification and some biochemical properties. *Journal of Basic Microbiology* 42:55–66.

Vinod, S. 2001. Enzymatic decolourisation of denims: A novel approach. *Colourage* 48:25–26.

Virk, S., Johri, B.N., and Singh, S.P. 1996. Production of polysaccharases by the wild type and potoclones of *Malbranchea cinnamomea*. *Indian Journal of Microbiology* 36:53–55.

Voordouw, G., Gaucher, G.M., and Roche, R.S. 1974. Physicochemical properties of thermomycolase, the thermostable, extracellular, serine protease of the fungus *Malbranchea pulchella*. *Canadian Journal of Biochemistry* 52:981–990.

Voorhorst, W.G.B., Eggen, R.I.K., Luesink, E.J., and De Vos, W.M. 1995. Characterization of the Cel B gene coding for β-glucosidase from the hyperthermophilic archean *Pyroccus furiosus* and its expression and site directed mutation in *Escherichia coli*. *Journal of Bacteriology* 177:7105–7111.

Waldrip, D.W., Padhye, A.A., Ajello, L., and Ajello, M. 1974. Isolation of *Dactylaria gallopava* from broiler-house litter. *Avian Diseases* 18:445–451.

Wallace, H.A.H. 1973. Fungi and other organisms associated with stored grain. In *Grain Storage: Part of a System*, ed. R.N. Sinha and W.E. Muir, 71–98. Westport, CT: AVI Publishing.

Wang, Y., Fu, Z., Huang, H., Zhang, H., Yao, B., Xiong, H., and Turunen, O. 2012. Improved thermal performance of *Thermomyces lanuginosus* GH11 xylanase by engineering of an N-terminal disulfide bridge. *Bioresource Technology* 112:275–279.

Wang, Y., Gao, X., Su, Q., Wu, W., and An, L. 2007. Cloning, expression, and enzyme characterization of an acid heat-stable phytase from *Aspergillus fumigatus* WY-2. *Current Microbiology* 55:65–70.

Widsten, P., and Kandelbauer, A. 2008. Laccase applications in the forest products industry: A review. *Enzyme and Microbial Technology* 42:293–307.

Wong, K.K., and Saddler, J.N. 1992. *Trichoderma* xylanases, their properties and application. *Critical Reviews in Biotechnology* 12:413–435.

Wyss, M., Pasamontes, L., Friedlein, A., Rémy, R., Tessier, M., Kronenberger, A., Middendorf, A., et al. 1999. Biophysical characterization of fungal phytases (*myo*-insoitol hexakisphosphate phosphohydrolases): Molecular size, glycosylation pattern, and engineering of proteolytic resistance. *Applied and Environmental Microbiology* 65:359–366.

Ximenes, E.A., Felix, C.R., and Ulhoa, C.J. 1996. Production of cellulases by *Aspergillus fumigatus* and characterization of one β-glucosidase. *Current Microbiology* 32:119–123.

Yaginuma, S., Asahi, A., Morishita, A., Hayashi, M., Tsujino, M., and Takada, M. 1989. Isolation and characterization of new thiol protease inhibitors estatins A and B. *Journal of Antibiotics (Tokyo)* 42:1362–1369.

Yoshioka, H., Nagato, N., Chavanich, S., Nilubol, N., and Hayashida, S. 1981. Purification and properties of thermostable xylanase from *Talaromyces byssochlamydoides* YH-50. *Agricultural and Biological Chemistry* 45:2425–2432.

Zamocky, M., Ludwig, R., Peterbauer, C., Hallberg, B.M., Divne, C., Nicholls, P., and Haltrich, D. 2006. Cellobiose dehydrogenase—A flavocytochrome from wood-degrading, phytopathogenic and saprotropic fungi. *Current Protein & Peptide Science* 7:255–280.

Zanphorlin, L.M., Cabral, H., Arantes, E., Assis, D., Juliano, L., Juilano, M.A., Da-Silva, R., Gomes, E., and Bonilla-Rodriguez, G.O. 2011. Purification and characterization of a new alkaline serine protease from the thermophilic fungus *Myceliophthora* sp. *Process Biochemistry* 46:2137–2143.

Zhai, M.M., Li, J., Jiang, C.X., Shi, Y.P., Di, D.L., Crews, P., and Wu, Q.X. 2016. The bioactive secondary metabolites from *Talaromyces* species. *Natural Products and Bioprospecting* 6:1–24.

Zhang, H., Woodams, E.E., and Hang, Y.D. 2011. Influence of pectinase treatment on fruit spirits from apple mash, juice and pomace. *Process Biochemistry* 186:1909–1913.

Zheng, Y.Y., Guo, X.H., Song, N.N., and Li, D.C. 2011. Thermophilic lipase from *Thermomyces lanuginosus*: Gene cloning, expression and characterization. *Journal of Molecular Catalysis B: Enzymatic* 69:127–132.

Zhou, J.H. 2000. Herbal sweetening and preservative composition comprising licorice extract and mogrosides obtained from plants belonging to cucurbitaceae and/or momordica. *U.S. Patent 6103240 (CA* 133 168393).

Zhu, Y.S., Wu, Y.Q., Chen, W., Tan, C., Gao, J.H., Fei, J.X., and Shih, C.N. 1982. Induction and regulation of cellulase synthesis in *Trichoderma pseudokoningii* mutants EA3-867 and N2-78. *Enzyme and Microbial Technology* 4:3–12.

Zimmermann, A.L.S., Terenzi, H.F., and Jorge, J.A. 1990. Purification and properties of an extracellular conidial trehalase from *Humicola grisea* var. *thermoidea. Biochimica et Biophysica Acta* 1036:41–46.

12 Future Perspectives and Conclusions

Although thermophilic fungi have existed in nature for millennia, they were cultured in the laboratory only in late nineteenth and early twentieth centuries. These fungi comprise a small assemblage among the Eukarya domain that possess a unique mechanism of growing at elevated temperatures up to 61°C. During the last five decades, many species of thermophilic fungi sporulating at 45°C have been reported. The first modern comprehensive account of the biology and classification of thermophilic fungi was published by Cooney and Emerson in 1964; it included the 11 species known at that time, with a few new to science. Since then, several thermophilic fungi have been discovered and documented in scientific literature. Notwithstanding their potential use in industrial processes, studies on thermophilic fungi have been neglected until recently. Further, their uncertain taxonomic affiliation puts them in a state of disarray, often leading to misidentification and confusion.

Thermophilic fungi have been playing a role in the economy of nature ever since they evolved on this earth. Their importance in the human economy has been realized from their ability to efficiently degrade organic matter, acting as biodeteriorants and natural scavengers; to produce extracellular and intracellular enzymes, organic acids, amino acids, antibiotics, phenolic compounds, polysaccharides, and sterols of biotechnological importance; and to produce nutritionally enriched feeds and single cell protein (SCPs), and as well as from their suitability as agents of bioconversion, for example, their role in the preparation of mushroom compost. The fungal protein, or "mycoprotein," is attracting the attention of food and feed scientists and protein engineers. *Chaetomium cellulolyticum* and *Sprotrichum pulverulentum* are the most widely used organisms for the upgrading of animal feed and producing SCP from lignocellulosic wastes. Similarly, investigations on the process of composting municipal solid wastes with thermophilic fungi, that is, *Chaetomium thermophile*, *Humicola lanuginosa*, *Mucor pusillus*, and *Thermoascus aurantiacus*, have revealed that the resulting compost is richer in N, P, and K.

On the industrial front, the use of thermophilic strains can be an effective solution to the maintenance of optimal temperature in industrial fermentation for the entire cultivation period. It is well known that thermophilic activities of microbes are generally associated with protein and enzyme thermostability. The advantages of the use of thermostable proteins and enzymes for conducting biotechnological processes at high temperature include a reduced risk of contamination with mesophilic microbes, a decrease in the viscosity of the culture medium, an increase in the bioavailability and solubility of organic compounds, and an increase in the diffusion coefficient of substrates and products, resulting in a higher rate of reactions. Further, their involvement in genetic manipulations is a much more recent development. Nevertheless, because of these and many more advantages, thermophilic fungi appear to be suitable candidates in biotechnological applications. Additionally, with the paradigm shift in industry as it moves from fossil fuels toward renewable resource utilization, the need for microbial biocatalysts is envisaged to increase, and undoubtedly there will be an unrelenting and increased need for thermostable selective biocatalysts in the near future. Thus, future perspectives relating to their diversity, taxonomy, phylogeny, genome-wide study, and biotechnology, entailing research on thermophilic fungi, are warranted.

12.1   DIVERSITY PERSPECTIVES

Of about 2 million kinds of living organisms on the earth, fungi encompass little more than 100,000 species. According to an estimate, less than 1% of the existing microbial population, including fungi, has been discovered. Among the isolated fungi, the proportion of thermophilic species is less than 0.1%. In the past few decades, thermophilic fungi have been isolated from a large variety of habitats, both natural and man-made. As noted by Whittaker (1975), there is an inverse relationship between biological diversity and the amount of adaptation required to survive in a specific habitat. Owing to their ubiquitous distribution, thermophilic fungi have been recovered from an array of habitats. Over the years, the majority of thermophilic fungi have been isolated from habitats such as manure, composts, stored grains, bird feathers and nests, industrial coal mine soils, beach sands, nuclear reactor effluents, Dead Sea valley soils, and desert soils. In these habitats, thermophiles may occur either as resting propagules or as active mycelia, depending on the availability of nutrients and favorable environmental conditions. Further, climate change entails increases in global temperature; therefore, a greater number of fungi that tolerate or prefer higher temperatures can be expected to evolve.

From the foregoing account, it is clear that the majority of thermophilic fungi occur in terrestrial habitats, depicting heterogeneity in terms of temperature and nutrient types and concentration. Efforts have been made to investigate their nutritional requirements, growth and metabolism, temperature relationships, pigmentation, protein induction, and enzymes, and compare these with those of their mesophilic counterparts to any find significant differences, which also helps to explain their unusual ability to grow at elevated temperature where most other organisms fail to grow. Nutritional studies play a very important role in understanding the close relationship and physiological behavior of two species of unrelated fungi. Until the 1980s, thermophilic fungi were thought to have complex or unusual nutritional requirements (Maheshwari et al., 2000). From this perspective, research efforts should be directed to their nutrient uptake systems, as well as their ability to utilize mixed substrates, in order to make their use in industries economically feasible.

Of special relevance to biodiversity-rich developing countries found in the tropical regions are the organisms that are able to survive at high temperatures, the thermophiles. Tropical countries represent a greatly underexploited pool of potentially useful biological material. Thus, it is very much warranted to focus our research on the isolation and characterization of thermophilic fungi from specific physiological growth niches. The implication of such studies has further increased with the establishment of a biodiversity action plan and new international policy on patenting. Bioprospecting of these microorganisms will further add to the existing knowledge database of the culture collections of microorganisms and enhance their exploitation, besides yielding additional information, leading to unraveling the mechanism of their survival at high temperatures. The culture-dependent methods do not display the whole spectrum of microbial diversity present in a particular ecosystem; therefore, culture-independent methods based on 16s rRNA gene sequence analysis must be exploited in order to recover the gene of interest. Molecular approaches, such as polymerase chain reaction (PCR) of the internal transcribed spacer (ITS) region, phospholipid fatty acid analysis (PLFA), denaturing gradient gel electrophoresis (DGGE) of PCR-amplified DNA fragments combined with the sequencing bands, terminal restriction fragment length polymorphism (T-RFLP) analysis, and more recently, next-generation sequencing, have given impetus to assessing the diversity of microbes in general and thermophilic fungi in particular.

The incredible scope of microbial diversity in human life and societal development accounts for bigger strides in biotechnological developments. However, one of the biggest problems in thermophilic fungal research is the lack of coordination between biotechnologists and fungal taxonomists. Joint research projects involving applied researchers and fungal taxonomists will play an important role in curtailing this problem. Further, as the use of industrially relevant enzymes from putative thermophilic fungi is economically feasible, albeit their structural and evolutionary aspects remain poorly understood. Thus, studies intending to describe their classification could provide information about the thermal stability of the molecular machinery of these organisms. The possible use of bioinformatic tools to explore uncultured species in order to harness their biocatalysts must be looked into. Analysis of genome sequences of certain thermophilic fungi has provided evidence for the presence of various enzymes of biotechnological interest. Further, metagenomic tools can help in the extrication of their diversity and function in various ecosystems. Concerted efforts should be made to further understand the diversity of thermophilic fungi and their biocatalysts using both culture-dependent and culture-independent approaches.

Considerable progress has been made in the search for thermophilic fungi, yet their true diversity has not been fully explored. The future challenges for tapping uncultured fungi using culture-independent molecular methods warrant that there are likely to be a high number of fungal species in various habitats, many of which might be thermophilic. Besides bioprospecting fungi in general and thermophilic fungi in particular, their conservation is of great concern, as they play significant role in human welfare. Several steps for their conservation have been suggested by Moore et al. (2001), including (1) conservation of their habitats, (2) *in situ* conservation of nonmycological reserves and ecological niches, and (3) *ex situ* conservation, particularly for saprophytic species growing in culture. In addition, the Slovak Republic has passed legislation to protect 52 species of fungi; it enables managers to prevent damage to their natural habitats (Lizon, 1999). Thus, research on microorganisms requires sound techniques for their stable preservation for the confirmation of results and future use.

12.2   TAXONOMIC PERSPECTIVES

The taxonomic classification and nomenclature of thermophilic fungi is in a state of disarray, often leading to misidentification and confusion (Mouchacca, 2000).

Presently, most of the known thermophilic fungi have been placed in the orders Sordariales, Eurotiales, Mucorales, and Onygenales (Berka et al., 2011). In addition, *Myriococcum thermophilum* is included as a mitosporic basidiomycete by the National Center for Biotechnology Information (NCBI), while Index Fungorum lists it as an agaricomycete. Further, assigning multiple names to a single taxon has added to the woes of fungal taxonomists.

In the recent past, strategies for maintaining a stable nomenclature have been in the forefront of fungal taxonomy, including proposals for the correct anamorph–teleomorph nomenclature, for subgeneric taxa, and for a list of accepted names (Pitt and Samson, 2007). Similar to other members of the Fungi kingdom, thermophilic fungi also face misleading taxonomic decisions. Mouchacca (2000) made an attempt to resolve taxonomic decisions and name changes for a number of thermophilic fungi. This group of fungi has received considerable attention due to their ability to produce thermostable enzymes with varied biotechnological applications. Therefore, the accurate naming of such fungi becomes a prerequisite in their industrial application, as their misidentification could lead to a chaotic state in the binomials of some thermophiles in published work. Further, it is also known that a few taxa encountered in ecological studies could have been misidentified, while some names reported in biotechnological work are obscured with uncertainties. Therefore, concerted efforts in resolving taxonomic issues pertaining to thermophilic fungi should be made with much ardor.

While describing and comparing new species with old names, taxonomists rely on DNA sequence data, besides morphological characters of the new fungus, and are governed by the International Code of Nomenclature for algae, fungi, and plants (ICN). The application of "one species, one name" is still in its infancy in mycology, primarily because many fungi are still only known in either their sexual (telomorph) or asexual (anamorph) state. In such circumstances, while describing new species, mycologists mostly recover and describe only of the states. This means that all legitimate names proposed for a species, regardless of what stage they are typified by, can serve as the correct name for that species.

In an effort to provide for fungi a natural system similar to that for plants and animals, the unification of fungal nomenclature was hard-pressed through the Amsterdam Declaration in 2012. Fungal taxonomists around the world are making enormous efforts to minimize the chaos resulting from the enforcement of one fungus, one name by generating fresh lists of valid fungal names that will become established in due course. In this context, however, Walter Gams (2016) is of a different opinion, stating that protected names are not the last words in fungal taxonomy, and mycologists cannot be forced to adopt a particular taxonomic system when they do not agree with it. The author further emphasizes that it is presently impossible to effectively squeeze all known species into recognized, available, and strictly monophyletic genera. A complete shift to unified nomenclature will entail the recombination of all included species into a single recognized genus for a particular group.

Furthermore, as suggested by Mouchacca (1997), strict restriction to nomenclatural rules governing citations of fungal binomials should be fundamental. It is known that many researchers use their own strains for the production of biocatalysts; this is, of course, welcome, as some strains of related isolates are highly industrious. Therefore, the authors of applied research dealing with thermophiles should necessarily follow such regulations in order to stabilize the names of strains used for applied research or in produced goods. This will definitely bring an end to the chaotic state prevailing especially in publications relating to fungal taxonomy and biotechnology.

One of the most controversial issues within any taxonomy and classification system is to come to an agreement about what a species is. To the majority of taxonomists, species is a pragmatic unit, conceived to identify those patterns of recurrence that occur in nature and that are thought to be units, and the basis for the construction of any classification. Post-Darwinism, "species" appeared as a scientific need for defining the basic unit of the biological order. Therefore, the species concept is an artifact of the mind, that is, an artificial concept that tries to embrace units of biological diversity. The question that is hampering taxonomy even today is whether this category reflects real discrete units in nature. Clearly, the current definition of species does not allow the classification of uncultured species. At most, one can classify something with the category *candidatus*. This category, with a provisional status in taxonomy, was created to accommodate such organisms that could be recognized by 16S rRNA sequences and a few other parameters. However, after the Amsterdam Declaration, it was widely accepted to encourage classification based on environmental sequences. Such data are usually obtained from metagenomic studies, which mostly include next-generation sequencing data (Hibbett and Taylor, 2013). However, developing community standards for sequence-based classification is the most difficult job. Currently, we have no reliable information regarding the extant of uncultivable fungi present in hot environments, in particular. Similarly, we do not have information on their ability to produce metabolites of biotechnological interest in comparison with their culturable counterparts.

The taxonomic knowledge of several fungal groups is still pretty inadequate and often does not yet allow decisions about the delimitation of natural taxa. The fungal world obviously remains alive in its native environment, awaiting discovery by scientists. However, due to the putative metabolic attributes of thermophilic fungi, several warranted nomenclatural and taxonomic decisions were not adopted by some fungal biotechnologists, creating a chaotic state in the cited binomials of some thermophiles in their published work (Mouchacca, 2000). Conversely, the adoption of gratuitous taxonomic decisions led to misunderstanding and misidentification of some thermophilic fungi. To overcome these uncertainties, fungal taxonomists working on thermophilic fungi, in particular, must fervently adopt the one-name, one-fungus concept, which may facilitate the study of fungal systematics by students and researchers.

12.3   PHYLOGENETIC AND GENOMIC PERSPECTIVES

The diagrammatic hypothesis about the evolutionary relationships of a group of organisms is referred to as phylogeny. An understanding of the diversity, systematics, and nomenclature of microbes is increasingly important in several branches of biological sciences. The molecular approach to phylogenetic analysis, pioneered by Carl Woese in the 1970s and leading to the three-domain model, has transformed the philosophy of microbial evolution. Presently, the classification and nomenclature of microorganisms is based on polyphasic taxonomy, an approach that includes phenotypic, phylogenetic, and genotypic characters. The phenetic system of classification is anchored in mutual similarity of their phenotypic characteristics. Organisms sharing many such characteristics form a single phenetic group or a taxon. This system of classification remained popular among microbial taxonomists for a long period of time. However, the major drawback of this system lays in its unrevealed evolutionary relationship. In contrast, the phylogenetic system of classification displays and compares organisms based on evolutionary relationships. This system is more practical, as it offers insights into the history of life on earth. Nevertheless, this system also remained ineffective for most of the twentieth century on account of the lack of information on the fossil record. It was only when small-subunit rRNA nucleotide sequences to assess evolutionary relationships among microorganisms were made available that the dawn of a new era revealing the origin and evolution of the majority of life-forms, including microorganisms, unfolded. Now, this approach is widely accepted for the classification of different varieties of organisms, and right now more than 0.5 million different 16S and 18S rRNA sequences are available in international public databases, such as GenBank and MycoBank.

The 16S and 18S rRNA sequences are considered the most conserved sequences and are helpful in tracing the evolutionary history of an organism. Similarly, genotypic classification sought to compare organisms on the basis of their genetic similarity. In this approach, individual genes or whole genomes can be compared and individuals displaying 70% homology are considered members of the same species. Techniques involving the study of DNA, RNA, and protein estimation have advanced our understanding concerning microbial evolution and taxonomy. Currently, the 18s rDNA gene-based systematic scheme appears to be justified as a starting point for puckering the ideal classification based on natural and evolutionary relationships.

As thermophiles are believed to share some characteristics with early life-forms, their further analysis would also play a critical role in the presumptions of these phylogenetic issues. Historically, the research and development efforts on thermophilic fungi have been focused on the identification and characterization of their enzymes from relatively few strains. The characterization of genes expressing these enzymes has largely been ignored for decades. This approach has produced advanced enzymes over time; however, recent studies based on genomics investigation has almost instantaneously yielded a diverse palette of novel, thermostable, high-efficiency gene products that can be combined and coordinated to improve existing enzyme concoctions or generate concoctions *de novo*.

Recently, researchers have become interested in developing aphylogenomics program to unravel the molecular basis of fungal thermophily (Bock et al., 2014) and to provide an empirical framework for linking genotype and phenotype to understand the molecular basis of fungal thermophily. The experimental findings of Bock et al. (2014) fit well with recent bioinformatics and deep sequencing studies, which also conclude that changes in protein primary structure lead to the thermostability of proteins and not to the differential expression of thermoinducible genes. As genome annotations of several thermophilic fungi are available, studies on the genomes of the remaining thermophiles will further demonstrate their complex molecular structure. Further, various biotechnologically important hydrolytic enzymes from thermophilic fungi need to be cloned and expressed in homologous or heterologous systems for large-scale applications in various industries.

12.4 BIOTECHNOLOGICAL PERSPECTIVES

One of the technical problems in industrial fermentation is the maintenance of temperature at an optimal level during the entire cultivation period. The use of thermophilic strains growing at elevated temperatures can be an effective solution to this problem. In view of their ability to produce thermostable enzymes, thermophilic fungal organisms are gaining importance in biotechnological industries requiring process operations at elevated temperatures. One of the major areas where thermophilic fungal strains have been investigated the world over is enzyme technology and protein engineering. On the industrial front, thermostable enzymes and proteins have been obtained from thermophilic fungi. The antifungal antibiotics obtained from thermophilic fungi include myriocin from *Myriococcum albomyces* and thermozymocidin from *Thermoascus aurantiacus*. Thermophilic species of *Mucor*, *Talaromyces*, and others have been reported to produce organic acids (lactic and citric acids) and extracellular phenolic compounds. Some thermophilic fungi are known to excrete various amino acids, like alanine, glutamic acid, and lysine.

It is now well recognized that thermophilic activities are in general associated with protein thermostability. Accordingly, proteins produced by thermophiles tend to be more thermostable than their mesophilic counterparts. Furthermore, these fungi grow in simple media containing carbon and nitrogen sources and mineral salts, rendering the industrial process more economical. Thermophilic fungi are capable of efficiently degrading organic materials by secreting extracellular thermostable enzymes, which are useful in the bioremediation of industrial wastes and effluents that are rich in oil, heavy metals, and anti-nutritional factors, such as phytic acid and polysaccharides. These fungi also synthesize several antimicrobial substances and biotechnologically useful miscellaneous enzymes. Some of these activities of thermophilic fungi are highlighted in Chapter 11.

Moreover, thermophilic fungi have effectively been used in the production of composts for the cultivation of various types of mushrooms. The effects of these fungi on the growth of mushroom mycelium and mushroom yield are significant in view of the fact that they decrease the concentration of ammonia in the compost, which otherwise would counteract the growth of the mushroom mycelium; immobilize nutrients in a form that is apparently available to the mushroom mycelium; and have a growth-promoting effect on the mushroom mycelium, such that a twofold increase in the yield of mushrooms has been obtained in the mushroom industry. The transfer of this technology to mushroom growers will further pave the way to enhancing mushroom production and help in alleviating protein deficiency in largely vegetarian populations in countries like India.

Bioremediation using fungal organisms is a rapidly growing field, primarily because of the rapid development of novel methods to qualitatively and quantitatively analyze the interactions occurring at the organismal level. In addition to this, the role of fungi in bioremediation and biomineralization also stems from the ability of their hyphae to traverse diverse substrates and their organic matter degradation, through either saprotrophy or parasitic and pathogenic interactions. Similar to bacteria, they also play a major role in the mobilization and immobilization of mineral and metal compounds. Numerous examples of induced biomineralization through fungal metabolic activity have been documented in the literature. Despite the fact that thermophilic fungi are able to produce an array of extracellular enzymes, such as phytases and laccases, that can be used in the biomineralization of a variety of different metals, studies on their role in this process are scanty. However, concerted efforts in highlighting the role of fungal processes from a diverse group of fungi, including thermophilic fungi, are needed to harness their true biotechnological potential.

Obtaining new biotechnological products from uncultivable microorganisms is quite interesting and a current topic of debate. Both basic research and biotechnological developments require routinely applicable tools for the functional analysis, expression, and manipulation of genes. These techniques include procedures for genetic transformation and selection of the transformants, well-characterized molecular markers, and expression signals. As thermophilic fungi colonize, multiply, and survive in habitats having elevated temperatures, they represent a formidable pool of bioactive compounds and are a strategic source for new and successful commercial products. Recent technological advances made in genomics, proteomics, and combinatorial chemistry show that nature continues to preserve compounds in its metagenome having the essence of bioactivity or function within the host and the environment. Bioprospecting of microbial genomes, such as thermophilic fungi, offers several advantages over their biocatalysts, besides being thermostable. However, studies on fungal distribution and mapping are challenging due to the lack of sufficient knowledge about their taxonomy and the lack of expert mycologists around the world. Nevertheless, the fungal world provides a fascinating and almost continual source of biological diversity, which is a rich source to exploit for human welfare.

REFERENCES

Berka, R.M., Grigoriev, I.V., Otillar, R., Salamov, A., Grimwood, J., Reid, I., Ishmael, N. et al. 2011. Comparative genomic analysis of the thermophilic biomass-degrading fungi *Myceliophthora thermophila* and *Thielavia terrestris. Nature Biotechnology* 29: 922–927.
Bock, T., Chen, W.H., Ori, A., Malik, N., Silva-Martin, N., Huerta-Cepas, J., Powell, S.T. et al. 2014. An integrated approach for genome annotation of the eukaryotic thermophile *Chaetomium thermophilum. Nucleic Acids Research* 42(22):13525–13533.
Cooney, D.G., and Emerson, R. 1964. *Thermophilic Fungi: An Account of Their Biology, Activities, and Classification.* San Francisco: W.H. Freeman and Co.
Gams, W. 2016. Recent changes in fungal nomenclature and their impact on naming of microfungi. In *Biology of Microfungi, Fungal Biology*, ed. D.W. Li. Cham, Switzerland: Springer International Publishing.
Hibbett, D.S., and Taylor, J.W. 2013. Fungal systematics: Is a new age of enlightenment at hand? *Nature Reviews Microbiology* 11:129–133.
Lizon, P. 1999. Current status and perspectives of conservation of fungi in Slovakia. In Abstracts XIII: Congress of European Mycologists, Madrid, Spain, September 21–25, 1999, 77.
Maheshwari, R., Bharadwaj, G., and Bhat, M.K. 2000. Thermophilic fungi: Their physiology and enzymes. *Microbiology and Molecular Biology Reviews* 64:461–488.

Moore, D., Nauta, M.M., Evans, S.E. and Rotheroe, M. 2001. Fungal conservation issues: Recognising the problem, finding solutions. In *Fungal Conservation*, ed. D. Moore, M.M. Nauta, and S.E. Evans. Cambridge: Cambridge University Press.

Mouchacca, J. 1997. Thermophilic fungi: Biodiversity and taxonomic status. *Cryptogamie Mycologie* 18:19–69.

Mouchacca, J. 2000. Thermophilic fungi and applied research: A synopsis of name changes and synonymies. *World Journal of Microbiology and Biotechnology* 16:881–888.

Pitt, J.I., and Samson, R.A. 2007. Nomenclatural considerations in naming species of *Aspergillus* and its teleomorphs. *Studies in Mycology* 59:67–70.

Whittaker, R. H. 1975. *Communities and Ecosystems*. New York: Macmillan.

Index

Page numbers followed by f and t indicate figures and tables, respectively.

A

Absidia blackesleana, 9, 292
Absidia corymbifera, 10, 32, 57, 58, 77, 277, 292
Absidia ramosa, 9, 10, 274, 292
Absidia spp., 9, 291–292
Accellerase 1000®, 271
Accurate identification of fungi, 187
ACDP, *see* Advisory Group on Dangerous Pathogens (ACDP)
Achaetomium macrosporum, 32, 77
Acquired thermotolerance, 45–46
Acremonium alabamense, 170
Acremonium albamensis, 7t, 58, 179, 274, 275, 277, 280
Acremonium thermophilum, 7t, 40, 61t, 170
Acrophialophora nainana, 77
Acrophialophora spp., 31
Actinomucor, 21, 290
Actinomucor spp., 21
Actinomycetes, 230
Adaptations
    homeoviscous *vs.* homeophasic, 46–47
    thermophilic, proteome and genome as determinants of, 38–40
    to thermophily, 33–34
α-D-glucuronidase, 281
Advisory Group on Dangerous Pathogens (ACDP), 111, 112
Aerial hyphae, 166
Agar-agar, 18t, 19t, 20t
Agaricus bisporus, 22, 221; *see also* Composting
    growth promotion of, by thermophilic fungi, 227–230, 228f, 229t
Agaricus subrufescens, 221
Agar plates, 17
Air, isolation from, 17
Alanines, 21, 38, 39
Alginates, 252
Allescheria fumigatus, 264
Allescheria terrestris, 64, 77, 264
American alligator (*Alligator mississipiensis*), 86
6-Aminopenicillanic acid, 21
Aminopeptidases, thermostable, 42
Ammonification, 213
Ammonium acetate, 57
Amsterdam Declaration in, 2012, 314

Amylases, 271–273
    in biotechnological processes, 20–21
    thermostable, 42
Anamorph (asexual forms), 185
Anglo-Dutch method of composting, 226; *see also* Mushroom compost
Animalia, 123
Animals nests, as natural habitat, 86–87
    alligator nesting material, 85–86
Anthus rufulus (pipit), 85
Antibiotics, antibacterial and antifungal properties, 21
Arabidopsis thaliana, 15
Arginines, 39
Arthrinium pterospermum, 7t, 61t, 175, 194t
Ascomycetes, 4t–7t, 123; *see also* Biodiversity and taxonomic descriptions
    identification of thermophilic fungi, 145–147
    thermophilic taxa, taxonomic descriptions of, 151–166
Aspergillus, 250
Aspergillus fischieri, 9, 292
Aspergillus flavus, 80, 253
Aspergillus fumigatus, 9, 10, 12, 21, 31, 32, 37t, 39, 41, 62, 70, 77, 78, 80, 253, 254, 255, 265, 274, 289, 292
    genomes of, 135–137, 136t
Aspergillus nidulans, 31, 32, 77, 278
Aspergillus niger, 42, 57, 64, 78, 253, 255
Aspergillus oryzae, 69, 280
Aspergillus spp., 9, 31, 59, 291–292
Aspergillus terreus, 78, 253
ATP sulfurylase, 287
Authentic, taxonomic definition, 189t
Azo dyes, 253
Azure B, 253

B

Bacillus brevis, 288
Bagassosis, 292
Barium, 252
Barotolerant organisms, 66
Basidiomycetes, 123
Beach sand, as natural habitat, 83–84
Bee-eater (*Merops superviliesus*), 85
β-fructofuranosidase, 285–286

β-glucosidase, 263, 265, 286–287
Bioaccumulation of heavy metals, 249–250
Bioaccumulation or biosorption, 242
Biocatalysts, 260–261
    enzymes as, 106
    potential industrial applications of thermophilic
        fungal, 260f
Biochemical oxygen demand (BOD), 16, 67
Bioconversion of lignocellulosic materials,
        211–213, 212f; see also Composting
Biodeteriorants, 260
Biodiversity and taxonomic descriptions
    identification of thermophilic fungi
        Ascomycota, 145–147
        Deuteromycetes (Anamorphic fungi),
            147–148
        Zygomycota, 145
    nomenclatural disagreement and synonymies,
        176–179
    overview, 143–144
    thermophilic taxa, taxonomic descriptions of
        Ascomycetes, 151–166
        Deuteromycetes (Anamorphic fungi),
            166–176
        Zygomycetes, 148–150
Biodiversity perspective, bioprospecting, 106–107
Bioethanol, 271
Biological efficiency (BE), 225
Biologically controlled mineralization (BCM),
        254–255
Biologically induced mineralization (BIM),
        254–255
Biological resource centers (BRCs), 111
Biological treatment, 246, 253
Biomass, recovery of heavy metals
        and regeneration of, 252
Biomineralization, 241–243, 254–256
    introduction, 241–243
Bioprospecting, 105–119
    action plan, 107
    biodiversity perspective, 106–107
    conservation of fungal diversity and, 110–118
        culture transportation, 116–117
        extinction culture technique, 107–108
        hazard-based classification
            of microorganisms, 111–113
        International Depository Authority (IDA),
            113–116
        Microbial Strain Data Network (MSDN), 111
        microroplet culture technique, 108
        organizations dealing with microbial
            cultures, 118
        premises before dispatch of culture,
            117–118
        world's major microbial culture collections,
            112t

culturable microbial diversity, 107–108
defined, 105
future perspectives, 118–119
overview, 105–106
the uncultivable, 108–110, 110f
Bioremediation, 241–254
    defined, 243
    heavy metals as environmental pollutants,
        243–245
    introduction, 241–243
    metals as a precious component of life,
        245–246
    strategies to control heavy metal
        contamination, 246–252
    thermophilic fungi in bioremediation, 252–254
Biosorbent for bioremediation
    immobilized, 252
Biosorbents, 252
    agro-industrial wastes as, 246
Biosorption, 243
    of heavy metals, 250–252
Biotechnological significance, 20–22
Biotechnology, 110
Bird nests and feathers, as natural habitat, 86–87
Black HFGR, 253
BLAST algorithm, 109
Bovine abortion, 70
BRCs, see Biological resource centers (BRCs)
Bromelain, 289
Budapest Treaty, guide to deposit of microorgan-
        isms under, 114, 115t

C

Calcarisporium spp., 31
Caldicellulosiruptor hydrothermalis, 40
Canariomyces thermophila, 4t, 60t, 166
Candida albicans, 289
Candidatus, 199
Capsicum annuum, 5t
Carbon sources and adaptations for mixed substrate
        utilization, complex, 67–68
Carboxypeptidases, thermostable, 42
Cardiac-related drugs, 287
Carpentaria acuminata, 5t
Cassia occidentalis, 6t
Cassia tora, 6t
CBS-KNAW COLLECTIONS, 118
Cell-bound enzymes produced by various
        thermophilic fungi, 283t
Cellobiose dehydrogenases (CDHs), 281–282
Cellodextrins, 263
Celluclast®, 271
Cellulases, 262–265, 271
    in biotechnological processes, 20–21
    thermostable, 42

Cellulose, 211
Centropus sinensis (crow pheasant), 85
Chaetomidium pingtungium, 4t, 60t, 164, 193t
Chaetomium brasiliense, 44
Chaetomium britannicum, 4t, 162
Chaetomium cellulolyticum, 9, 21, 22, 58, 290, 291, 311
Chaetomium globossum, 9, 38, 39, 40, 264, 282, 292
Chaetomium mesopotamicum, 5t, 162
Chaetomium olivicolor, 40
Chaetomium senegalensis, 5t, 32, 60t, 162
Chaetomium spp., 10, 31
Chaetomium thermophile, 9, 12, 21, 22, 42, 57, 62, 67, 277, 282, 311
    var. *coprophile*, 5t, 10, 60t, 66, 70, 159, 264, 276, 287
    var. *dissitum*, 5t, 10, 60t, 77, 159, 254, 264, 276
    var. *thermophile*, 44
Chaetomium thermophilum, 33, 38, 39, 40, 273, 279, 291
    genome of, 136t, 138
    var. *coprophilum*, 265
Chaetomium virginicum, 5t, 162
Chaperones, 38
Chemical precipitation, 246, 247–248
Chemoheterotrophs, 55
Chemoorganoheterotrophs, 55
Chemoorganotrophs, 55
Chimeric enzymes, 291
Chromatin, HSPs and, 38
Chrysosporium keratinophilum, 10, 292
Chrysosporium pruinosum, 264
Chrysosporium spp., 31
Chrysosporium thermophilum, 9, 292
Chrysosporium tropicum, 7t, 168f, 171, 171f
Citric acid, 56–57
Citrus peel, 253
Clade/subclade, taxonomic definition, 189t
Classification system; *see also* Taxonomy
    binomial system, 122–123
    described, 122–124
    examples, 123
    genotypic, 122
    genus, 122
    history, 121
    phenetic system, 121
    phylogenetic system, 121–122
    species, 122
    taxonomic ranks, 122–124
ClustalW, 132
Cluster/group, taxonomic definition, 189t
C:N and C:P ratio, initial, 213–214
Coagulation and flocculation, 248
Coal mine soils, 81
Coal spoil tips, as natural habitat, 86

Co-composting, 230–231
Code of practice, for IDA, 114
Colony-forming units (CFU), calculation, 16, 78, 220
Composting
    bioconversion of lignocellulosic materials, 211–213, 212f
    co-composting, 230–231
    defined, 209
    growth promotion of *Agaricus bisporus* by thermophilic fungi, 227–230, 228f, 229t
    hydrolytic enzymes of thermophiles in, 222–223
    mushroom compost, ecology of thermophilic fungi in, 217–222, 218t–219t, 221t
    mushroom composting methods
        Anglo-Dutch method, 226
        INRA method, 226–227
        long method of composting, 223–225, 224f
        short method of composting, 225–226
    overview, 209–211
    phases of, 93
    physicochemical aspects of composts
        C:N and C:P ratio, initial, 213–214
        initial pH value of compost, 216–217
        moisture content, 215
        pile size, 216
        temperature, 215, 216f
    primary goals of, 210
    prospects, 231–232
Composts
    defined, 93
    as man-made habitat, 93–94, 95t
    mushroom compost, thermophilous fungi reported from (global), 95t
Composts, physicochemical aspects of
    C:N and C:P ratio, initial, 213–214
    initial pH value of compost, 216–217
    moisture content, 215
    pile size, 216
    temperature, 215, 216f
Concanavalin A-Sepharose chromatography, 280
Conflict over name change, 186–188, 189t
Conservation of fungal diversity, bioprospecting and, 110–118
    culture transportation, 116–117
    extinction culture technique, 107–108
    hazard-based classification of microorganisms, 111–113
    International Depository Authority (IDA), 113–116
    Microbial Strain Data Network (MSDN), 111
    microroplet culture technique, 108
    organizations dealing with microbial cultures, 118
    premises before dispatch of culture, 117–118
    world's major microbial culture collections, 112t

Coonemeria aegyptiaca, 5t, 60t, 154, 192t
Coonemeria crustacea, 5t, 152, 192t
Coonemeria verrucosa, 60t, 193t
Copper, effect on human health, 246
Coprinopsis spp., 10
Corvus splendens (crow), 85
Corynascus sepedonium, 5t, 165
Corynascus thermophilus, 5t, 165
Crow (*Corvus splendens*), 85
Crow pheasant (*Centropus sinensis*), 85
Cryptic species, taxonomic definition, 189t
Culturable microbial diversity, bioprospecting,
        107–108
Culture-dependent methods, 108–109
Culture-independent approaches, 199
Culture media, for isolation of thermophilic fungi,
        15–16, 17–20
Culture transportation, bioprospecting and,
        116–117
Cytosol, 282, 285
Czapek–Dox agar medium, 18t

D

Dactylaria gallopava, 10, 292
Dactylomyces thermophilus, 5t, 40, 77, 154, 193t
Darwin's prophecy, 107
Dead Sea valley soil, 82, 83f
DEAE-cellulose, 280
Debaromyces hansenii, 65
Denaturing gradient gel electrophoresis (DGGE),
        109–110, 313
Desert soils, 80–81
Deuteromycetes (Anamorphic fungi), 7t–8t, 123;
        see also Biodiversity and taxonomic
            descriptions
    about, 179
    identification of thermophilic fungi, 147–148
    thermophilic taxa, taxonomic descriptions of,
        166–176
Dextrose, 19t
Dextrose-peptone-yeast extract agar medium, 18t
D-Glucosyltransferase or transglucosidase, 282
Diauxic growth, 67
Dictyostelium spp., 38
Dilution plate technique, 16
Dinitrosalicylic acid (DNS), 22
Dipterocarp spp., 8t
Disaccharide trehalose, thermoprotective, 43
DL-alanine, 57
DL-leucine, 57
DL-phenylalanine, 57
DNA damage
    due to metal toxicity, 250
DNase, 282
Drosophila buskii, 36

Drosophila melanogaster, 15
Dual nomenclature, 185
Dulcitol, 56, 57
Duncan's Multiple Range Test (DMRT), 229, 229t
Dye colors, 253
Dynamic hypothesis of thermophily, 69

E

EC, *see* European Community (EC)
ECCO (European Culture Collections
        Organization), 114
EC Directive on Biological Agents (93/88/EEC),
        111–112
Ecological relationships, origin and, 30–33
Ecosystems
    characteristics, 75
EFB, *see* European Federation of Biotechnology
        (EFB)
Elaeis guineensis, 7t
Electrochemical processes, 249
Electrodeposition, 246, 249
Electrodialysis, 248
Electron transport chain (ETC), 64
Emericella nidulans, 39, 62
Endoglucanase or carboxymethyl cellulase,
        262–263
Enterococcus faecalis, vancomycin-resistant, 289
Environmental factors on growth, 59–67
    effect of oxygen, 64
    effect of pH, 62–64
    effect of solutes and water activity, 64–65
    hydrostatic pressure, 65–66
    relative humidity, 66–67
    temperature, effect, 59–62
Environmentally controlled composting (ECC),
        227
Environmental nucleic acid sequences (ENAS)
        to, 199
Enzymes
    as biocatalysts, 106
    from thermophilic fungi, 262
    thermostable, 20
Epitype, taxonomic definition, 189t
Escherichia coli, 36, 275, 287, 291
European Community (EC), 111
European Culture Collections Organization
        (ECCO), 114
European Federation of Biotechnology (EFB), 111
Eurotiomycetidae, 39
Eurythermal fungi, defined, 11
Exocellulases, 211
Exoglucanase or avicelase, 263
Exopolysaccharides, heavy metals accumulation
        by, 250
Extinction culture technique, 107–108

Extracellular thermostable enzymes produced by,
        261–265, 271–282
    amylases, 271–273
    cellobiose dehydrogenases (CDHs), 281–282
    cellulases, 262–265, 271
    D-Glucosyltransferase or transglucosidase, 282
    α-D-glucuronidase, 281
    DNase, 282
    glucoamylase, 273
    laccases, 280–281
    lipases, 276–277
    pectinases, 278–279
    phosphatases, 280
    phytases, 279–280
    proteases, 277–278
    xylanases, 273–276
Extremophiles, 75

F

"Farmer's lung" disease, 70
Ficin, 289
Ficus spp., 7t
Filamentous fungi, 57, 83–84, 209
Flow cytometry, 108
Foeniculum vulgare, 5t
Freundlich model, 251
Fungal biomass, 250
Fungal biomineralization, 256
Fungal sugar or mycose, 284–285
Fungal taxonomists, taxonomic definition, 197
Fungi, 123
Fusarium graminearum, 209
Fusarium oxysporum, 36, 37
Future perspectives, 311–318
    biotechnological perspectives, 317–318
    diversity perspectives, 312–313
    phylogenetic and genomic perspectives,
        315–317
    taxonomic perspectives, 313–315

G

GBlocks, 132–133
GenBank, 109, 134, 316
Generally recognized as safe (GRAS), 272
Genes, thermotolerance (*THTA*), 41
Genomes
    as determinants of thermophilic adaptation,
        38–40
    thermophilic fungi, 135–138, 136t
Genome size, reduction in, 40–41
Genozymes Research Project, 40
Geographical distribution, thermophilic fungi,
        4t–8t
Geomycology, 254

Geothermal soils, 81–82, 81f
Geotrichum spp., 31
Glucoamylase, 273
Glucose, culture media for isolation, 18t, 19t
Glucose-6-phosphate dehydrogenase,
        thermostable, 42
Glucuronic acid, 281
Glutamic acid, 21
Glycines, 38, 39, 57
Glycoside hydrolases (GHs), 263–264
GPCA, *see* Great plate count anomaly (GPCA)
Great plate count anomaly (GPCA), 108
Growth, thermophilic fungi, 57–59
    environmental factors on, 59–67
        temperature, effect, 59–62
Guayule (*Parthenium argentatum*), 91
    retting, as man-made habitat, 91

H

Habitat diversity, 75–95
    man-made habitats, 88–95
        composts, 93–94, 95t
        hay, 88–89
        manure, 90
        municipal waste, 92–93
        nuclear reactor effluents, 89–90
        retting guayule, 91
        stored grains, 91–92
        stored peat, 90–91
        wood chip piles, 89
    natural habitats, 76–88
        beach sand, 83–84
        coal spoil tips, 86
        hot springs, 86–88
        nesting material of birds and animals,
            84–86
        soil, 76–83
    overview, 75–76
Habitats, thermophilic fungi, 4t–8t
    relationships, 13–15
Hansenula polymorpha, 45
Hay, as man-made habitat, 88–89
Hazard
    classification of microorganisms based on,
        111–113
Heat shock proteins (HSPs), 36–38
    defined, 36
    molecular weight, 37–38
    OGT, 36
    synthesis, 36, 37, 45
Heavy metal contamination, strategies to control,
        246–254
    bioaccumulation of heavy metals, 249–250
    biosorption of heavy metals, 250–252
    conventional treatment techniques, 247–249

immobilized biosorbent for bioremediation, 252
recovery of metals and regeneration of biomass,
    252
Heavy metals
    as environmental pollutants, 243–245
    on plants, animals, and humans, 244t
    in various industrial effluents, predominant,
        242t
Hemicelluloses, 273–274
Hemoflavoprotein, 281–282
High-fructose corn syrup (HFCS), 273
Homeophasic adaptations, homeoviscous *vs.*,
    46–47
Homeoviscous adaptations, homeophasic *vs.*,
    46–47
Hordeum vulgare, 6t
Hot springs, as natural habitat, 86–88
House sparrow (*Passer domesticus*), 85
Human health, effect of heavy metals on, 244–246
Humicola brevis, 253
Humicola grisea, 32, 58, 77, 174f, 265
    var. *thermoidea*, 10, 44, 254, 264, 273, 285
Humicola hyalothermophila, 40
Humicola insolens, 10, 58, 64, 67, 173f, 174f, 223,
        225, 253, 264, 272, 275, 279, 281–282,
        288
    var. *thermoidea*, 37t
Humicola lanuginosa, 9, 10, 21, 31, 42, 57, 77,
        254, 272, 276, 277, 279, 292, 311
Humicola lanuginose, 273, 274
Humicola nigriscens, 280
Humicola spp., 21, 42, 59
Humicola stellata, 77, 175, 194t, 272, 279
Humidified chamber technique, 16–17
Hygromycin-resistant gene (*hph*), 41
Hypothesis, thermophilism in fungi, 34–45
    HSPs, 36–38
        defined, 36
        molecular weight, 37–38
        OGT, 36
        synthesis, 36, 37
    lipid solubilization, 44–45
    macromolecular thermostability, 42–43
    overview, 34–35
    protein thermostability and stabilization, 35–36
        sequence-based mechanism, 36
        structure-based mechanism, 35–36
    proteome and genome as determinants
        of adaptation, 38–40
    rapid turnover of essential metabolites, 41–42
    reduction, in genome size, 40–41
    thermotolerance genes, 41
    ultrastructural thermostability and
        pigmentation, 43–44

I

ICBN, *see* International Code of Botanical
        Nomenclature (ICBN)
IDA, *see* International Depository Authority (IDA)
Immobile biopolymers, 256
Immobilized biosorbent for bioremediation, 252
Importance value index (IVI), 78
Index Fungorum, 134, 314
Industrial fermentation, 250, 260
Industrial laccases, 280–281
INRA method of composting, 226–227; *see also*
        Mushroom compost
Internal transcribed spacer (ITS), 65, 222
    sequencing, 10, 87
Internal transcribed spacer (ITS) region, 313
    PCR of, 109
International Air Transport Association (IATA)
        Dangerous Goods Regulations (DGR),
        116
International Code of Botanical Nomenclature
        (ICBN), 123, 185
International Code of Nomenclature for algae,
        fungi, and plants (ICN), 187, 314
International Depository Authority (IDA), 113
    code of practice for, 114
    distribution, and the biological material
        accepted, 114
    future development of network worldwide,
        115–116
    guide to deposit of microorganisms under the
        Budapest Treaty, 114, 115t
    responsibilities of, 113
Intracellular or cell-associated thermostable
        enzymes, 282–288
    ATP sulfurylase, 287
    β-Glycosidases, 286–287
    invertase, 285–286
    lipoamide dehydrogenase, 288
    protein disulfide isomerase, 288
    trehalase, 284–285
Invertase, 285–286
Ion exchange, 241, 246–251
Isolation, of thermophilic fungi, 14, 15–20
    culture media for, 17–20
    overview, 14, 15–16
    techniques, 16–17
        from air, 17
        dilution plate technique, 16
        humidified chamber technique, 16–17
        paired petri plate technique, 17
        Waksman's direct inoculation method, 17
        Warcup's soil plate method, 16
IVYWREL, amino acid composition, 38, 39

K

Kitchen garbage composting, 215

L

Laccases, 280–281
Langmuir and Freundlich equations, 251
Larch wood xylan, 281
L-Asparagine, 18t
Last universal common ancestor (LUCA), 39
Leptochloa fusca, 265
Lichtheimia, 70
Lignin, 211
Lignocellulosic materials, 93
   bioconversion of, 211–213, 212f; *see also*
      Composting
Lineage, taxonomic definition, 189t
Linnaeus, Carolus (binomial nomenclature of), 185
Linoleic acid, 44, 46
Lipases, 276–277
   in biotechnological processes, 20–21
   thermostable, 42
Lipid solubilization, 44–45
Lipoamide dehydrogenase, 288
Litchi sinensis, 5t
Long method of composting (LMC), 223–225,
      224f; *see also* Mushroom compost
L-serine, 57
Lugol's iodine solution, 272
Lysine, 21

M

Macromolecular thermostability, 42–43
Magnesium nitrate, 57
Maillard reaction, 216
Malassezia furfur, 65
Malate dehydrogenase, thermostable, 42
Malbranchaea pulchella var. *sulfurea*, 69
Malbranchea cinnamomea, 7t, 37t, 43, 63–64,
      169f, 170, 195t, 273
Malbranchea connamomea, 61t
Malbranchea flava, 10
Malbranchea pulchella, 31, 41, 77, 80, 264, 275,
      277, 288
   var. *sulfurea*, 22, 67, 77, 88, 272, 285, 288, 291
Malbranchins A and B, 21
Malt extract, 19t
Maltodextrins, 282
Man-made habitats, 88–95; *see also* Habitat
      diversity
   composts, 93–94, 95t
   hay, 88–89

manure, 90
   municipal waste, 92–93
   nuclear reactor effluents, 89–90
   retting guayule, 91
   stored grains, 91–92
   stored peat, 90–91
   wood chip piles, 89
Mannitol, 56
Manure, as man-made habitat, 90
Martin's rose bengal agar medium, 19t–20t
Matrix-assisted laser desorption/ionization time-of-
      flight mass spectrometry (MALDI-TOF
      MS), 278
MDA-MB-231 human breast carcinoma cell line,
      290
Megapodiidae, 30
Melanocarpus albomyces, 5t, 22, 40, 60t, 67, 165,
      254, 274, 275, 279
Melanocarpus thermophilus, 6t, 60t, 166, 193t
Membrane filtration, 248
Merops superviliesus (bee-eater), 85
Mesophiles, 10
Mesophilic fungi, 57, 58
Mesophilic microorganisms in bioremediation,
      application of, 242–243
Metabolism, thermophilic fungi, 57–59
Metabolites, essential, rapid turnover of, 41–42
Metagenomic tools, 313
Metal toxicity, 245
Michaelis–Menten kinetics, 68
Microbeads, 108
Microbe-mediated bioremediation, 242; *see also*
      Bioremediation
Microbial Strain Data Network (MSDN), 111, 118
The Microbial Underground, 118
Microorganisms
   hazard-based classification of, 111–113
Microroplet culture technique, 108
Miehein, 21
Mineral acids, 252
Modified Czapek–Dox medium, 18t–19t
Moisture content, in composting process, 215
Molds, defined, 55
Molecular phylogeny, of thermophilic fungi,
      125–129, 127f–128f, 129f
Molecular weight, HSPs, 37–38
Mortierella spp., 31
MSDN, *see* Microbial Strain Data Network
      (MSDN)
Mucor, 70, 252
Mucor circinelloides, 70
Mucor miehei, 69, 77, 265, 272, 277, 278, 279
Mucormycoses, 70

Mucor pusillus, 9, 10, 12, 21, 31, 42, 67, 77, 254,
    265, 272, 274, 276, 277, 278, 279,
    291–292, 311
Mucor spp., 21, 31
Municipal compost, 213
Municipal waste, as man-made habitat, 92–93
Mushroom
    composting, 4t, 5t, 9, 20, 21, 57
    growth of, 22
    mycelium, 22
Mushroom compost; *see also* Composting
    ecology of thermophilic fungi in, 217–222,
        218t–219t, 221t
    production methods
        Anglo-Dutch method, 226
        INRA method, 226–227
        long method of composting, 223–225, 224f
        short method of composting, 225–226
    thermophilous fungi from (global), 95t
Myceliophthora fergusii, 7t, 32, 37t, 61t, 172, 193t
Myceliophthora guttulata, 196t
Myceliophthora heterothallica, 196t
Myceliophthora hinnulea, 7t, 37t, 40, 61t, 172
Myceliophthora spp., 10
Myceliophthora thermophila, 8t, 37t, 39, 45, 46,
        61t, 172, 193t, 253, 273, 279, 280, 289,
        290
    genomes of, 136t, 137
Myceliophthora thermophilum, 275
Mycelium–mineral aggregates, 255
MycoBank, 134, 316
Myconet, 134
Myriocin, 21, 289, 317
Myriococcum albomyces, 21, 77, 264, 289, 317
Myriococcum thermophilum, 314
Myrothecium gramineum, 255
Myrotheciurn verrucaria, 42

N

Name change and synonymies
    conflict over name change, 186–188, 189t
    onefungus, one-name regime (Amsterdam
        Declaration), 189–190
    overview, 185–186
    taxonomies and name changes, 190–198,
        192t–196t
    uncultured species, classification of, 199–200
    unwarranted taxonomies, 200–201
National Center for Biotechnology Information
        (NCBI), 314
    nucleotide database, 109
Natural habitats, 76–88; *see also* Habitat diversity
    beach sand, 83–84
    coal spoil tips, 86
    hot springs, 86–88

nesting material of birds and animals, 84–86
soil, 76–83
    coal mine soils, 81
    Dead Sea valley soil, 82, 83f
    desert soils, 80–81
    distribution of thermophilic and
        thermotolerant fungi in North India, 79t
    geothermal soils, 81–82, 81f
    physical and biological parameters of, 78t
Natural scavengers, 260
Neotype, taxonomic definition, 189t
Nests
    alligator nesting material, 85–86
    of animals, 84
    of birds, 84
    material of, 84
Neurospora crassa, 40, 282, 284–285, 286
Neurospora sitophila, 66
Next-generation sequencing, 109
Nitrobacter, 213
Nitrosomonas, 213
Nomenclatural disagreement and synonymies,
        176–179; *see also* Biodiversity and
        taxonomic descriptions
North India
    distribution of thermophilic and thermotolerant
        fungi in soils of, 79t
Notoamide E, 289
NPC, *see* Nuclear pore complex (NPC)
Nuclear magnetic resonance (NMR), 289
Nuclear pore complex (NPC), 138
Nuclear reactor effluents, as man-made habitat,
        89–90
Nutrient transport, 68
Nutritionally enriched feeds, 311
Nutritional requirements, 56–57
    thermophilic fungi, 56–57

O

Olekminskite, 255
Operational taxonomic units (OTU), 126, 200
Optimum growth temperature (OGT), 36, 38,
        39–40
Orange M2R, 253
Organization for Economic Cooperation and
        Development (OECD), 111
Organomineralization sensu strict, 255
Origin, of thermophily in fungi, 29–47
    acquired thermotolerance, 45–46
    adaptations to thermophily, 33–34
    ecological relationships and, 30–33
    homeoviscous *vs.* homeophasic adaptations,
        46–47
    hypothesis, 34–45
        HSPs, 36–38

lipid solubilization, 44–45
macromolecular thermostability, 42–43
overview, 34–35
protein thermostability and stabilization,
    35–36
proteome and genome as determinants
    of adaptation, 38–40
rapid turnover of essential metabolites,
    41–42
reduction, in genome size, 40–41
thermotolerance genes, 41
ultrastructural thermostability and
    pigmentation, 43–44
overview, 29–30
OTUs, *see* Operational taxonomic units (OTUs)
Oxalic acid, 56–57

P

Paecilomyces thermophila, 274
Paecilomyces varioti, 9, 10, 292
Paired petri plate technique, 17
Papain, 289
Paper industries
    thermophilic fungi in, 253
Papulospora thermophila, 8t, 61t, 166, 279
Parthenium argentatum, 12
Parthenium argentatum (guayule), *see* Guayule
    (*Parthenium argentatum*)
Passer domesticus (house sparrow), 85
PCR, *see* Polymerase chain reaction (PCR)
Pectinases, 278–279
Pectinases, in biotechnological processes, 20–21
Penicillin G, 21
Penicillium, 250
Penicillium chrysogenum, 41, 69, 84, 287
Penicillium dupontii, 9, 10, 12, 41, 42, 57, 64, 67,
    69, 277, 279, 286, 287, 292
Penicillium notatum, 41, 69
Penicillium sp., 77
Penicillium spp., 12, 31, 32
Peptone, 18t, 19t, 57
Peritonitis, 293
Pestalotiopsis sp., 255
Phialophora lignicola, 10, 292
Phosphatases, 280
Phosphate and carbonate salts of calcium, 254
Phosphatidic acids (PAs), 44–45
Phosphatidylcholines (PCs), 44–45
Phosphatidylethanolamines (PEs), 44–45
Phospholipid-derived fatty acid (PLFA), 221
Phospholipid fatty acid analysis (PLFA), 109, 313
pH value of compost, 216–217
Phycomycetes, 123
Phyllosilicate mineral, 255
Phylogenetic analysis, 124

molecular phylogeny of thermophilic fungi,
    125–129, 127f–128f, 129f
phylogenetic trees, constructing, 129–133,
    131f–132f
systematics, 134–135
thermophilic fungal genomes, 135–138, 136t
Phylogenetic trees, constructing, 129–133,
    131f–132f
Phylogeny; *see also* Phylogenetic analysis
defined, 124
described, 124–125
Physical separation, 247
Physicochemical processes for treatment of
    wastewater containing heavy metal,
    247t
Physiology, 55–70
complex carbon sources and adaptations for
    mixed substrate utilization, 67–68
environmental factors on growth, 59–67
    effect of light, 66
    effect of oxygen, 64
    effect of pH, 62–64
    effect of solutes and water activity, 64–65
    hydrostatic pressure, 65–66
    relative humidity, 66–67
    temperature, effect, 59–62
nutrient transport, 68
nutritional requirements, 56–57
overview, 55
protein breakdown and turnover, 68–69
virulence, 69–70
Phytases, 279–280
Phytic acid, 279, 317
Pichia pastoris, 271, 275–276, 277
Piezophile, 65
Pigmentation, 43–44
Pile size, in composting process, 216, 216f
Pinus spp., 5t
Pïnus taeda, 7t
Pipit (*Anthus rufulus*), 85
PKS-NRPS, 289
Plantae, 123
Poikilophilic fungi, defined, 11
Poikilotrophic fungi, defined, 11
Poiklothermic fungi, defined, 11
Polygalacturonic acid, 278
Polymerase chain reaction (PCR), 82, 109, 125,
    199, 313
Polyphasic taxonomy, 121
Polysaccharides, 317
Preechinulin, 289
Prime lethal event, concept, 34
Prostrate hyphae, 166
Proteases, 277–278
    in biotechnological processes, 20–21
    thermostable, 42

Protein, thermostability and stabilization, 35–36
  sequence-based mechanism, 36
  structure-based mechanism, 35–36
Protein breakdown and turnover, 68–69
Protein disulfide isomerase, 288
Protein inactivation
  due to metal toxicity, 250
Protein turnover, 69
Proteome, as determinants of thermophilic
          adaptation, 38–40
Protolog, taxonomic definition, 189t
Pseudomonas, 264
Psychrophiles, 10
Psychrophilic fungi, 57
Pyrococcus furiosus, 35–36, 287

R

Radionuclides, 252
Rapid amplification of cDNA ends (RACE)
          amplification, 277
Rapid-resynthesis hypothesis, 69
Rapid turnover, of essential metabolites, 41–42
Rapid-turnover theory, 68
Rasamsonia byssochlamydoides, 60t, 158, 196t
Rasamsonia composticola, 61t, 156
Rasamsonia emersonii, 60t, 156, 196t
Reactive Blue MR, 253
Red M8B, 253
Reduction, in genome size, 40–41
Remazol Brilliant Blue R, 253
Remersonia thermophila, 8t, 40, 61t, 169, 195t
Representative, taxonomic definition, 189t
Requirements
    thermophilic fungi
      nutritional, 56–57
Reverse osmosis, 248
Reverse transcriptase polymerase chain reaction
          (RT-PCR), 277
Rhizoctonia spp., 10
Rhizomucor miehei, 4t, 10, 37t, 39, 40, 41, 44, 60t,
          62, 63, 148, 149f, 192t, 274, 275, 276,
          279
    genomes of, 136t, 137
Rhizomucor nainitalensis, 4t, 60t, 77, 148
Rhizomucor pusillus, 4t, 12, 37t, 44, 60t, 63, 88,
          148, 151f, 192t, 220, 274, 279
Rhizomucor spp., 21
Rhizomucor tauricus, 45, 46, 177
Rhizopus, 252
Rhizopus arrhizus, 70, 254, 276
Rhizopus caespitosus, 70
Rhizopus homothallicus, 70

Rhizopus microspores var. *chinensis*, 37t
Rhizopus microsporus, 4t, 32, 60t, 62, 70, 77, 150
    var. *rhizopodiformis*, 4t, 32, 37t, 43, 58, 60t,
          77, 150, 265, 280
Rhizopus schipperae, 70
Rhizopus sp., 253
Rhizopus spp., 31
Ribonucleases, thermostable, 42
Rose bengal, 18t, 20t
16S rRNA gene–based hylogeny, 125
Rubredoxin, 35

S

Sabouraud dextrose agar medium, 19t
Saccharomyces cerevisiae, 36, 43, 252
Scolecobasidium spp., 31
Scopulariopsis fusca, 9, 292
Scytalidium indonesicum, 8t, 61t, 175
Scytalidium thermophile, 227
Scytalidium thermophilum, 8t, 10, 22, 37t, 56, 57,
          61t, 94, 172, 174, 179, 194t, 214, 220,
          221, 222, 225, 254, 273, 278, 279, 280,
          285
Semilogarithmic growth curve, 58
Sequence-based protein thermostabilization, 36
Serpula himantioides, 255
Short method of composting, 225–226; *see also*
          Mushroom compost
Sibling species, taxonomic definition, 189t
Sillucin, 21
Single cell proteins (SCPs), 9, 311
Sodium dodecyl sulfate–polyacrylamide gel
          electrophoresis (SDS-PAGE), 45, 273
Sodium glycerophosphate, 19t
Soil(s)
    microbes, interaction of metals and their
          compounds with, 245
    as natural habitat, 76–83
      coal mine soils, 81
      Dead Sea valley soil, 82, 83f
      desert soils, 80–81
      distribution of thermophilic and
          thermotolerant fungi in North India, 79t
      geothermal soils, 81–82, 81f
      physical and biological parameters of, 78t
    plate method, Warcup, 16
Solid-state fermentation (SSF), 253, 262
Solvent extraction, 246
Spezyme®, 271
Sphaerospora spp., 31
Sporotrichum pulverulentum, 21, 290
Sporotrichum sp., 77

Sporotrichum spp., 21, 31, 32
Sporotrichum thermophile, 10, 21, 57, 58, 265, 274–275, 277, 281, 287, 290
Sporotrichum thermophilum, 291
Sportotrichum thermophile, 264
Sprotrichum pulverulentum, 9, 311
Stabilization, protein
    sequence-based, 36
    structure-based, 35–36
Staphylococcus aureus, 289
Stenothermal fungi, defined, 11
Stilbella thermophila, 40
Stored grains, as man-made habitat, 91–92
Stored peat, as man-made habitat, 90–91
Streptomycin sulfate, 19t, 20t
Streptopenicillin, 18t, 20t
Strontianite, 255
Strontium, 252
Structure-based protein thermostabilization, 35–36
Structure-guided recombination approach (SGRA), 291
Submerged fermentation (SmF), 253
Succinic acid, supplementation, 56
Sun-heated soils
    thermophilic and thermotolerant fungi isolation from, 77
Synthesis, HSPs, 36, 37
Synthetic dyes, 253
Systematics; *see also* Taxonomy
    defined, 121
    phylogeny and, 134–135

T

Talaromyces byssochlamydoides, 6t, 10, 274
Talaromyces duponti, 6t, 43, 64, 67, 158, 282
Talaromyces emersonii, 6t, 10, 22, 37t, 40, 57, 62, 64, 77, 252, 265, 272, 290–291
Talaromyces leycettanus, 279
Talaromyces species, 178
Talaromyces spp., 21, 31
Talaromyces thermophilus, 6t, 10, 12, 21, 39, 46, 60t, 64, 77, 159, 212, 254, 264, 276, 279, 280, 289
Taxon (plural taxa), 122
Taxonomies and name changes, 190–198, 192t–196t
Taxonomy; *see also* Classification system
    classification and taxonomic ranks, 122–124
    defined, 121
    future prospects, 139
    genotypic, 122
    phenetic system, 121
    phylogenetic analysis

molecular phylogeny of thermophilic fungi, 125–129, 127f–128f, 129f
    phylogenetic trees, constructing, 129–133, 131f–132f
    systematics, 134–135
    thermophilic fungal genomes, 135–138, 136t
    phylogenetic system, 121–122
    phylogeny, defined, 124–125
    polyphasic, 121
T-Coffee, 132
Teleomorph (sexual forms), 185
Temperature
    in composting process, 215, 216f
    on growth of microorganisms, 59–62
Terminal restriction fragment length polymorphism (T-RFLP) analysis, 109, 313
Thermoascus, 177
Thermoascus aegyptiacus, 41
Thermoascus aurantiacus, 6t, 9, 10, 12, 21, 31, 32, 43, 56, 57, 58, 60t, 62, 64, 77, 88, 151, 254, 265, 274, 276, 279, 281, 289, 290, 291–292, 311, 317
    var. *levispora*, 10
Thermoascus crustaceous, 293
Thermoascus crustaceus, 37t, 275, 276
Thermoascus spp., 59
Thermoascus taitungiacus, 37t
Thermofilia, 15
Thermoidium sulfureum, 12
Thermomucor indicaeseudaticae, 274
Thermomucor indicae-seudaticae, 57, 253, 273, 279
Thermomyces dupontii, 196t
Thermomyces ibadanensis, 8t, 9, 61t, 174, 176, 221, 292
Thermomyces lanuginosus, 8t, 9, 10, 12, 21, 31, 32, 39, 41, 42, 43, 44, 45, 46, 56, 57, 58, 61t, 62, 63, 64, 67–68, 70, 77, 88, 138, 174, 175, 179, 194t, 220, 221, 222, 254, 265, 272, 273, 274, 275, 280, 284–285, 286, 287, 291–292
Thermomyces stellatus, 174
Thermophiles, 10
    hydrolytic enzymes of, 222–223
Thermophilic fungal genomes, 135–138, 136t
Thermophilic fungi
    bioprospecting of, *see* Bioprospecting
    biotechnological significance, 20–22
    classification of, *see* Classification system
    defined, 10–11
    discovery, 75
    distribution in North Indian soils, 79t

geographical distribution, 4t–8t
habitat diversity, *see* Habitat diversity
habitat(s), 4t–8t
    relationships, 13–15
historical background, 11–13
isolation, 15–20
    from air, 17
    culture media for, 17–20
    dilution plate technique, 16
    humidified chamber technique, 16–17
    overview, 15–16
    paired petri plate technique, 17
    techniques, 16–17
    Waksman's direct inoculation method, 17
    Warcup's soil plate method, 16
isolation from sun-heated soils of south-central
    Indiana, 77
molecular phylogeny of, 125–129, 127f–128f,
    129f
origin of thermophily, *see* Origin, of thermophily
    in fungi
overview, 3–10
physiology, *see* Physiology
reported from mushroom compost (global), 95t
*Thermophilic Fungi: An Account of Their Biology,
    Activities, and Classification*, 75
Thermophilic fungi, biocatalysis of, 259–293
    bioactive compounds from, 288–290
    detrimental activities, 291–293
    extracellular thermostable enzymes produced
        by, 261–265, 271–282
        amylases, 271–273
        cellobiose dehydrogenases (CDHs),
            281–282
        cellulases, 262–265, 271
        D-Glucosyltransferase or transglucosidase,
            282
        α-D-glucuronidase, 281
        DNase, 282
        glucoamylase, 273
        laccases, 280–281
        lipases, 276–277
        pectinases, 278–279
        phosphatases, 280
        phytases, 279–280
        proteases, 277–278
        xylanases, 273–276
    intracellular or cell-associated thermostable
        enzymes produced by, 282–288
        ATP sulfurylase, 287
        β-Glycosidases, 286–287
        invertase, 285–286
        lipoamide dehydrogenase, 288

    protein disulfide isomerase, 288
    trehalase, 284–285
single-cell protein production, 290
tools for genetic recombination, 290–291
Thermophilic fungi in bioremediation, 252–254
Thermophilic hyphomycete, defined, 11–13
Thermophilism, defined, 30
Thermophilous fungi, defined, 11
Thermophymatospora fibuligera, 41
Thermostability
    macromolecular, 42–43
    protein, 35–36
        sequence-based mechanism, 36
        structure-based mechanism, 35–36
    ultrastructural, 43–44
Thermostable cell-free enzymes, 266t–270t
Thermostable enzymes, 20
Thermotoga maritima, 36
Thermotolerance, acquired, 45–46
Thermotolerance genes (*THTA*), 41
Thermotolerant fungi, 37t, 75
    defined, 11
    distribution in North Indian soils, 79t
    habitat diversity, *see* Habitat diversity
    isolation from sun-heated soils of south-central
        Indiana, 77
    in mushroom compost, 218t–219t
Thermozymes, defined, 15
Thermozymocidin, 21, 289, 317
Thermus thermophilus, 35
Theromduric fungi, defined, 11
Thielavia australiensis, 6t, 164
Thielavia heterothallica, 33, 38, 40, 77
Thielavia minor, 7t, 32, 58, 77, 164, 277
Thielavia sepedonium, 77
Thielavia sp., 77
Thielavia spp., 31
Thielavia terrestris, 33, 38, 39, 40, 274, 275
    genomes of, 136t, 137
Thielavia terricola, 7t, 164
Thielavia thermophile, 264
Thiourea, 57
TM-417, 282
Torula thermophila, 21, 22, 32, 57, 58, 77, 264,
    272
*Trade-Related Aspects of Property Intellectual
    Rights* (TRIPS) agreement, 113
Transitional thermophile, defined, 11
Tree of Life, 134
Trehalase, 284–285
Trehalose
    synthesis, 43, 44, 45, 46
    thermoprotective disaccharide, 46

Tricarboxylic acid (TCA), 56
Trichoderma reesei, 39, 280, 291
Trichoderma viride, 264
Triticum spp., 5t
Tritirdium spp., 31
Turning process, 223

U

Ubiquitin, 38
Ultrafiltration, 246
Ultrastructural thermostability and pigmentation,
    43–44
Uncultivable microbial diversity, bioprospecting,
    108–110, 110f
Uncultured species, classification of, 199–200;
    see also Name change and synonymies
United Nations Educational, Scientific, and
    Cultural Organization (UNESCO), 111
Unwarranted taxonomies, 200–201; *see also* Name
    change and synonymies
Uranium, biosorption of, 252
Urease-producing fungi, 255
U.S. Public Health Service (USPHS), 111

V

Vancomycin-resistant *Enterococcus faecalis*, 289
Ventilative heat management, 226
Viable but nonculturable (VBNC) microbes, 108
Vioxanthin, 21
Virulence, 69–70
Voucher, taxonomic definition, 189t

W

Waksman's direct inoculation method, 17

Warcup's soil plate method, 16
WFCC, *see* World Federation for Culture
    Collections (WFCC)
Wheat bran, 253
Wheat straw composts, 220, 223
WHO, *see* World Health Organization (WHO)
WIPO (World Intellectual Property Organization),
    114
Wood chip piles, as man-made habitat, 89
World Data Centre for Microorganisms, 118
World Federation for Culture Collections (WFCC),
    115, 118
World Health Organization (WHO), 111
World Intellectual Property Organization (WIPO),
    114
Xylanases, 273–276
    in biotechnological processes, 20–21

Y

Yeast extract
    Difco-powdered, 18t
    starch agar medium, 18t
Yeast glucose agar, 18t
Yellow M4G, 253
Yh-50, 274
YM3-4, 289
Young compost, 226, 228f

Z

Zinc, 245
Zygomycetes, 4t; *see also* Biodiversity
    and taxonomic descriptions
    identification of thermophilic fungi, 145
    thermophilic taxa, taxonomic descriptions
    of, 148–159

T - #0112 - 111024 - C352 - 234/156/16 - PB - 9780367571894 - Gloss Lamination